Geography, Resources, and Environment

Geography, Resources, and Environment

Volume I

Selected Writings of Gilbert F. White

Edited by

Robert W. Kates
and Ian Burton

The University of Chicago Press
Chicago and London

Robert W. Kates is professor of geography and Research Professor at Clark University and holds a MacArthur Prize fellowship. **Ian Burton** is professor of geography and former director of the Institute for Environmental Studies, University of Toronto, and vice-chairman of the International Federation of Institutes for Advanced Study. Kates and Burton, who have published widely, are coauthors, with Gilbert F. White, of *The Environment as Hazard* (1978) and coeditors of *Readings in Resource Management and Conservation* (1965).

The University of Chicago Press, Chicago 60637
The University of Chicago Press, Ltd., London
© 1986 by The University of Chicago
All rights reserved. Published 1986
Printed in the United States of America
95 94 93 92 91 90 89 88 87 86   5 4 3 2 1

Library of Congress Cataloging-in-Publication Data

White, Gilbert Fowler, 1911–
  Selected writings of Gilbert F. White.

  (Geography, resources, and environment; v.1)
  Bibliography: p.
  Includes index.
  1. Water resources development—Addresses, essays, lectures.   2. Anthropo-geography—Addresses, essays, lectures.   I. Kates, Robert William.   II.   Burton, Ian.
III. Title.   IV.   Series.
GF75.G46 vol. 1.   333.7 s   [333.91′15]   85-13987
[HD1691]
ISBN   0-226-42574-6
ISBN   0-226-42575-4 (pbk.)

Contents

Illustrations

Maps

Tables

Introduction

Gilbert F. White is the outstanding geographer in the man-land tradition, in the study of natural resources and hazards, and the study of human environment. The twenty-nine papers and the accompanying brief excerpts from twenty-four other works have been selected from a body of work at least four times larger that spans a period of fifty years of professional work as civil servant, scientist, and educator. Their publication, along with the companion volume *Themes from the Work of Gilbert F. White,* is designed to honor that lifetime of work.

Brief introductions guide the reader across the expanse of scholarly work and public service. Each attempts to place the selection in historical perspective, identify the author's intentions, and assay its significance. The latter judgment is ours, but the former two are informed by White's recollections in a series of conversations with Anne White and Robert Kates in October 1981. The times, places, and opportunities of White's rich and varied life are interwoven throughout.

Limited by constraints of publication, we have tried to represent the substance, style, and spirit of his professional life. Water as resource or hazard is central to fourteen papers—too little or too much in arid lands and humid floodplains, households and factories, river basins and metropolitan areas. In other papers, water often serves as an example to illustrate broader issues of resource, hazard, and environment. Four papers differ by focusing on issues of geographic education and professional life.

Overall, the selections and excerpts span twelve somewhat distinct themes that we have distilled from White's work for the purposes of inviting the essays in volume II. Each is datable to his first published work on the topic: water supply (1935); floodplain management (1936); natural resources (1948); international resources and environment (1949); arid lands (1954); river basins (1957); decision making (1958); geographic education (1958); weather and climate modification (1966); environmental perception (1966); environmental hazards (1968); and energy (1974). In general, each selection includes an early paper on one of the themes and the accompanying excerpt is taken from a related paper to enlarge upon the theme or to offer a counterpoint to it.

xii Introduction

These papers make three seminal contributions to geographic science: the concept of human adjustment, its broad theoretical and practical range, and the role of perception and decision making in resource management. The choice of human adjustment, as a range of alternative actions intended either to control nature or to adapt to nature, was initially developed for floods in the 1930s. Thus it predates by two decades the wide diffusion of concepts of adaptation and feedback within systems in the social and biological sciences.

In the mid-fifties, White provided a common approach for analyzing and comparing resource use in activities both large and small, public and private, by conceiving of a central role for "managers" in resource use. Unlike the spatial tradition in geography during the same period, his approach did not simply transfer normative economic analyses to the resource domain. Instead, White moved quickly beyond utilitarian analysis to the boundaries of psychology and economics, contributing geographic field study and experience to the rapidly developing, interdisciplinary study of decision processes. In so doing he found himself a leader in the newly developing geography of perception, the world inside people's minds.

The subject of his work is specifically water, more generally resources, hazard, and environment. The key actors are managers, the process is decision making, the goal is human adjustment. Within this framework is an immense body of fact—case studies, initial estimates, key numbers, and observations. Included in this volume are selections from or summaries of major empirical studies: the measurement of trends in urban occupance on the floodplains of the United States between 1936 and 1955; the intensive study of the choice of adjustment in the floodplain of Big Creek in La Follette, Tennessee, in 1961; the interdisciplinary exploration of attitudes toward water in 1965; the broad-based study of domestic water use in East Africa in 1967; and the worldwide collaborative study of natural hazards between 1969 and 1972.

Throughout this body of work is the coherent, repeated style of White's research: goal-oriented, persistent, expanding over time, enlarging in space, scope, and collaboration, but modest in claims and presentation. There is only one research paper in this volume for which the dominant motivation was curiosity. All the other papers had a functional motivation—to answer a question of social interest, to identify the choices open to individuals and society, to influence the selection of a public policy. Given this goal orientation, research for White is defined by management problems, not disciplinary directions. In Selection 21 he writes:

> One of the common and commonly destructive questions
> about research runs, "But is it Geography?" I would like
> to see us substitute "Is it significant?" and "Are you com-
> petent to deal with it?"

Significant research often requires the competence of several specialists. White gradually enlarged his circle of research collaborators,
beginning in flood studies with engineers and economists, and adding
psychologists, sociologists, biologists, geophysical scientists, and
medical doctors.

His commitment is long-term; the themes of his earliest work on
drought, water supply, and floods persist to this day. To illustrate
this, we present at least two selections for these themes, and several
selections for his major commitment to the study of floods. Over
time the scale of his research expanded from the local and the particular to the national and the worldwide, culminating in a global
analysis. Within any one study there are almost always multiple
tiers—in-depth case studies embedded in a broader matrix of like
situations. Similarly the scope of his research expanded—droughts
and floods led to the investigation of seventeen natural hazards,
extreme events led to other forms of environmental concern—culminating in an overview illustrated by the massive and path-breaking
assessment edited by White with Martin Holdgate and Mohammed
Kassas of *The World Environment, 1972–1982* (1982b).*

This long-term commitment and expanded scale permit the retrospective view. White is always looking backward, harvesting the
experience of the natural experiment, the staff of the social science
life. He looks forward but cautiously so. His breadth of hindsight
is panoramic, his foresight is measured in glimpses.

His hindsight is complemented, his foresight augmented, by many
of the essayists in volume II. For most of the twelve themes, the
authors, often beginning with the date of White's first published
paper in that area, assess what has been achieved in the past, what
needs to be done in the future, and how professional inquiry links
with public policy. As White does in his own writing, they focus on
the problem area and the collective effort, not on White's work.

White's books and papers are inevitably slim, his writing sparse,
concise, elegantly summarized. He is a stranger to polemic, hyper-
bole, and theoretical speculation. Subject and style are infused with

---

* The dates in parentheses refer to the publications of Gilbert F. White listed in the
Bibliography at the end of this volume.

a spirit of strong values and an activist philosophy. Underlying his pragmatic social involvement are deeply held views on the sanctity of human life, the stewardship of nature, and the liberating qualities of education and science. These form an anthropocentric vision of human life sustained by a nature transformed and transformable by human action. Thus the responsibilities of respecting and understanding nature, sustaining life, and learning and teaching what one knows are united in a single moral imperative.

To base one's worldly actions upon such respect, sustenance, and learning is special, but also to be efficacious in such actions is unique. Caring deeply about values shared only superficially by others in the society, White nonetheless became a mover and a shaker in the areas of his competence.

In Anne White's words:

> As I think about Gilbert's life, the image of the juggler with balls in the air does fit. The balls are discrete, all spinning around gaily. If one does particularly well it can be kept in orbit a long time; if one falls, it can be kicked aside and another one added. But the important aspect is that, spinning around, they form a pleasing pattern as a whole. More and more I see how Gilbert feels his work life has been very much of a piece, fitting together, as he moved from land use in England to the most complex interactions between humans and the world around them over the whole globe.
>
> What makes him keep at this? I think it is all tied together by his Quaker faith in the ability of humans to marshal their inner resources to deal competently and lovingly with the outer world and with their fellow human beings. Then there is the real fun he gets out of the exchange of ideas with others, and the challenge of making real friends, not just coworkers, out of those others. And not least, there is his innate and humble desire to leave the world a bit better place than he found it.

Robert W. Kates
Ian Burton

I Shortage of Public Water Supplies in the United States During 1934

This major excerpt from Gilbert F. White's first published paper is a straightforward report on a national survey of the effect of the 1934 drought on public water supplies. It contains elements of form, function, and findings that would continue to characterize his professional work for the next fifty years.

All of White's writing is functional, designed to answer a question and/or deliver a message. The drought of 1934, as we now know from tree-ring data, was the greatest to that date in United States settlement experience. There was intense public interest in the drought and much media discussion of its impact on and threat to public welfare. In the area of public policy, there was need for emergency action to meet the perceived crisis and for a longer-term New Deal water program. For the staff of the fledgling National Resources Board, some data, even modest data, would surely help the policy decision. With the help of the American Water Works Association, a survey was undertaken. Preparation of the report was assigned to the junior member of the group, a twenty-two-year-old ABD[1] from Chicago, who was a research assistant to Harlan Barrows. It seemed natural for him to publish the findings in the journal of the group that assisted in the collection of the data.

The sparse, concise, hyperbole-free White style is evident. The purpose and audience are clearly targeted—to feed back information to those who helped in the study and to use an obvious journal of record for those desiring access to the data.

Surprisingly, given its modest intent, the paper documents three principles which follow in the research traditions initiated by White and his mentor Harlan Barrows and which would be repeated in much of White's subsequent work and in the work of others.[2] First, droughts are "acts of God," but drought losses are largely acts of

Condensed from *Journal of the American Water Works Association* 27, no. 7 (July 1935): 841–54. © 1935, The American Water Works Association. Reprinted by permission. The data used in this study are taken chiefly from letters from the respective state sanitary engineers and from the planning board staffs in Ohio and Texas.

man. That principle is well documented in the survey. The greatest
drought in history actually affected the public water supply of only
2 percent of the American people and created difficulties only in
areas already vulnerable either because of prevailing physical problems of supply or the underprovision of capacity. Second, there are
few truly national (and even fewer global) problems. Most environmental problems are regional problems, with considerable difference
between regions. It is the geographer's duty to carefully document
the regional difference and suggest ameliorative action matched specifically to the needs of a place. Finally, even "quick and dirty"
data, limited by issues of definition, accuracy, and coverage, can be
better than ignorance in making policy decisions.

Although precipitation was deficient to a greater extent than ever
before during the spring and summer of 1934 in many parts of the
United States west of the Appalachians, most of the public water
supplies of the nation proved to be adequate. The abnormally small
runoff and the recession of the shallow groundwater table, which
caused severe shortage of water for domestic use and for livestock
in numerous rural areas, resulted in general and acute shortage of
public supplies in only a few sections of the country.

The conditions resulting from the drought focused attention on
certain areas in which unfavorable geologic and hydrologic conditions hinder the development of supplies that are adequate during
very dry years, and on specific municipalities which have failed to
make efficient provision for the use of sources that are readily
available.

In an effort to discover the areas of especially difficult environmental conditions, and the municipalities having deficient collecting
systems, the Water Resources Section of the National Resources
Board last autumn requested the aid of the American Water Works
Association in determining the public supplies which suffered shortage during 1934.

Sources of Data

Tables 1.1 and 1.2 incorporate the returns from letters sent to all
state sanitary engineers by the Association, and information from

Table 1.1 Public water supplies experiencing shortage during 1934

STATE	POPULATION CLASSES 250 to 1,000		1,000 to 5,000		5,000 to 10,000		More than 10,000		TOTAL		PERCENTAGE OF TOTAL POPULATION SERVED BY PUBLIC WATER SUPPLIES*
	Number	Population	Number	Population	Number	Population	Number	Population	Number	Population	
Arizona	1	847	5	18,172					6	19,019	9.4
Arkansas			1	3,234	1	5,182	2	132,526	4	140,942	30.8
California	2	1,300	4	6,353	1	8,434	1	10,439	8	26,526	5.5
Colorado	2	1,182	2	4,124	1	8,665			5	13,971	2.2
Georgia	1	538	2	2,513					3	3,051	0.3
Idaho			6	15,487			1	18,337	7	33,824	17.9
Illinois			1	2,162					1	2,162	0.04
Indiana	1	877	4	6,334	1	5,709	4	96,114	10	109,034	5.5
Iowa	2	1,445	7	11,855	3	19,719			12	33,019	2.5
Kansas	6	3,736	9	20,763			1	12,243	16	36,742	3.9
Michigan					1	9,514			1	9,514	0.3
Minnesota			2	5,186	2	15,710			4	20,896	1.4
Missouri			19	44,081	2	11,402	1	13,875	22	69,358	3.4
Nebraska			1	4,930					1	4,930	0.8
Nevada					Insufficient data						
New Mexico	1	698	1	4,739					2	5,437	3.6
New York	11	7,060	12	26,231	2	13,602	2	332,570	27	379,463	3.1
North Carolina	2	1,500	6	8,815					8	10,315	1.1
North Dakota	13	7,963	6	9,213	2	14,638	2	45,731	23	77,545	43.5
Ohio	8	5,808	18	31,798	5	39,931	3	110,572	34	188,109	3.9
Oklahoma	14	8,137	12	26,738	3	20,459	2	40,498	31	95,832	9.1
Pennsylvania	3	2,250	7	15,800			1	19,300	11	37,350	0.6
South Dakota	1	690					1	12,000	2	12,690	5.1
Tennessee	2	1,469							2	1,469	0.1
Texas	9	6,471	2	5,658	10	24,319			21	36,448	0.1
Utah	65	32,243	33	61,622	1	5,282	1	130,948	100	230,095	60.7
Vermont	1	297	5	9,962			1	10,008	7	20,267	12.4
West Virginia			1	1,192					1	1,192	0.2
Wisconsin					1	7,394			1	7,394	0.4
Wyoming					Insufficient data						
Total	145	84,511	166	346,962	36	209,960	23	985,161	370	1,626,594	2.1

*Total population served by public water supplies taken from tabulation in *Journal of American Water Works*, January, 1935.

special reports prepared by several state planning board consultants and staff members. The state returns are not all comparable. The Kansas towns are only those in which water relief work was done by the state relief administration, the North Dakota towns are only those which appealed for aid to the state geologist, and no entries are made for Nevada in view of the fact that all public supplies there were reported as depleted and that no supporting information was furnished. In several states, notably Wyoming, the authorities consulted, lacking requisite facilities for collecting precise information on the adequacy or inadequacy of supplies, submitted only rough estimates. It was also impossible to check the sources of supply in

Table 1.2 Public water supplies experiencing shortage during 1934 classified by size of population and source

State		250 to 1,000				1,000, to 5,000			
		Shallow wells	Deep wells	Springs	Surface	Shallow wells	Deep wells	Springs	Surface
Alabama	None								
Arizona	6			1 847				1 3,891	4 14,281
Arkansas	4								1 3,234
California	8				2 1,300				4 6,353
Colorado	5				2 1,182				2 4,124
Connecticut	None								
Delaware	None								
Florida	None								
Georgia	3		1 538				1 1,178	1 1,335	
Idaho	7					2 4,016		1 3,235	3 8,236
Illinois	1								1 2,162
Indiana	10	1 877				2 2,764	2 3,570		
Iowa	12				2 1,445				7 11,855
Kansas	16	1 723	4 2,714			5 10,606			4 10,157
Kentucky	None								
Louisiana	None								
Maine	None								
Maryland	None								
Massachusetts	None								
Michigan	1								
Minnesota	4								2 5,186
Missouri	22					2 4,321		2 5,932	15 33,828
Montana	None								
Nebraska	1								1 4,930
Nevada	?								
New Hampshire	None								
New Jersey	None								
New Mexico	2				1 698				1 4,719
New York	27			7 4,474	4 2,586			2 2,720	10 23,511
North Carolina	8				2 1,500	2 2,677	1 1,730		3 4,408
North Dakota	23	6 3,540	3 1,450	1 750	2 1,570	2 2,513	4 6,700		
Ohio	34	6 4,195	1 830		1 783	11 17,868	2 2,140		5 11,790
Oklahoma	31	11 5,876			3 2,261	8 13,767			4 12,951
Oregon	None								
Pennsylvania	11	1 950			2 1,300	2 2,300		1 1,800	4 11,700
Rhode Island	None								
South Carolina	None								
South Dakota	2				1 690				
Tennessee	2	1 515		1 981					
Texas	21	1 813	1 544		7 5,114	1 3,780	1 1,878		
Utah	100			53 25,556	12 6,687	2 2,959		23 46,210	8 12,453
Vermont	7			1 297				3 4,652	2 5,310
Virginia	None								
Washington	None								
West Virginia	1						1 1,192		
Wisconsin	1								
Wyoming	?								
D.C.	None								
Total	370	28 17,489	10 6,076	64 32,905	41 27,116	39 67,571	12 18,388	34 69,775	81 191,188

5,000 to 10,000				More than 10,000				
Shallow wells	Deep wells	Springs	Surface	Shallow wells	Deep wells	Springs	Surface	Source not Known
√3.9			1 5,182				2 132,526	
			1 8,434				1 10,439	
			1 8,665					
				1 18,337				
	1 5,709				1 12,795		3 83,319	
			3 19,719					
							1 12,243	1 299
		1 9,514						
			2 15,710					
			2 11,402				1 13,875	
			2 13,602				2 332,570	
	2 14,638						2 45,731	1 653
4 31,175			1 8,756	1 31,820			2 78,752	
			3 20,459	1 26,399			1 14,099	
							1 19,300	
							1 12,000	
			10 24,319					
		1 5,282					1 130,948	
							1 10,008	
1 7,394								
5 38,569	3 20,347	2 14,796	26 136,248	3 76,556	1 12,795		19 895,810	2 952

all cases. There appears to have been some difference in opinion from state to state as to the meaning of "serious shortage." In the final compilation an effort was made to include only those supplies which failed completely or which were in such danger of failing that special emergency measures were taken. Wherever possible, supplies reported as presenting minor control problems of taste, odor, and the like, or from which the use of water for sprinkling purposes was restricted for short periods were excluded.

Notwithstanding certain variations in the data obtained, it is believed that the returns furnish a moderately dependable basis for making generalizations with respect to the distribution of shortage in the United States as a whole.

Problem Areas

The conditions of precipitation, groundwater, and runoff in 1934 seem to have caused acute problems of water supply in eight major areas and in a number of lesser areas. Within each of these problem areas the difficulties encountered in securing adequate supplies in both normal and drought years are somewhat similar. Throughout each, ground and surface water conditions are roughly similar, and the major groundwater provinces as delimited by O. E. Meinzer[3] have been used to distinguish most of the boundaries. In some the shortages affected only a small proportion of the population served by public supplies, while in others the shortage was felt by all or nearly all of the urban population.

The outstanding characteristics of each problem area are summarized briefly in the following paragraphs. . . .

Supplies Having Chronic Shortage

The shortages shown in Tennessee, North Carolina, Georgia, Michigan, and Wisconsin seem to have been the consequence of a failure to maintain wells or to build suitable impounding works rather than the consequence of drought. In the opinion of the respective state sanitary engineers, the shortages were due to inadequate plants that deserve corrective work if satisfactory service is to be given in most years.

It is probable that an extension of the drought to other parts of the country in future years would emphasize the existence of other problem areas, and it is possible that a continuation of the recent drought might cause widespread and serious shortage in areas that experienced only slight difficulties during 1934. However, from the standpoint of quantity, the events of last year have demonstrated for a large portion of the country the general adequacy of supplies and, at the same time, the existence of special problem areas.

Summary

1. Notwithstanding the severe and widespread nature of the drought of 1934, relatively few public water supplies in the United States experienced shortage. Approximately two percent of the total population served by public water supplies was affected.

2. The chief areas in which there were serious shortages of water were: (a) the Lower Colorado Basin; (b) the Colorado Plateaus; (c) the Great Basin; (d) the Southwest Permian and Pennsylvanian area; (e) the Kansan Drift area; (f) the Dakota Drift area; (g) the Red River Valley of the North; and (h) the Indiana-Ohio-New York Drift area.

3. Minor shortages due in large measure to the drought occurred on the upper coastal plain of Texas, the Snake River plains, along the east slope of the Rocky Mountains, in western Kansas, and in Vermont.

4. Shortages also were experienced by certain supplies in Wisconsin, Michigan, New York, Tennessee, Georgia, and North Carolina which have chronic deficiency.

5. The public supplies of communities ranging in population from 250 to 5,000 constituted the greater proportion of those adversely affected. In this class the sources were about evenly divided among shallow wells, springs, and surface sources. In communities having populations of more than 5,000 each, surface waters were the major sources of supply in all but thirteen cases.

Conclusions

1. From the standpoint of quantity of water, the public water supplies in many areas affected by the drought appear to have been adequate during 1934. This reflects general credit on the engineers and geologists responsible for the location and design of collecting systems and on the managers who maintained their services without

marked deficiencies in quantity during drought. There are, however, a number of areas in which the public has faced shortage problems.

2. In two of the major problem areas—the Red River of the North and the Southwest Pennsylvanian and Permian areas—the environmental conditions make difficult the development of public water supplies which are inadequate in normal years, let alone in abnormal years of drought. Careful geological and engineering surveys involving groups of communities as well as individual communities will probably be necessary if satisfactory supplies are to be secured throughout those areas.

3. In the remaining six major problem areas the shortages, although severe in 1934, seem for the most part to be preventable in the future by the completion of work now in progress or by the construction of relatively small additional works. Satisfactory solution may require the improvement of storage facilities in accordance with lessons learned from the precipitation, runoff, and evaporation extremes of past years as modified by the 1934 experiences. It may further require an extensive exploration of groundwater resources, and in some cases it may necessitate a regional rather than a local approach to water supply and sanitation problems.

Resources and Needs: Assessment of the World Water Situation

This paper has described a condition where the global stock of water is fixed, and postulated an accelerating future demand. It is tempting in such circumstances to extrapolate demand curves for future times when they might outstrip supply. But there is little help in such an exercise.

To be sure, water demand has already exceeded supply in some areas. And there is no question that demand will have to be curbed in some instances in the near future, unless available supplies can be increased and water management radically improved.

Rather than extrapolating from present data on the presumption that current conditions will persist, it might prove more salutary and realistic to focus attention on alternative and improved methods that will correct current carelessness or profligate practice. The crucial question is how to implement effective and socially acceptable demand management procedures before they are dictated by shortages.

Further, the basic data about water supply and its rational use is inadequate for large sectors of the land surface. Decisions about future management for such areas are riddled with uncertainty and frustrated by large margins of error in data derived from inadequate observation networks and equally inadequate modes of analysis. Also, the gap between scientific knowledge and its application is vast and widening in most parts of the world.

The opportunities for radical improvement in the socioeconomic, financial, technological, administrative and legal conditions that influence present circumstances are immense. It should therefore not be taken for granted that any sector of the world's population need drink contaminated water, that industry need continue its present pattern of largely unregulated water use and discharge, that agriculture cannot alter its current pattern of irrigation loss and misuse of water, or that productive soils need be destroyed and aquifers exhausted beyond our ability to replenish them in our own lifetime.

(1978a), 53–54

Notes

1. "All but dissertation" requirements have been completed for the degree of doctor of philosophy.
2. See Wolman and Wolman, vol. II, chap. 1.
3. O. E. Meinzer, "The Occurrence of Groundwater in the United States." U.S.G.S. *Water-Supply Paper,* No. 489.

2   Human Adjustment to Floods

Human Adjustment to Floods may well have been the most influential dissertation in U.S. geography. It was begun in 1938, four years after White, having completed all the formal requirements for the doctorate at the University of Chicago, went to Washington—ostensibly to assist Harlan Barrows for six weeks in preparing the Mississippi Report of the Mississippi Valley Committee. In 1934 and 1935 White reviewed flood control project proposals for the Public Works Administration. Later, as secretary of the National Resources Board's Land and Water Committees, he reviewed numerous projects under the Flood Control Acts of 1936 and 1938.

In all of these projects the stated purpose was, by the construction of engineering works, to reduce the future toll of lives lost and property damaged. But would the net effect of constructing dams, levees, and channel diversions actually be to reduce the future toll of damages, or would such construction increase the possibility of damages as people built on the floodplains in anticipation of, or in response to, flood control projects?

A series of papers followed that explored issues of floodplain economic analysis (1936a; 1937b; 1939), the need for land management and regulation (1936b; 1937a; 1940), and methods for forecasting effects (1937b and 1939). Gradually the list of issues expanded into a generalized conception, what Wesley Calef describes in his introduction to the printed dissertation as "nothing less than a comprehensive theory of the geographic approach to the problem of dealing with floods."[1]

Human Adjustment to Floods is rooted in White's professional work, his view of the human relationship with nature, and the heady sense of new beginnings and opportunities that permeated Washington.[2] The actual writing of it, however, was undertaken to meet the pragmatic requirements of the academic license the doctorate grants; and in the time-honored way of dissertations, it was completed at a forced pace, the last footnotes in place just days before White boarded a boat in 1942 to Lisbon—the first step in a journey

From *Human Adjustment to Floods,* Department of Geography Research Paper no. 29 (Chicago: The University of Chicago, 1945), chaps. 1 and 5.

to occupied France and refugee service with the American Friends Service Committee.

As a doctoral candidate White fussed over drafts with faculty reviewers, but the dissertation was not intended only for them. It was to be published first in a limited Department of Geography edition in 1942, then reprinted in 1945 as a paper in the department's new research series. It quickly found its way into the hands of scholars, scientists, planners, and administrators, in a number of cases because White gave them copies. Two influential books on flood policy in the early fifties made direct use of substantial sections of his arguments: *Muddy Waters* by Arthur Maass[3] and *Floods* by William G. Hoyt and Walter B. Langbein[4]. This work still serves as a remarkable blueprint for the enormous changes in the attitudes and policies of floodplain adjustment that were to take place over the next forty years.

---

1. A Comprehensive View of the Flood Problem

The Flood Problem in the United States

Every year receding flood waters in one or more sections of the United States expose muddy plains where people were poorly prepared to meet the overflow. Small-town shopkeepers digging their goods out of Ohio River silt; Alabama farmers collecting their scattered and broken possessions; and New England manufacturers taking inventory in water-soaked warehouses, testify to the dislocating effects of floods and to the unsatisfactory adjustment which man has made to them in many valleys. For the most part, floods in the United States leave in their wake a dreary scene of impaired health, damaged property, and disrupted economic life.

The effects of floods are not everywhere disastrous, however, or even disturbing to the economy. Each year ebbing flood waters also reveal plains in which a relatively satisfactory arrangement of human occupance has taken place. Pittsburgh merchants returning to stores which, because of adequate preparations, suffered only minor losses; Montana ranchers appraising the increased yields of hay to be obtained because of fresh deposits of moisture; and New Orleans citizens carrying on their business behind a levee withstanding a flood crest high above the streets, illustrate wise adjustments to flood hazard.

It has become common in scientific as well as popular literature to consider floods as great natural adversaries which man seeks persistently to overpower. According to this view, floods always are watery marauders which do no good, and against which society wages a bitter battle. The price of victory is the cost of engineering works necessary to confine the flood crest; the price of defeat is a continuing chain of flood disasters. This simple and prevailing view neglects in large measure the possible feasibility of other forms of adjustment, of which the Pittsburgh and Montana cases are examples.

Floods are "acts of God," but flood losses are largely acts of man. Human encroachment upon the floodplains of rivers accounts for the high annual toll of flood losses. Although in a few drainage areas the frequency and magnitude of floods have increased as a result of exploitative use of the up-stream lands, the flood menace elsewhere has changed but little while man has moved into the natural paths of flooded rivers or has restricted the channels so as to heighten normal flood crests. Moreover, floods may be beneficial as well as harmful, and even where they are completely harmful there are remedies other than physical structures built to afford protection. Recognizing these facts, floodplain occupance cannot be considered realistically as a matter solely of man against the marauder.

Dealing with floods in all their capricious and violent aspects is a problem in part of adjusting human occupance to the floodplain environment so as to utilize most effectively the natural resources of the plain, and, at the same time, of applying feasible and practicable measures for minimizing the detrimental impacts of floods. This problem in the United States involves at least 35 million acres of land known to be subject to flood. A large part of that land is not cultivated, but the cultivated portions are among the more productive agricultural resources of the nation. Of the 59 cities in the United States having a population of more than 150,000 in 1940, 19 or more suffer at times from high water. Eight of them (Springfield, Hartford, Pittsburgh, Cincinnati, Louisville, Kansas City, Denver, and Los Angeles) have serious flood hazards in highly important sections. In addition, two cities—Dayton and New Orleans—occupy land which has been protected fully from flood. Although most of the densely settled floodplains are in the Northeastern Manufacturing Belt and along the Lower Mississippi River, economically important encroachments have been made upon floodplains in all sections of the United States. For the nation as a whole, the mean annual property loss resulting from floods certainly is more than $75 million and probably exceeds $95 million. The toll in human life is approximately 83 deaths annually. For the heavy damages to health and to productive activity no measuring units are available.

Purpose and Method of Analysis

Because of the great diversity in flood conditions and in floodplains and their occupance, it is impracticable to formulate more than a few generalizations with respect to flood problems in the United States. Solutions to such problems can be developed effectively only be examining the environmental and social conditions in each locality having a flood problem. No attempt is made in this dissertation to make that kind of an examination, locality by locality. It is believed, however, that specific local problems could be appraised more fully, and that better solutions could be found for them if a broader and essentially geographical approach to the flood problem were to be adopted. Such an approach would take account of all relevant factors affecting the use of floodplains, would consider all feasible adjustments to the conditions involved, and would be practical in application.

The remainder of this chapter outlines the points of view which have dominated public action in dealing with the flood problem in the United States, and suggests a more nearly comprehensive approach meeting the foregoing requirements as to breadth and practicability. Succeeding chapters attempt to show the validity and implications of that approach. Chapter 2 defines the concepts of flood, floodplain, and floodplain occupance. Chapter 3 points out the chief factors—natural and social—which have been important in the occupation of American floodplains. The range of human adjustment to the flood hazard is described in chapter 4. Finally chapter 5 states the conclusions of the investigation and suggests ways of applying them to public policy affecting the flood problem and to geographical research.

These findings are the results of an examination of the available literature on flood problems in the United States, comprising chiefly the reports of the U.S. Corps of Engineers and the U.S. Department of Agriculture on their flood-control surveys; reports of state and

municipal engineering surveys; bulletins on floods prepared by the
U.S. Geological Survey and the U.S. Weather Bureau; geographical
studies of floodplains; and relevant statements in technical and trade
journals. They also reflect a large body of unpublished material
which the author was privileged to review while associated with the
National Resources Planning Board and its predecessors. As the
major findings began to take shape from the review of the literature
on floods, they were tested by reconnaissance studies of floodplains
selected for their diversity of conditions and lying within the Potomac, Delaware, Upper Ohio, and Los Angeles basins.

Three Public Approaches to the Flood Problem

Public action with respect to floods in the United States has emerged
from three streams of thought, each reflecting a distinct social technique, and each fostered by a separate professional group.

The engineer has approached the problem by inquiring "Is flood
protection warranted?" He has utilized levees, dams, floodways,
channel improvements, and similar engineering devices to curb flood
flows. The public welfare official has sought to determine "How
best to alleviate flood distress?" He has relied upon soup kitchens,
rescue boats, emergency grants, rehabilitation loans and like measures to cushion the social effects of flood. The property owner has
been aided somewhat by the meteorologist who, asking "When will
the next flood occur and how high will it be," has made forecasts
that enable public officials and property owners alike to evacuate
some of their goods and to prepare in other ways for the on-coming
flood. Each approach has helped to reduce flood losses and to increase the utility of floodplain resources. Each has developed fruitful
methods of coping with floods. These three approaches, either singly
or in combination, do not point, however, to solutions of the flood
problem which promise maximum use of all floodplains with minimum social costs. . . .

Summary of Prevailing National Policy

The policy declared by the Congress in the Flood Control Act of
1936, as amended, represents one segment only of the total national
policy relating to the flood problem. Taking into account all phases

of public action and inaction, the policy in essence is one of protecting the occupants of floodplains against floods, of aiding them when they suffer flood losses, and of encouraging more intensive use of floodplains. By providing plans and all or at least half of the cost of protective works, the federal government, under the policy established in 1936 and 1938, reduces the flood hazard for the present occupants and stimulates new occupants to venture into some floodplains that otherwise might have remained unsettled or sparsely settled. Even though no protection is provided or planned, the federal forecasting system tends to encourage continued use of floodplains by reducing the expectancy of loss and discomfort from flood disasters. Public relief is now so widespread that the threat of flood, while not pleasant, has lost many of its ominous qualities. If a community wishes to relocate outside of a floodplain, federal help is given to the extent that funds might otherwise have been expended on local-protective works, but if a floodplain occupant wishes to rehabilitate a relatively profitable business or desirable residence in the old location after a flood he may obtain federal aid for that purpose.

At the same time, the occupants are themselves concerned in an important degree with reducing flood losses by emergency removal and by changes in land use and structures. Except in so far as the forecasting system promotes emergency removal, the prevailing public policy is largely neutral; it neither encourages nor discourages such activities.

Obviously, the floodplains of the United States will not be permanently evacuated and returned to nature merely because of the annual bill for their occupancy, which now approaches $95 million. Neither will they be occupied as intensely as consistent with other relevant physical and cultural conditions solely because, irrespective of cost, suitable engineering and land-use devices can be developed to curb or prevent floods. No general rule can be established as to the most satisfactory arrangement of land occupance in relation to local stream regimen and floodplain conditions. In some instances, profound modifications in the stream regimen or channel have been necessary, and in other instances the cultural forms and patterns have been adjusted delicately to the earlier landscape. By and large, a fairly harmonious combination has been developed. Wherever the adjustments are not satisfactory, as attested by crippling flood losses, wherever a regressive occupance obtains, or wherever the floodplain resources are not used as fully as practicable, a readjustment may be in the public interest. This, it has been shown, is the central flood problem: how best to readjust land occupance and floodplain phenomena in harmonious relationship.

Outline of a Geographical Approach

From the three converging streams of public action with respect to the flood problem, and from corollary fields of action, such as land-use planning, we may draw an approach to this problem more comprehensive than any one of them. It is a view which considers all possible alternatives for reducing or preventing flood losses; one which assesses the suitability of flood-protective works along with measures to abate floods, to evacuate people and property before them, to minimize their damaging effects, to repair the losses caused by them, and to build up financial reserves against their coming. It is a view which takes account of all relevant benefits and costs. It analyzes the factors affecting the success of possible uses of a floodplain. It seeks to find a use of the floodplain which yields maximum returns to society with minimum social costs, and it promotes that use.

Unless the major factors affecting floodplain use are appraised, there can be no assurance that the recommended use is beneficial. Unless all possible forms of adjustment to floods are canvassed, the less expensive ones cannot be selected with certainty. Unless the analysis leads to practicable forms of readjustment, there is little purpose in examining these possibilities.

Analyses of this character have not been made in the past, and even the need for them has been stated only in general terms.

Marsh, while primarily interested in the prevention and protection phases of the flood problem, appears to have recognized these propositions in his discussion of floods in 1898.[5] McGee called attention in 1891 to several possible adjustments and noted with a tinge of pessimism that, "As population has increased, men have not only failed to devise means for suppressing or for escaping this evil, but have, with singular short-sightedness, rushed into its chosen paths."[6] Semple in 1911 described several types of riverine adjustments but did not analyze the problem of reducing flood losses.[7] Russell merely noted some of the factors affecting the occupance of floodplains.[8] J. Russell Smith called attention to the need for a different attack upon the Mississippi River problem following the flood of 1927[9] and various editorial writers[10] and public agencies[11] suggested after the

1936 and 1937 floods that a broader approach was desirable, but their suggestions have not found wide acceptance in practice. Today there are no studies or programs which meet the requirements outlined above.

This geographical approach to the flood problem appears to be more nearly national in scope, and more nearly sound from a social standpoint than the approaches which dominate prevailing public policy. The remainder of the dissertation states the evidence in support of this approach and shows its implications in public policy and in geographical research. . . .

5. Conclusions

In the light of the preceding evidence relating to possible adjustments to floods and to factors affecting those adjustments, the geographical approach to the flood problem which was outlined briefly in chapter 1 may now be reconsidered with a view to making it more specific and to indicating its relationship to prevailing public policy and geographical research.

Essentials of a Sound Approach to the Flood Problem

If the resources of the floodplains of the United States are to be used in the public good so as to yield maximum returns to the nation with minimum possible social costs, it seems clear that action affecting their continued occupance will be based upon four essentials.

1. *It will take account of all possible adjustments which might be made to the flood hazard.* At least eight forms of adjustment have been tried successfully. Singly or in combination, they offer lines of readjustment where present occupance has been unsuccessful, or where the flood hazard is unduly costly under existing conditions.

Land elevation provides a permanent means of escape from floods at relatively high construction costs. It is impracticable for densely-settled areas, but may be suited to new urban developments, to strategic sections of highways and railways, and to isolated residential, commercial, and manufactural occupance in sparsely-settled floodplains.

Flood abatement by means of erosion control, forest-fire control, forest planting, and relating methods of land improvement and management in areas upstream from a floodplain affords the possibility of reducing the magnitude of floods in a few sections. The complexity of the hydrologic factors involved and the scarcity of detailed field observations make it impracticable to generalize as to the opportunities for reducing the frequency of floods, checking the movement of debris, curbing bank erosion, and reducing highway erosion by these means in all parts of the country, but in a few areas such opportunities have been shown to exist. Experience to date suggests that land-management measures are the most feasible remedy for flood losses in a few places, and that such measures are a corollary of engineering works wherever heavy flows of debris threaten the life of the works. Their major benefits accrue to the owners and operators of the land on which the improvements are made, however, and such programs in the interest of reducing flood losses should be considered also as a part of integrated programming for all relevant phases of rural land use.

Flood protection by levees and floodwalls, channel improvements, diversions, and reservoirs is the most reliable and, in many instances, the easiest means of reducing flood losses. In balancing the costs of protective works against the benefits expected from them, it should be recognized that expenditures greater than engineering costs may be involved in the disturbance of urban land use through levee and floodwall construction, of the populations and industries of reservoir areas, and of the occupants of floodway areas, and in deleterious effects which the works may have upon erosion and sedimentation processes in the floodplain and stream channel involved. Levees and floodwalls carry a special disadvantage; if overtopped by a flow greater than the design flood, the maximum loss occurs. The other three types of works result in only partial loss when their designed capacity is exceeded. In the light of meager evidence it seems possible that reservoirs and channel improvements, unless supplemented by land use measures, may induce or promote further encroachment upon a floodplain, and so may increase rather than decrease mean annual losses. Sound evaluation of flood-protective works requires appraisal of these possible costs, but also appraisal on a consistent basis of all benefits, in addition to those resulting from prevention of flood loss, which are involved in enhancing the productivity of the area to be protected, in improving public facilities through the protective works themselves, in training laborers on the projects, and in stimulating better land use. Just as amelioration of flood conditions through land-use practices can be evaluated effectively only in conjunction with broader programs for land improvement and management, so flood-protective works, if fully effective, must be planned with an eye to the needs and possibilities of water control and use in the same drainage area for other purposes, such as navigation, irrigation, power, and pollution abatement.

Emergency measures may reduce greatly the impact of floods if there are accurate, timely forecasts of their occurrence and height, if efficient plans for emergency action have been prepared, and if the persons affected know the plans sufficiently to act promptly. Properly-organized emergency removal measures have proved effective in evacuating large populations with a minimum of discomfort and distress, and with no serious effects upon public health. They also have been used and can be used to prevent losses of movable property, such as furniture, store furnishings, stored goods, and machinery, items which in urban areas account for a large proportion of the total property losses. Flood-fighting measures, ranging from emergency levees and bulkheads to coating parts of immovable machinery with protective oil, can serve well to decrease damages. Rescheduling operations by manufacturers, transportation companies, and public utility agencies can assist materially in maintaining essential services and in minimizing production losses. It is believed that in most urban areas the mean annual flood losses could be reduced at least 15 percent, and, under favorable circumstances, as much as 50 percent by these emergency measures. They have been adopted by only some of the public utilities and large manufacturers, and they are generally not practiced or even understood by small property owners and by occupants of upper flood zones.

Structural adjustments may be used to good advantage to prevent, or reduce losses of valuable property, interruption of essential public services, and scouring of farm land. Without attempting to provide protection for an entire area subject to floods, changes in building design, building layout, communication lines, street grades, and the like can be made while previous flood losses are being repaired, and can be executed as a part of regular replacement and maintenance operations. Such measures, in conjunction with emergency flood-fighting and rescheduling measures, can minimize, or even eliminate, public-utility interruptions in urban areas, and they can reduce materially losses to buildings and lands in rural areas.

Land use readjustment can largely prevent those losses in agricultural areas which accrue to property and crops that do not depend upon special advantages of floodplain location, and also can curb unsound urban occupance of undeveloped land. The chief deterrent, particularly in urban areas, lies in obtaining group action by land owners in readjusting the uses. Such readjustment therefore depends for its effectiveness upon public subsidy of urban relocation, public subsidy of property abandonment, public acquisition of land, and public land-use regulation. Zoning has been an effective means of preventing further impairment of channel capacity through human encroachment, and it promises aid in promoting improved land use in floodplains.

Public relief will remain a necessity in cushioning the social impacts of floods so long as other adjustments are not adopted. It has come to be well organized under federal auspices.

Insurance against flood losses has failed under private management in the United States, but it is a measure which probably would be practicable if national coverage and guarantees against catastrophic losses were to be provided during the early years of operation. Once in operation, it would allow systematic indemnification of losses, and an inspection service which would promote the adoption in unprotected areas of emergency measures and of structural and land-use readjustments.

2. *In comparing possible adjustments for a given area, the benefits and costs of each adjustment will be evaluated on a consistent basis which recognizes all costs of appropriate remedial action, and which considers benefits in terms of the welfare of the entire community affected.* Such comparison involves various costs and benefits to which precise monetary values cannot be assigned, but which must be given judicious weight in the comparison in order to prevent the unwarranted dominance of other items in deciding upon desirable lines of action. It is believed that there has been a general tendency to place undue emphasis upon hazards to life and health, and to assume without much foundation that production losses from flood were large in proportion to property losses. Because these components of costs and benefits are complex and difficult to measure, caution should be exercised in any attempt to express the feasibility of a given adjustment as a simple ratio of measured costs to measured benefits. Such a ratio may be more misleading than helpful unless all available data have been properly evaluated.

3. *Any action will seek to take full account of all factors affecting the success of the occupance which is possible under the various adjustments or readjustments.* In so doing, it will recognize: (a) that location upon a floodplain may be essential to certain types of land occupance, such as generating stations, because of special factors

of slope and contour, soil, surface water, groundwater, and corridor facilities; (b) that for many types of land occupance, some floodplains afford marked advantages of location which do not, however, necessarily outweigh the disadvantages of flood hazard; (c) that much occupance has developed on floodplains in association with other occupance depending upon floodplain features, but without itself being related directly to such features; (d) that the present occupance of many floodplains reflects an earlier adjustment to advantageous factors which, because of technological changes, are no longer significant; (e) that many alluvial floodplains afford opportunities for maintaining a permanent agriculture on soil free from erosion and subject to replenishment by natural means.

4. *Any action will promote adjustments or readjustments that favor the type or types of land occupance most likely to contribute to effective use of floodplain resources.* Some considerations that might affect a choice in terms of this principle of action are the following: (a) all possible adjustments except those in land use and insurance tend to favor the preservation of existing land occupance; (b) public relief favors further encroachment upon a floodplain by bearing part of the costs of such encroachment; (c) effective emergency measures also may favor encroachment in less degree by reducing the hazard of flood loss; (d) encroachment upon floodplains is likely to continue so long as the riparian doctrine is not modified by public regulation; (e) insurance and structural adjustments, by requiring a property owner to make some payment for the advantages of floodplain location which he enjoys, stimulate the abandonment or movement of occupance that is not profitable; (f) flood abatement, flood protection, and public relief, by placing upon public agencies the major burden for reduction of losses, encourage the occupants of floodplains to seek those adjustments at public expense even though other adjustments at private expense might be less costly and more effective from the standpoint of the nation.

Present Public Policy

Prevailing public policy in the United States falls short of the foregoing four essentials in several respects. Under present legislative directives the federal government's concern with reducing flood losses is limited to flood protection, flood abatement by land management, certain types of emergency measures, public relief, and relocation of a community if it can be accomplished at a cost less than that of protection. Surveys of the flood problem by federal agencies are directed at flood protection and flood abatement primarily, and a large and outstanding engineering organization has been developed for that purpose. The forecasting system maintained by the Weather Bureau is unsatisfactory in its coverage of small drainage areas and its dissemination of forecasts for large drainage areas. Hence opportunities for stimulating emergency measures, structural adjustments, land-use changes, and insurance are largely lost. So long as present surveying and forecasting methods are not expanded, the potentialities of such adjustments will remain undeveloped.

In estimating the limit of feasibility for flood-protective works, no consistent basis is in use for evaluating benefits and costs. Unless some agreement is reached among the agencies responsible for flood investigations or unless a suitable congressional policy is adopted, important classes of costs and benefits will be overlooked or will be compared in a misleading manner.

None of the present survey procedures for flood protection takes adequate account of the factors which, in floodplain occupance, have been advantageous or disadvantageous for the present uses. As a consequence, it cannot be said with confidence that present federal activities have the net effect of promoting sound land use. Public relief, flood abatement, and flood protection favor the retention of present occupance, whether desirable or undesirable. With little or no attention to the desirability of such occupance, present policy helps to stabilize uneconomic occupance. Works for protection and abatement minimize the flood hazard at public expense for the most part, and in the process provide increments in land value to landowners, some of whom occupy the floodplain on a highly speculative basis. Heavy losses by small property owners are largely subsidized by the government. Further encroachment is thereby tacitly encouraged, and Congress has not yet seen fit to require regulation of encroachments as a condition of federal participation in flood protection. In effect, the national treasury bears a large part of the costs of those who prefer to live on floodplains, and does so without inquiring as to whether or not such plains afford any pronounced advantages for such occupance.

In areas where neither protection nor prevention is found to be economically feasible, the federal government assumes no responsibility for fostering types of adjustment other than public relief. It goes to great lengths to provide protection from floods if the cost-benefit ratio is favorable, but contents itself with merely helping to relieve and rehabilitate flood sufferers in areas where the ratio is unfavorable. Inasmuch as the cost-benefit ratio is no index of the economic vigor of a community, protection is given to some towns that are definitely decadent, while it is withheld from some areas where development, though recent, is highly promising. Indeed, the more vigorous types of occupance which are highly dependent upon floodplain locations, such as power plants, water works and railways, have been leaders in developing emergency measures and structural adjustments, quite independently of publicly-financed protection measures. Their successful activities have attracted, in many instances, subsidiary residential and commercial occupance which has lacked the foresight and skill to make such adjustments, and which has been more dependent, therefore, upon public aid.

On the whole, present policy fosters an increasing dependence by individuals and local governments upon the federal government for leadership and financial support in dealing with the flood problem. While encouraging solicitation of further federal aid and the establishment of types of occupance requiring such aid, the policy does not help or stimulate beneficiaries to explore the possibilities of making other adjustments with a view to promoting the most effective use of floodplain resources.

Needed Geographical Contributions

In arriving at these conclusions concerning the characteristics of a sound approach to the flood problem, a general theory as to the factors affecting human adjustment to floods has been stated tentatively. This, it is believed, deserves more detailed examination and experimental application in the light of field studies. Meanwhile, it seems to provide a useful frame of reference for two lines of geographical research which are needed as contributions to the solution of the flood problem in the United States.

First, the development of new and improved adjustments to floods would be promoted by intensive studies of adjustments in representative areas here and in other countries, studies designed to identify the conspicuously successful and unsuccessful adjustments for each important type of floodplain and of floodplain occupance. Second, it would be helpful to have studies of the present and prospective importance of the various factors affecting adjustments to floods in areas where protective works are under consideration and particularly in areas where protection has not been deemed to be justified.

If the floodplains of the United States are to be developed progressively so as to utilize as fully as practicable the advantages afforded by them, and to minimize their disadvantages, it will be necessary to adopt a broad geographical approach of the type outlined in the preceding pages. That approach will demand an integration of engineering, geographic, economic, and related techniques. The solutions will not involve a single line of public or private action but will call for a combination of all eight types of adjustments, judiciously selected with a view to the most effective use of floodplains.

A Unified National Program for Managing Flood Losses

The nation needs a broader and more unified national program for managing flood losses. Flood protection has been immensely helpful in many parts of the country—and must be continued. Beyond this, additional tools and integrated policies are required to promote sound and economic development of the floodplains.

Despite substantial efforts, flood losses are mounting and uneconomic uses of the nation's floodplains are inadvertently encouraged. The country is faced with a continuing sequence of losses, protection and more losses. While flood protection of existing property should receive public support, supplemental measures should assure that future developments in the floodplains yield benefits in excess of their costs to the nation. This would require a new set of initiatives by established federal agencies with the aid of state agencies to stimulate and support sound planning at the local government and citizen level.

(1966d), 1

Notes

1. Gilbert F. White, *Human Adjustment to Floods,* Department of Geography Research Paper no. 29 (Chicago: University of Chicago, 1945), chap. 8, vii.

2. For a detailed accounting of these beginnings and opportunities, see below, vol. II, chap. 2 (Platt).

3. Arthur Maass, *Muddy Waters: The Army Engineers and the Nation's Rivers* (Cambridge: Harvard University Press, 1951).

4. William G. Hoyt and Walter B. Langbein, *Floods* (Princeton: Princeton University Press, 1954).

5. George P. Marsh, *The Earth as Modified by Human Action— A Last Revision of "Man and Nature"* (New York: Charles Scribners Sons, 1898), 472–74, 498.

6. W. J. McGee, "The Flood Plains of Rivers," *Forum* 11 (1891):221–34.

7. Ellen Churchill Semple, *Influence of Geographic Environment* (New York: Henry Holt and Company, 1911), 322–27, 363–70.

8. I. C. Russell, *Rivers of North America* (New York: G. P. Putman's Sons, 1898), 114.

9. J. Russell Smith, "Plan or Perish," *Survey* 58 (1927):370–77.

10. "A Modest Proposal for Flood Control," *New Republic,* 19 May 1937, 34.

11. New York, Division of State Planning, *A Common Sense View of the Flood Problem,* Bulletin no. 28 (Albany, May, 1937). A similar view is taken by Allen Hazen in *Flood Flows: A Study of Frequencies and Magnitudes* (New York: John Wiley and Sons, 1930), 2–3, 177–79.

3 A New Stage in Resources History

Change and stasis are recurrent themes in art and science—one favors evolutionary progress and the other favors a state of equilibrium. White employs both in his thinking and writing, but overall he is committed to change. There is an essential optimism in his perception of time. Looking backward, it provides insight by measuring a resource-use trend, a social practice, a pattern of human occupance. Looking forward, it provides opportunities to change an undesirable trend, innovate a useful behavior, develop human occupance consonant with nature. Most prized of all, for him, is the "turning point," or hinge of history, when opportunity is maximized.

This modest paper, given as a speech to Michigan business executives, but also published to reach soil conservationists as well, illustrates White's tempered and temporal stance. It is contemporaneous with Vogt's *The Road to Survival*,[1] and predates slightly Osborn's *The Limits of the Earth*[2] and Brown's *The Challenge of Man's Future*,[3] but specifically eschews such neo-Malthusian statements. Instead it begins with the documentation of past trends and future perspectives found in three landmark conferences and commissions: the United Nations Scientific Conference on Conservation of Resources held at Lake Success, New York, in 1949; the President's Water Resources Policy Commission (PWRPC) of 1950; and the President's Materials Policy Commission of 1951.[4] White had intimate knowledge of all three activities, and personal experience with two as a member of the PWRPC and chairman of the river basin development discussion at Lake Success, but he does not indicate his role in the paper. The documentation stands by itself.

This turning point in resources history is characterized in typical White style in a single key paragraph. The pattern of full water development, frozen as of 1950, was substantially completed by 1970, as White had predicted in 1953. Materials demand has turned sharply upward, real production costs after decades of decrease have leveled off, and the United States is now a net importer of materials

Condensed from *Journal of Soil and Water Conservation* 8, no. 5 (September 1953): 228–32, 248. © 1953, Soil Conservation Society of America.

with new responsibilities in developing nations. At the same time there are major opportunities to consider nonrenewable materials in their entirety and to administer renewable resources for many purposes. This emphasis on a new stage of rational, economic use of resources at home and humane assistance abroad (even admidst the intemperate cold war designations) was to emerge as the dominant theme of resource management for the next twenty years.

The past fifteen years have seen a momentous change in the resources situation in the United States. Domestic resources development has entered a crucial stage in many areas. Long-run trends in the production of basic materials have been reversed. The United States position in the world market has altered rapidly and radically. New ideas as to resources use and management have begun to replace ideas which have prevailed for half a century.

The combination of these events—some the result of advancing technology, some the result of a chaotic world situation, some the product of widely increasing pressure of population upon the land— gives the United States a new prospect for the wise development and use of its natural resources. This picture is extremely complex, but a few major elements can be seen clearly. At the risk of oversimplifying, it is possible to point out a number of the events that have combined to create this new prospect. Most of the relevant facts are presented in a series of reports issued during recent years; the President's Materials Policy Commission (Paley Report),[5] the President's Water Resources Policy Commission,[6] and the proceedings of the United Nations Scientific Conference on Conservation of Resources.[7] These draw together a great deal of previously dispersed thinking and statistics. Some of their estimates are rough guesses at best, and many of the recommendations in them are highly controversial, but there appears to be wide agreement on a few of the facts.

From them an outline of future development begins to emerge which has a special relevance for representatives of business, industry, and finance. In the new picture business groups have, it would seem, a more direct stake and a clearer responsibility for action than in earlier decades. No doubt the very fact that this conference is taking place this week indicates a recognition by business groups of changes that have already occurred.

The prospect which we now face is not as discouraging as that presented in some of the recent conservation literature which ominously maintains that the United States itself already is overpopulated in terms of the available resources, and that other densely settled areas will continue to be explosively over-populated unless drastic measures are taken to curb population. At the same time the evidence does not seem to support the assertion by some business leaders that given freedom of research and enterprise, and stable investment possibilities, there is no need to fear any shortage of materials. It seems neither as hopeless nor as hopeful as these views would suggest.

Freezing the Pattern of Water Development

One recommendation in the report of the President's Water Resources Policy Commission was that new water projects in the United States should be undertaken only as they are clearly shown to be in conformity with drainage basin plans that have been restudied and approved. The Commission found that the cost of federal projects under construction or authorized was equal to the entire cost of all federal projects constructed in the preceeding one hundred and sixty years. Projects planned but not yet authorized would have a minimum cost at least four times as great.

Many of these projects had been planned without adequate basic data on water, land, and mineral resources. Most had taken shape while the experience from the pioneering efforts on the Columbia, the Missouri and the Tennessee was still undigested. It seemed desirable to take a fresh look at the schemes for development of other basins before letting them reach the construction stage.

There is a sobering finality in river basin development: once a major construction plan is undertaken, little can be done to change the pattern of water use which it imposes upon the surrounding area. A dam cannot be moved and there are relatively few suitable damsites. Irrigation projects cannot be shifted merely because the soil turns out to be unproductive. Once a commitment is made to protect a city from floods it is extremely difficult to withdraw from doing whatever is essential, regardless of costs.

In its river development the United States has entered into a period which is similar to the great transcontinental railroad building period which followed the Civil War. Heavy investment is being made in new dams, power plants, navigation facilities, and levees which when completed will fix the main outlines of economic development for decades to follow. When these works are done, the pattern of water development in the United States will be frozen as surely as the pattern of transcontinental freight traffic was frozen by the turn of the century. There still will be opportunity for competing facilities, and for extensions and changes in the pattern, but the major outlines will remain.

Only a few basins such as the Tennessee are fully developed with river-regulating works. There still is time to make changes in the present plans of the other basins if it seems desirable. The time is short. Each year sees new plans authorized and new commitments made. At the present rate of construction the next twenty years will see the major part of the job done. Students entering into business and professional life this summer will witness in their active lifetimes the completion of large-scale water development as we now envisage it. Whether or not that development is sound will depend upon review and appraisal during the years immediately ahead.

For example, the present federal policy for dealing with flood damages lays heavy stress upon engineering works to curb and delay flood flows. Minor attention is given to land treatment upstream, and little or no attention to such alternative means of reducing flood losses as emergency evacuation, relocation, readjustments in land use, and zoning. If these alternatives are ignored for the next few years the federal commitment to protect people who own property or live in the major floodplains will be so heavy that there will be little point in considering cheaper or more effective action.

Accelerated Increases in Demand

The past decade also has brought tremendous increases in the demand for certain raw materials. In the Paley Report an effort is made to forecast the consumption of raw materials in the decade 1970–80, choosing a period of approximately 25 years in which the current defense production problems hopefully will become less dominant and in which many technological shifts probably can be forecast.

If it is assumed that by 1975 the total population will increase to 193 million and that total national output will be twice the 1950 level, and if it is assumed that the relative prices of various materials remain unchanged from 1950, projections can be made as to the demand for basic raw materials. These projections by the Commission show that considerably less than a doubling of total materials input will be needed to support a doubling of national output. The Commission believes that a 50–60 percent increase in total materials production will support a doubled national production.

The demand for all materials is expected to rise, but the rate of increase for some materials will far exceed others. The Commission published estimates as to prospective changes which are highly controversial and have been widely challenged, but a few of them may be quoted here to indicate the magnitude of development anticipated by the Commission. By comparison with 1950 consumption, tin demand is expected to increase 18 percent and iron ore 54 percent. At the other extreme, among the minerals, petroleum is expected to increase more than 100 percent and aluminum more than 291 percent. For products of the land, excepting forests, an increase of 40 percent in demand is anticipated. Timber product demand is expected to increase by about 10 percent with major changes taking place among timber uses. Obviously, a great claim is likely to be made upon all of our natural resources during the decades immediately ahead.

Water supply illustrates one of the materials which has been considered abundant but which may be in short supply over large areas by 1975 unless prompt measures are taken. Consumptive use of water for municipal, industrial and irrigation purposes has mounted at a rapid rate and promises to continue to rise. Approximately 185 billion gallons were used daily for those purposes in 1950. Within the next three decades this use is expected to mount to 350 billion gallons daily. The greater part of that 90 percent increase is expected to take place in the industrial field. There the trend is not alone to use more water but to make more exacting requirements of water free from organic and inorganic impurities, as shown in the recent report by the Conservation Foundation.[8] Demands upon groundwater have tripled since 1935. This trend will continue so long as pollution of both surface and underground sources is not curbed more effectively. It must be asked, however, whether the surface and ground supplies can meet the prospective demand. It is common knowledge that there now are large sections of the United States in which certain types of industrial development are impracticable because of the inadequacy of existing water systems.

In the case of water, as with the other less abundant materials, it is difficult to offer any simple solution for meeting the forecast demand. The size of known reserves, the cost of substitutes and improvements, the possibility of technological improvements, and the availability of the same or substitute materials outside the United States must be taken into account.

Net Producer to Net Consumer

In 1950 the United States, with 9.5 percent of the free world's population and 8 percent of its land area, was consuming half the volume of the free world's production of materials. It had arrived at this position by making huge increases in its use of materials: in the preceding 50 years bituminous coal consumption had increased 2½ times, iron ore 3½ times, zinc 4 times, and petroleum 30 times, as compared with 1900. Minerals as a group increased much more rapidly than either agricultural or forest products.

The effect of this extraordinary growth in materials use was to change the United States trade relations with the rest of the world in at least two important respects. *First,* the country, after having had a production surplus of 15 percent in 1900, had developed a production deficit of 9 percent in mid-century. After having been a net world exporter of copper, petroleum, zinc and lumber, during the late 30s or 40s the United States became a net world importer of those and other key materials.

Second, the United States was using up its known reserves of such materials at a much higher rate than other countries in the free world.

The result of this great period of use of natural resources was to make the United States the dominant user in the free world market, to change it from being an exporter to being an importer, and to use up its reserves more rapidly than the other countries. Paley Commission projections of possible production and demand for 1975 suggest that this shift will continue, and that by that time the annual production deficit will be in the neighborhood of 20 percent.

Reversal of Real Costs

Even though the materials consumption sky-rocketed during the first half of the century, the real cost of producing most of the materials declined. That is, the hours of human work and the volume of capital required to bring a unit of material into use decreased. The Paley Commission tried to estimate these "real costs," expressing them so as to eliminate the major effects of changes in current dollar values. Thus, labor time required to mine a ton of bituminous coal declined from three hours and twenty minutes in 1903 to one hour and twenty minutes in 1950.

This decline in most of the materials industries was made possible by new technology, by heavy investments of capital and energy, and by drawing upon the cheaper sources of supply. In general, the supplies of minerals and land most fully exploited were those which were available at the lowest cost. All through the early decades of the century the cost of materials in relation to general price levels tended to diminish. In 1900, "each dollar of raw materials cost supported $4.20 worth of finished goods and services; by 1950 the raw materials dollar was supporting $7.80 worth (after discounting for changes in the general price level)."[9] We were skimming the cream and it was both rich and cheap.

Then in the 1940s, the trend appears to have reversed. The metals, fuels and agricultural industries began to experience a change which long before had set in with the forest products industries: real costs began to rise or to level off. The chemical industry was the only major one in which real costs continued to decline rapidly. Part of the cause for this otherwise general reversal in trend is laid by the Commission at the door of slowness in readjusting to changes in market conditions following the war. Much of the cause lies in the increasing demand for production from dwindling and high cost supplies.

If, as seems likely, this upward trend continues, we will witness a strategic change in the natural resources situation in the United States. The fact that forest products have suffered from rising real costs since the beginning of the century may help explain the heavy emphasis which foresters in both government and private industry long have placed upon conservation practices.

A New World Position

In seeking to find the right course of action in meeting the prospective demands in each resources field, there is a wide range of choice. In general, the Paley Commission seeks to suggest ways consistent with the methods of free enterprise, the profit motive, and the price system, but it also suggests helps and restraints from government to keep the system working well. Some of the measures considered include promoting more mineral exploration, preventing waste in use, developing lower-quality reserves, encouraging sustained-yield management of renewable resources, and perfecting substitute materials. Possible ways of reducing or shifting demand also are reviewed. Another major opportunity lies in the direction of imports from overseas producers.

Here, account should be taken of a wide shift in the United States position. At the same time that the United States has attained a position of world political dominance it has become more dependent than ever before upon other countries for raw materials, and it has assumed heavy moral reponsibilities for resources development in other countries. Part of this new responsibility stems from American investment in producing facilities such as the oil refineries in Saudi Arabia. Private investment thus far has been relatively small because of the uncertainty of world markets and because of the political instability of many of the nations involved.

The larger part of the new moral responsibility attaches to government support of both United Nations and bilateral technical assistance work in underdeveloped countries. Those nations are undergoing profound economic and social revolutions leading to radically different forms of both production and consumption. Much of the Point Four work relates to the spread of technology for resources use, and involves the United States in efforts to improve production methods, open up new resources, promote the public health, and introduce new consuming habits.

Often public support of technical assistance in underdeveloped areas is justified either as a necessity of military strategy or as an essential feature in the program to restrict and turn back communism. Both arguments now find rather wide public support.

There are, however, two other grounds on which vigorous American action in that direction may be expected even though there happily might be an unexpected relaxation of the international political tensions and in the accompanying military preparations. So long as the United States faces an increasing dependence upon overseas areas for basic materials—and self-sufficiency seems both economically and physically impossible—it will be involved in the maintenance and stability of those areas. Much more basically, so long as the United States wishes to act in a spirit of Christian and democratic friendship toward its fellow nations it must be prepared to share in their efforts to improve their own positions.

This commitment has far-reaching implications for our current resources policies. For example, the newly independent countries of Southeast Asia are joined in a program for mutual economic development known as the Colombo Plan. Expenditures of approximately five and one-half billion dollars over a six-year period are contemplated to bring 13 million acres of land under cultivation, to increase the production of food grains by 6 million tons, and to increase electric generating capacity by 1,100,000 kilowatts.[10] It is not certain that these results will be obtained. At best, with a completely successful program, the net will be to maintain the status quo in the Southeast Asian area. Population is increasing so rapidly that improvements of this size will only serve to support the new population at the present low standards. Serious famines will remain a hazard. The situation is not hopeless, but drastic efforts on the one hand to curb population and on the other hand to build industrial and crop production wil be required for a long time to come. As a minimum this will demand contributions of American equipment and personnel. As a maximum it may demand foodstuffs and producers' goods to help carry the area over crisis periods.

Industrial Analysis

Out of the recent studies that have been made of resources problems has come appreciation for a relatively new kind of thinking as to lines of constructive action. More and more, the problems of an industry are being viewed as a whole, with attention to ways of coordinating different groups of operators.

The forest products industry is a case in point. From the Paley Report it may be learned that the principal problem in looking to meeting future timber demands is in promoting sustained-yield management by the small landowners who hold more than half of the total commercial timber area but who for the most part manage poorly if at all. The large private owners and the public agencies are responsible for the best management. Increasingly, the proper management of the large tracts requires cooperation between public and large private owners, and for the welfare of the industry as a whole the smaller owners must somehow be related to the large operations if supplies are to be used cheaply and efficiently. In dealing with forests, the Paley Report summarizes the results of previous surveys, but it pioneered in dealing with some of the other materials industries in a unified fashion.

It is interesting that the report dealt with energy sources as a group, recognizing the high degree of dependence there is among the coal, oil, gas and hydrosectors of the industry in meeting total energy needs. The Paley Report is bound to stand out as a landmark in thinking about resources problems because of its relatively successful effort to deal with total demand and total supply for each of the basic raw materials. A new type of analysis is established for industries as a whole.

Multiple-use Administration

The past twenty years have marked very wide recognition of the multiple uses which resources may have. In Michigan there has been an effort over a much longer period of time to show the close inter-relation of agriculture, forestry, wildlife management, recreation and minerals use in areas such as those adjoining this meeting place. Increasingly, within local areas resources use is seen as multiple-use.

There have been numerous attempts to translate this concept of multiple-use into administration of government or cooperative enterprises for resources management. By and large, the federal government has failed in its attempts. With the exception of the Tennessee Valley Authority, which has no clear local roots and which operates largely independently of other federal agencies, the tendency has been for each department to go its own way. Efforts have been made by individual departments, such as Agriculture and Interior, to co-ordinate their own bureaus' activities within certain regions, and there is an informal inter-agency committee on river basin development, but no thoroughly satisfactory devices have been found thus far.

The soil conservation districts which have been organized under state law with local responsibility probably have proved the most effective agencies in helping private operators deal intelligently with resources problems. A number of watershed associations, of which the Brandywine is the most publicized, are being organized and are beginning to find their halting way in coordinating public and private work. In the water field there has been strong support for local administration of new federal projects but very little interest in the suggestions made by the President's Water Resources Policy Commission that local beneficiaries should bear a larger share of the construction cost. Sharing of cost and sharing of responsibility go hand in hand.

It appears that new administrative organization is likely to unfold in the years immediately ahead in order to assure the kind of wise multiple-use which now is an ideal rather than a reality. The local efforts, such as the soil conservation districts, emphasize authority resting in local owners, township commissioners, county boards and the like. The new emphasis on coordinated industrial programs in the Paley Report attacks the problem from the standpoint of national goals and responsibilities. Both are necessary and are antidotes for undue emphasis on the other. The local administrative agency does embody, however, a point of view which seems absolutely essential as this country looks forward to enlarged production in a radically different situation than that which prevailed 20 years ago. This is the view that our central concern should be with the quality of individual and family life enjoyed by the people who use the resources.

Emerging Responsibilities of Private Enterprise

These events add up to a formidable and challenging combination of circumstances. They suggest that this country is at a major turning point in its resources history. The pattern of its full water development is beginning to freeze. Demand for many basic materials, including such presumably abundant materials as water, is turning sharply upward. From being a net producer of raw materials the United States has become a net consumer. At the same time the real costs of production have, after a long period of persistent decline, begun to level off or increase. In its new position as a world leader the United States has assumed heavy responsibilities for promoting and cushioning the shocks of profound changes in resources use in underdeveloped and heavily populated countries overseas. A new concept of the coordination of demand and supply factors in basic materials industries has begun to take hold. New administrative devices to deal with genuine multiple-purpose use of resources are being tried on a broad scale.

It could be fairly said two decades ago that while a representative of a large business enterprise might feel a responsibility as a citizen to prevent undue waste in resources use he could show very little reason for committing his business to a conservation effort. The future was too promising: demand did not exceed supply; costs were decreasing; overseas supplies offered opportunities rather than responsibilities. It was left chiefly to the government and to civic groups to deal with local emergencies of resources exhaustion or to attempt to probe the future.

This has changed. Today, a good case can be and is being made for direct participation in conservation work by business. Rather than working against government it means working with and through local, state and federal agencies. . . .

As the Rich Grow Richer

We can regard the current outlook as unique in human history in terms of the opportunities it offers. Only in the last decade has technology reached a point where we can confidently assert that if present knowledge were properly applied the whole human race, even if greatly enlarged, could be fed, clothed, and housed at minimum standards. Only in the past five years have the techniques of birth control permitted us to think of practicable methods of family planning being available at choice to all the world's population—and the techniques promise further rapid change. The United Nations has survived, albeit badly scarred, the political batterings of more than twenty years, and now provides a framework through its specialized agencies for multilateral efforts at both economic development and environmental preservation. Worldwide, simultaneous distribution of good news, bad news, violence, and love is close to being reality as the radio set and TV reach the remotest communities. Notwithstanding widespread outbreaks of tribal and national violence, there is official expression and popular support, particularly among young people, for nonviolent methods on an unprecedented scale. For the first time in human history, it can be said that the technical methods and rudimentary social organization for dealing with resource use and preservation in the face of rapidly expanding population, for at least a few more decades, are in hand.

Can the leading nations, whatever their size, grasp the opportunities presented by this situation with the same enthusiasm that the United States and the Soviet Union have embraced the opportunity to get a man to the moon and back? Were expenditures of the order of 8 billion dollars annually—double the current U.S. outlays for its space exploration—to be authorized over a suitably slow development period, the effects might be momentous. But more important than money would be the sense of political and intellectual commitment in such an undertaking. Given its immense complexity, it would require not only international activity on a whole order of magnitude greater than the United Nations Development Programme and the major bilateral programs, but an enlistment and encouragement of imaginative innovation and investigation by countless individuals and private groups. It would take more than an international survival year or an international development decade. This unified effort seems essential, unlikely, and not quite impossible.

Commencement Address, Earlham College,
Richmond, Indiana, January 20, 1969, 10–11

Notes

1. William Vogt, *The Road to Survival* (New York: William Sloane Associates, 1948).
2. Fairfield Osborn, *The Limits of the Earth* (Boston: Little, Brown and Company, 1953).
3. Harrison Brown, *The Challenge of Man's Future* (New York: Viking Press, 1954).
4. For these and other integrative studies of natural resources, see below, vol. II, chap. 3 (Clawson).
5. *Resources for Freedom,* a report by the President's Materials Policy Commission, 5 vols. (Washington, D.C.: Government Printing Office, 1952).
6. *A Water Policy for the American People,* a report by the President's Water Resources Policy Commission, 3 vols. (Washington, D.C.: Government Printing Office, 1950).
7. *Proceedings of the United Nations Scientific Conference on the Conservation and Utilization of Resources,* 7 vols. and index vol. (New York: United Nations, 1950).
8. *Water in Industry* (New York: National Association of Manufacturers and the Conservation Foundation, 1950).
9. *Resources for Freedom,* vol. 1, 7.
10. "Building Leadership for Peace," *New York Herald-Tribune Forum* (1952), 15.

4 A Perspective of River Basin Development

Dividing the world has never been an easy task, but there has always been one easy way to define a region—an irregular line can be drawn anywhere in the world on the basis of the direction a drop of water flows. In selection 3 of this volume, White spoke of the "sobering finality in river basin development." Today, at least in the United States, that sobering finality is a reality. It is now nine years since the last major dam structure has been authorized, a comparatively less known hydrological equivalent to the current de facto nuclear reactor moratorium.

White's career paralleled the rise of intense interest in the river basin as a unit for integrated land and water planning. During his years at the National Resources Planning Board, this interest spread beyond the Tennessee Valley to other reigons. It was widely supported by state planning agencies and less so by the competitive federal agencies. While President of Haverford College, through service on the Hoover Commission task forces in 1949, and as vice-chairman of the President's Commission on Water Resources Policy in 1950, White monitored the postwar surge in large dam construction.

Written soon after White went to the University of Chicago, and while serving as chairman of the United Nations Committee on Integrated River Basin Development, this paper taps a reflective vein. In the longest paper in this volume, White explores the historical roots of river basin development and how the concept spread across state and national boundaries. The piece foreshadows his influential work in the 1960s on man-made lakes in developing countries (see selection no. 17). The paper concludes with another familiar plea of his, the need for continuing appraisal—audit—of this and the other social and technological experiments in resource development.[1]

Reprinted from *Law and Contemporary Problems* 22, no. 2 (Spring 1957): 157–84. © 1957, Duke University School of Law. The author is indebted to Edward A. Ackerman for helpful criticism of the manuscript.

I. Introduction

The river systems of the world flow today with only a small proportion of their total volume harnessed and applied for human good. With the exception of a few small drainage basins in arid regions, the water of no stream has been fully regulated or used. There are physical limits to such regulation and use, but the degree to which those limits are approached is related to conditions which are partly technological, partly economic, partly political, and partly ethical.

Using the concept of integrated river basin development, each major network of streams draining the land masses of the earth may be viewed as the backbone for a possible planned use of a unified system of multiple-purpose and related projects to promote regional growth. This view of river basin development has come, during the past sixty years, to be employed rather widely as a technical tool for achieving social change. It has found imaginative support, and it appears to be on the threshold of wider application. How much further it wisely may be applied would seem to depend, in part, upon the sharpening of our knowledge as to its utility and implications as a tool. Like any tool, it is not inherently good. Its value must be judged in terms of the growth and changes it can effect and upon its flexibility and precision.

The concept of river basin development is used here to mean three component ideas having separate roots in western civilization but coming to be associated with each other in present-day theory and practice.

A. Limits and Degrees of Development

Although the Nile and Tigris-Euphrates Valleys cradled the early civilizations of the eastern Mediterranean and still support their basic irrigation agriculture, and although the flows of the Rhine, the Ohio, and the Thames long have been essential to the industrialized populations along their banks, it would be inaccurate to regard any of them as having attained a high degree of development by contemporary standards of river regulation. Development implies at least two physical changes in stream flow. One is the regulation of flow by storage, diversion, or land management so that the water is available when and where needed, rather than as dictated by natural fluctuations over days, seasons, and years. The other is the use of the water to maximize returns from other resource use. Under this definition, no stream can be considered fully developed unless, over long periods, its flow has been so regulated as completely to serve whatever purposes can, on grounds of social needs and economic growth, be shown to be important to the society involved. Only a few drainage basins are free from annual periods of water deficiency.[2] The ideally regulated stream would fluctuate in its main channels only to meet fluctuating human demands, the natural variations having been evened out.

The engineering means—dams, diversion canals, water-spreading devices—are available to carry out such regulation in most, but not all, situations. In some areas, because of special conditions, such as unstable foundations in deltas, no engineering solution is known to problems of controlling river flow.

Whether or not regulation is possible under prevailing technology, however, the physical limits to such regulation may be estimated with modest accuracy. For example, the total amount of electric power which a stream is capable of generating may be calculated: the amount of power is a function of volume of natural flow, fall, and regulation. It is possible, as well, to calculate the total acreage of land which may be irrigated from a stream, if fully regulated, taking into account the consumptive use made by different assumed crops and cropping practices. The consumptive use of water by animal and human populations also may be computed, but such uses are only a small proportion of the water withdrawn for rural and urban purposes. Typically, more than 90 percent of water employed in a municipal system is used nonconsumptively—that is, returned to a stream or aquifer, but with some change in chemical quality. The theoretical limit to the amount of nonconsumptive use for urban purposes is very high, so long as appropriate means of purification, waste treatment, and recycling are assumed to be available. Similar considerations apply to the use of water for navigation, wildlife, and recreational purposes, although in those cases, the loss of water by evaporation from reservoir and stream surfaces or by transpiration from plants may be heavy.

Every stream, accordingly, has a definite maximum potential output of hydroelectric power and maximum possible consumptive use of water. But the use which is made within those limits depends upon the complex interworking of factors affecting the particular technologies and uses which will be employed. For example, the decision to generate electric power at a given site is influenced by the engineering feasibility of alternative designs, costs of construction, alternative means of generating power, opportunities for marketing, availability of capital, investment alternatives, availability of labor force, accessibility of construction materials, and organization to plan and carry out the project. These and other factors will be discussed elsewhere in this symposium. It is important here only to note that the degree to which the ideals of full regulation and full use are approached is a function of those factors.

Thus, the actual accomplishment in the Nile or the Tigris-Euphrates or the Rhine falls far short of full regulation or full use. Several lower right-bank tributaries of the Rhine probably are more nearly regulated and used than any other humid or subhumid drainage area. A population of approximately 4,700,000 gains its livelihood from mining, heavy industry, and associated activities in a drainage area of 5,437 square miles. The waters of the Emscher, Lippe, Ruhr, and Wupper are handled by an intricate system of works, including storage reservoirs, diversion tunnels, collection sewers, city and industrial waste-treatment plants, channel improvements, pumping plants, channel dams, and hydroelectric installations. Water from one channel reservoir is pumped back to the upper reservoir at night to serve again during the daytime, and in periods of low flow, the water may be used several times. The responsible authorities are, however, looking forward to much more intensive development.[3] The physical limits of development are being approached, but they have not been reached.

The exceptions—the streams with more nearly complete regulation and use—are those, such as the upper Salt, in Arizona, where the storage structure in an area of heavy and chronic water deficiency halts all flood flows and fails to yield enough reliable supply to meet irrigation commitments.

What is true on a small scale for these highly developed tributaries is true on a vast scale for the major drainage basins of the world. Even the Tennessee is not completely developed. While its main navigation channel has been assured and an intricate system of reservoir regulation has been perfected, it still could have a flood high enough to damage low-lying portions of Chattanooga, and the planned management of some tributaries has only begun. The great rivers typically have flood peaks which reach the sea unused, leaving behind a muddy trail of damage. Their low flows still hinder economic life in times of unusual drought. In many of them, fruitful uses are precluded by lack of regulation or by unsatisfactory bacterial or chemical quality. The gap between what is technically attainable and what exists is everywhere large, although smaller in a few streams of Europe, Asia, and the United States than elsewhere. There seems little doubt that in every basin of more than 2000 square miles drainage area and in many smaller ones, there is the physical possibility of evening out flow by further storage, of decreasing the pollution of waters, and of readjusting upstream land use so as to reduce unnecessary soil loss and make wiser use of water. The social feasibility of such water and land management is, of course, a separate problem.

Before generalizing further about either the idea or the practice of river development, however, it is in order to offer a word of caution as to rivers themselves.

B. No Two Rivers Are the Same

If there is any conclusion that springs from a comparative study of river systems, it is that no two are the same. Each river is distinctive in characteristics of basin and flow. And rare are the streams that, regardless of size, are homogeneous within their own drainage areas.

The essential elements in a stream system are the river channels, the soils, and aquifers by which water reaches the channels, and the flowing water itself. At any one time, the channel section, the contributing slopes and aquifers, and the stream flow bear definite, but not fully measured, relationships to each other, and these relationships change as the volume and quality of water in the stream change. For a true picture of a river, it thus is necessary to describe not only its condition at a given time, but its changes from day to day, from season to season, and from year to year. It is possible to measure, for example, for any stream:

channel cross section at representative points,
channel gradient for the entire stream and by reaches,
length of channel,
angle of junction of tributaries,
density of tributaries,
area of the entire drainage basin and tributary basins,
shape of the entire drainage basin and tributary basins,
flow of water,
mineral and biological content of water,
slope and soil condition of tributary land surface, and
slope, permeability and thickness of contributing aquifers.

When any one of these characteristics is examined, an amazing range is found among the streams, and no two are found to be

precisely the same. Adequate explanation of differences in flow behavior under different conditions of land use, for example, is complicated and is not entirely practicable in our present state of
knowledge. Knowledge of origin and flow of sediment in suspension
is even less complete.

The point here is that streams are unique combinations of natural
features whose processes follow principles for the most part known.
They cannot be regarded as interchangeable, and while they may
be grouped into broad classes according to their combinations of
characteristics, the planning of their development always involves
a new, adventurous exploration for each stream, revealing differences in flow, channel, sediment, and chemical quality.

II. Central Ideas in River Basin Development

The concept of integrated river basin development, as it has come
to be used by many scientists, engineers, and statesmen around
the world today, seems to consist in three associated ideas. These
are the ideas of the multiple-purpose storage project, the basin-
wide program, and comprehensive regional development. They took
shape over more than half a century, forming side by side, each
drawing stimulus from a different set of conditions, but not clearly
combining into single programs in the United States until the middle 1930s. They still are far from finding full expression in many
areas of water and land resources activity. Their combination is
more an ideal than a reality, but it is an ideal which recurs in
differing form so frequently and widely and which commands such
warm enthusiasm as a symbol in public thinking that it should be
reckoned with as a unit.

Other ideas have entered into the arena of public action, sometimes figuring prominently, and these also should be considered. The
idea of articulated land and water programs is one such line of
thought. The idea of unified basin administration is another. These
are less persistent and influential in the work undertaken thus far,
however, and appear to have played a secondary role.[4] This review
deals primarily with the United States, but it outlines some of the
more influential changes in other countries.

One hundred years ago, Humphreys and Abbot, in making the
first monumental survey of flood problems in the lower Mississippi
Valley, could be comfortable in dismissing remedial reservoir work

and in limiting their detailed recommendations to the main stem of the river.[5] Engineers for irrigation works in Egypt or India could feel warranted in putting forward single projects to serve single purposes.[6] Whenever man was dealing with rivers, he was touching them at a particular point for a particular purpose and, with the exception of a few far-seeing men, like Powell and Willcocks, rarely dreamed of laboring with the whole river for multiple purposes.

By 1900, however, three new ideas had begun to emerge and to receive discussion. They were slow in finding acceptance, and it was several decades before all had been translated into action in an appropriate scale. Today, they still await a full and thorough demonstration in a single area.

A. Multiple-purpose Storage

Hoover Dam, on the Colorado River, is a conspicuous example of a form of engineering design which uses a single structure to store water for multiple purposes.[7] A concrete gravity dam, with a height of 726 feet and a mass of 3,250,000 cubic yards, it blocks the flow of the Colorado in the Black Canyon, providing sufficient storage and accompanying structure to assure that the flow can be fully controlled below that point to serve four major uses. The stored water may be diverted downstream to irrigate farms in the Imperial and Coachella Valleys of California. It also is diverted to serve the residential, agricultural, commercial, and manufacturing needs of Southern California to the extent of 1,500 cubic feet per minute. The water released through the dams is used to generate electric power, which is marketed as falling water to the states of Arizona and California and to the city of Los Angeles. A part of the reservoir storage, approximately 9,500,000 acre-feet, is reserved for the control of peak floods, so as to prevent them from traveling on downstream, where damage, chiefly to the Imperial Valley, might result from the river going over its banks. In addition to these four objects of irrigation, municipal water supply, hydroelectric power, and flood control, the dam was designed to help maintain a flow of water for navigation in the lower channel. Navigation has only theoretical importance; the theory, however, being a crucial one, inasmuch as the improvement of navigation formed the ostensible constitutional peg upon which federal action in undertaking a project of this magnitude first was hung.

Hoover Dam was authorized under the Boulder Canyon Project Act of 1928,[8] upon the basis of an interstate compact first concluded by the states of the Colorado basin in 1922[9] for the allocation of the waters of the basin, but it was not finally ratified with reservations by the requisite number of states, Arizona refraining, until 1929. Its design had begun in preliminary form on the drafting tables of the Bureau of Reclamation engineers ten years earlier, when the first intensive studies were made of the possibility of harnessing the river, and they applied to a monumental problem the fruits of their experience with many smaller, less complicated river projects in the western states.

When Hoover Dam was authorized, there were in the world few conspicuous examples of a large multiple-purpose storage project. There were numerous projects at which two or three purposes were served, most of them not involving storage. The storage projects were chiefly in western Europe and western United States. In the Ruhr-Westphalian area, there was a series of seventeen smaller dams constructed around the turn of the century to store water for industrial and municipal supply, to generate power, and to reduce floods.[10] Mill dams had provided water for industrial processing as well as to turn water wheels, and in some instances, they stored water for municipal purposes, but they typically were single-purpose and small in height.[11] With the coming of long-distance electricity transmission in 1891, electric power began to be linked with other purposes. Thus, in the United States, the Bureau of Reclamation combined electric power and irrigation storage at the Minidoka Dam on the Snake River in 1909 and at the Roosevelt Dam on the Salt River in 1911.[12] And the Hetch Hetchy water supply, with its controversial use of National Park lands to aid in furnishing municipal water and power for San Francisco, had commanded public attention from 1913 until construction began in 1924.[13]

Nowhere in South America, Asia, or Africa was there a major multiple-purpose storage project. This is not to say, of course, that there were no combinations of two or more purposes in water storage or that the concept of multiple use was not current in public thinking. The irrigation barrages constructed under British supervision in India as a measure of famine prevention in drought areas had provided for some use of water for domestic purposes, a use ever common where water is conducted through a dry area, and their irrigation canals had been used for navigation where suitable. Engineers, using the new knowledge of electricity, had sought to install power turbines where water fell through conduits to supply municipal needs. And a few river canalization schemes had installed generators in connection with navigation locks, thus capturing the energy otherwise lost. In eastern Asia, too, works on the lower stretches of streams combined navigation or flood control with irrigation.

A different form of multiple-purpose project, not involving storage, however, had already taken shape in the lower reaches of a few navigable streams, such as the Seine, the Po, and the Mississippi, where, by the construction of cutoffs, deepened channels, or training works, it was practicable both to improve the all-season navigability of channels and to hasten the flow of floods, thus reducing the area and depth of flooding. Likewise, the Dutch and English engineers, operating along their coasts and in the English Fens, had, as early as the seventeenth century, devised schemes which would dispose of flood waters and reclaim land for agriculture through subsidiary drainage. Strictly speaking, drainage goes hand in hand with flood control wherever the latter is practiced by channel regulation, and the Yazoo and the Chao Phya provide excellent examples of this.

In all these countries, the vision of multiple-purpose development was far ahead of the practice. By 1890, Sir Wilfred Willcocks was dreaming of dual-purpose dams on the Nile—indeed, a system of them—and viewed his design for the low Aswan structure as the precursor for what might follow on a large scale.[14] A German engineer, Mattern, using the work of Intze and others, was pleading in 1902 for multiple-purpose projects as the key to more intensive water use.[15] Similar thinking is also to be found in the report of the United States Inland Waterways Commission in 1908, which, while deeply concerned with problems of public-private responsibilities for water transport and water power and with coordination of federal activities, recognized the possibilities of combining irrigation, navigation, power, flood control, water supply, and related purpose in the same programs.[16] And writing in 1915, the Professor of Civil Engineering at Harvard was saying about water power:[17]

> This power arises from the water flowing in a stream, but this stream affords other uses, as for irrigation, water supply, and navigation. For the proper conservation of the water in a stream, all these four uses must be considered. Its development for one purpose must, so far as possible, be consistent with its development for the others. All should be developed so as to be productive of the greatest total good. The four uses above referred to are inseparably connected;

particularly so are the two uses of a river for water power and for purposes of navigation, in cases where both of these uses are economically practicable.

In view of the early articulate championing of multiple-purpose approaches, it may be asked why a project such as Hoover Dam did not materialize sooner. Part of the explanation rests upon the perfection of engineering techniques. Multiple-purpose storage design generally means larger structures, and these require refinements of foundation treatment and mass design in order to carry the larger load. Concrete was not adapted to dams of large height until the early 1900s, and then it remained for Arrowrock Dam in 1915, with a height of 354 feet, and, by way of culmination, Hoover Dam to perfect the means of pouring large blocks as much as sixty feet across.[18] The method of rolling raw earth into great structures, such as at Fort Peck on the Missouri, was not developed until later years, for while it is cheaper in materials, it depends for its economy upon large-scale earth-moving machinery and for exact knowledge of methods of handling soil foundation materials, which did not become available until the 1920s.[19] Until some of those new techniques were in use, the building of large earth dams, particularly on streams in semiarid or arid regions, was a hazardous occupation.

Other factors than dam design and construction also were at work. Perhaps most important, improvement in methods of long-distance transmission of electricity was rapid between 1900 and 1925, so that a line of 100 miles, considered uneconomic by operating companies in 1910, seemed short by comparison with the 200-mile line which the Southern California groups were prepared to build to Hoover Dam in 1929. Power clearly was the most readily vendible of all the products of multiple-purpose dams: municipal water supply might warrant heavy investment, but it was not as easily distributed and sold. Once cheap transmission became practicable, the active market for power for residential and industrial use grew at breathtaking speed. For many projects, the combination of vendible power made attractive an enterprise that otherwise would have commanded little attention if solely for irrigation or flood control.[20] Electricity generation played a major role at Hoover Dam, in the pioneer French plans for the Rhône in 1933, and for the larger multiple-purpose structures in India, such as the Bahkra Dam on the upper Sutlej.

The private power companies in the United States, however, vigorously opposed the incorporation of power-generating facilities in public water-storage facilities. Threat to their monopolistic position under public franchise was one consideration in their opposition, which also stemmed from a deep concern with the constitutionality of direct government action in power generation. The moment that it was proposed to combine power generation with a purpose such as navigation, which was the exclusive function of the federal government, complications arose, because the private companies obviously could not be expected to take direct responsibility for navigation or flood control or noninterest-bearing reclamation projects. On their side, the advocates of public power opposed any grant of authority to private companies to develop power in single-purpose projects on streams under public control. The General Dam Act of 1906 was the first broad legislative effort in the United States to deal with the conditions in which nonfederal development might take place on navigable waters and gave special attention to preventing interference with navigation and fish movement.[21] There followed a long controversy culminating in the Federal Water Power Act of 1920, with its much more precise regulation of nonfederal development.[22] The issue was both federal and state, having been drawn, for example, in 1908 in New York, when the State Water Supply Commission proposed public construction of a dam on the Sacandaga River to reduce floods and produce electric power.[23]

Opposition also was supported by inertia in design and by the widely held view that on grounds of design safety and economy, it often was impracticable to combine power with certain other purposes. Thus, an examination of a dozen civil engineering textbooks published during the first three decades of the century shows little or no attention to multiple-purpose design. And a special committee of the American Society of Civil Engineers, in its progress report on Mississippi River flood problems, in 1916 stated:[24]

> There is a popular delusion that the same reservoir can be utilized simultaneously to reduce floods, increase the low-water discharge of a stream, and increase the water-power that can be developed therefrom, but ordinarily its utilization for any one of these purposes precludes its efficient use for either of the others. . . . Your committee, however, does not intend to condemn in toto the utilization of reservoirs for more than one purpose. In fact, it believes that the practical solution of the flood problem in some valleys will be found in permitting corporations to build reservoirs in which a portion of the stored water can be utilized to a limited extent for power purposes and the remainder for flood prevention.

One of the standard texts on waterpower engineering was still taking a somewhat similar position ten years later.[25] The single-purpose approach persisted. Perhaps the extreme public demonstration was in the case of the Miami Conservancy District, where, under a new Ohio state conservancy law following the tragic flood of 1913,[26] the District constructed a system of channel improvements and five detention reservoirs to prevent any repetition of peak floods. The reservoirs were designed to detain water only long enough to permit the outflow to be limited to bank-full capacity downstream. On each dam was placed a plaque reading as follows:[27]

> The Dams
> of the Miami Conservancy district are for
> Flood Prevention Purposes
> Their use for power development
> or for storage
> would be a menace to
> the cities below

For twenty years after 1908, there were recurring federal authorizations of water-resources studies, in which the surveying agency was directed or permitted to consider other uses in addition to the major purpose in view. Thus, the Bureau of Reclamation was authorized to investigate power and later municipal water supply,[28] the Corps of Engineers was authorized to consider power in connection with navigation and then flood control,[29] and the Federal Power Commission was to take those other uses into account in reviewing applications for power-site permits.[30] These did not specify the kind of report which was expected, however, and the agencies regarded them more as permissive than directive. Accordingly, water-pollution control was neglected throughout those decades; and although attention was given to preservation of fish life at dams early in the period, as the size of the projects increased and competition between federal agencies also increased, the interests of recreation, wildlife habitat preservation, and esthetic enjoyment of wilderness lost out, and the conservation groups had to fight hard to gain any serious consideration from the engineers.[31]

The period of emergency public works in the United States in 1933–39, however, gave occasion for broad application of the multiple-purpose project idea across the country. Aided by substantial grants and loans in the interest of relieving unemployment, a whole series of new dams combining two or more purposes was constructed. They were favored, in principal, by the reviewing agencies; they offered the most widely distributed direct benefits; and they lent themselves particularly to public agencies, the only ones which could qualify for the public subsidy. In addition to the Tennessee Valley Authority, the Bureau of Reclamation launched the Grand Coulee, Central Valley of California, and Colorado-Big Thompson projects; the Corps of Engineers launched the Bonneville, Fort Peck, and upper Ohio projects; and various state agencies, such as the Nebraska power and irrigation authorities and the Muskingum Conservancy District in Ohio, obtained federal help. The reports of the Mississippi Valley Committee and National Resources Board and their successors gave heavy attention to multiple-purpose storage.[32]

By 1939, multiple-purpose projects were the order of the day: the idea was accepted and considered practical, and, with the lag in private construction owing to depression conditions, the single-purpose storage project was no longer dominant.

Single-purpose projects were not abandoned, however; they only gave way to multiple-purpose projects in relative weight. Irrigation storage on the upper Rio Grande, flood-control dams on the Yazoo and Muskingum, navigation dams on the upper Mississippi, power dams on the Wisconsin were examples of structures that were considered best suited to serving a single aim. They have continued in relatively diminishing number, often a source of controversy that probably has had fullest recent public expression in the dispute over the development of privately financed power dams in Hell's Canyon, on the Snake. In 1953, the chief of project planning for the Bureau of Reclamation could say, ". . . multiple-purpose projects are now thoroughly accepted, and an engineer would be considered remiss if he did not consider all possible uses, in connection with the planning of any irrigation project."[33]

B. The Basin-wide Program

While the idea of multiple-purpose storage was gathering force, there was a complementary, but not corollary, formation of the idea of basin-wide development. Again, Willcocks, Powell, and others had seen that if regulation of stream flow was to be achieved fully, it could only be by harnessing the flow of an entire drainage basin, and this meant designing control works with a regard to all other works—existing or possible—in the basin. Willcocks had made tentative plans for dealing with the Nile and Tigris-Euphrates basins as unified systems and was aware of, but not discouraged by, the political complications in store.[34] Powell, observing from his studies of the arid and semiarid regions that each stream presented problems peculiar to itself, proceeded to appraise the irrigation possibilities of each basin separately.[35] And President Theodore Roosevelt, in transmitting the Inland Waterways Commission preliminary report, could say, "Each river system from its headwaters in the forest to its mouth on the coast, is a unit and should be treated as such."[36]

The first major basin in the United States in which this idea was incorporated in a complete design was the Miami basin. There, as already noted, the program was strictly a single-purpose, flood-control effort. New York City, also, had undertaken to deal with entire tributary drainages of the Hudson River in single-purpose development of new water supply at the turn of the century, and while there had been no thought of covering larger areas, the concept of planning for the complete subbasin had been established.[37] Prior to that time, relatively complete development of small streams for mechanical water power had taken place, but not in accord with a single plan.[38]

The same situation had prevailed in a more acute form along the alluvial valley of the Mississippi, where the multiplication of levee districts and drainage districts without clear relation to a basin-wide program had resulted in direct conflict and oftentimes serious injury among the various works, leading to increased federal participation.[39] It is significant that A. E. Morgan and his associates came directly out of efforts to plan drainage projects in the alluvial valley and had seen the folly of attempting to deal with large flood flows by taking small bites. The Miami, they were determined, was to be handled as a unit.

In the same year that major works were undertaken in the Miami basin, a National Waterways Commission was authorized to prepare a comprehensive plan for the development of the nation's water resources for navigation and every useful purpose and to make recommendations for carrying out such work.[40] War conditions deferred its appointment, however, and it never came to grips with the problems.

In France, the desirability of treating basins as units was recognized in law in 1919,[41] and in Germany and Italy, the concept was accepted by engineers as necessary to effective planning.

But even after the Miami program had been demonstrated, it was a long time before the idea caught on elsewhere in the United States. Clearly, it played a major role in the discussions leading to final authorization of Hoover Dam: the 1922 compact recognized that the waters of the entire Colorado basin would need to be allocated under one agreement and that such allocation might lay the basis for designing works to control and use those waters. It was not sufficiently strong, however, to require that the final design of the first dam should be shown as a part of a system for the full basin. Nevertheless, the idea was finding progressively wider support in engineering circles, and threatening water shortages in western streams lent weight to it.

The great Mississippi flood of 1927 brought this thinking to focus. That catastrophe dramatized the inadequacy of Corps of Engineers' plans, which had sought to control flooding and maintain navigation channels along the main stem without planning works for the tributaries. Only the year before, the Chief of Engineers had reported that all was well with the levee and channel works.[42] A national debate ensued on the efficacy of levees versus cutoffs versus dams versus forests. Many extreme and poorly grounded assertions were made, and out of the heat and confusion, there emerged a few policy agreements. In the River and Harbor Act of 1927, Congress authorized comprehensive examinations and surveys by the Corps of Engineers to formulate, ". . . general plans for the most effective improvement of [navigable streams and their tributaries] for the purposes of navigation and the prosecution of such improvement in combination with the most efficient development of the potential water power, the control of floods, and the needs of irrigation."[43]

This was the second time in American history that a public agency had been directed to make comprehensive studies of navigation, flood control, irrigation, and power for complete drainage basins, but the first to have effect. The "308 reports," as they were known from their areas having been described in House Document no. 308,[44] became the point of departure for river basin development in the United States. Under the authority of the 1927 act, the possibility of multiple-purpose projects for at least four major purposes was recognized, and the comprehension of all parts of the basin in a single report was required. In doing so, it brought together federal concerns that had grown up along four separate lines of navigation, flood control, irrigation, and power policy.

Of the many basin programs that emerged in later years out of 308 reports—the Columbia, the Missouri, the upper Ohio—none commanded more attention than the Tennessee. After the 1927 authorization, the Corps of Engineers decided to concentrate its studies in the early years on one pilot basin. They selected the Tennessee,

in part because it seemed to lend itself to unified planning, in part
because it was endowed with a relatively generous amount of basic
hydrologic data, and in part because the government was confronted
with the difficult decision of what to do with Wilson Dam, a hydroelectric plant at Muscle Shoals which had been constructed during World War I to provide power for manufacture of nitrogen. It
was the Corps of Engineers which prepared the first plan for the
Tennessee, and when the Tennessee Valley Authority Act was passed
by Congress in 1933,[45] the Corps plan was the available one.[46] The
new Authority set up its own engineering staff and asked the Bureau
of Reclamation to aid in revising the plans so as to design a series
of high dams having large hydroelectric output, rather than the moderately low dams designed by the Corps to serve navigation primarily
and to produce relatively small amounts of power. Thus, the Tennessee basin was the first to be studied with a view to designing a
single, unified program, and it was the first in which such a program
was authorized for construction. As the TVA worked out its revised
designs over the years, it developed a program under which there
are twenty-seven dams serving navigation, flood control, and hydroelectric power, operated so as to regulate flow throughout the
main stem and major tributaries and to contribute to reduction of
flood flows in the lower Mississippi. From the standpoint of historical
evolution and of popular regard, the TVA may be considered the
prototype for unified basin-wide programs of multiple-purpose projects. It was intended to demonstrate the feasibility of such programs,
and it clearly has done so.

Throughout the late 1930s, the National Resources Committee and
its successors promoted thinking among both federal and state groups
on the meaning of drainage basin plans by drawing them together to
assess needs and to attempt to agree upon unified programs of investigation and construction.[47] Several significant basin-wide programs already were on the drafting boards about the time that the
TVA took shape. One of these was the Rhône plan, which had its
inception in the device of French engineers to combine further navigation improvements with power generation.[48] In Spain, a national
survey of water resources had recommended unequivocally in 1933
for the treatment of rivers as units.[49] The Central Valley plan in
California had been proposed as early as 1921 and was authorized
by the state in 1933, but awaited federal financing.[50]

There is little doubt that all the integrated systems of multiple-
purpose projects attempted since the 1930s were influenced in some
degree by the TVA, Columbia River and Central Valley works. Prob-

ably more important than the direct connections, of which there are many, has been the fact of accomplishment, the patent demonstration that what engineers around the world had been describing as possible and feasible could be built and operated in a relatively short time.[51]

C. Comprehensive Regional Development

The third of the major ideas in unified river development is more difficult to describe than the other two, because it has not been fully realized in any part of the earth. There is no prototype, no sterling demonstration of the idea; only partial, incomplete ventures in a direction that is, thus far, obscure. Implicit in the Hoover Dam project and in the TVA was an aim that, while assumed in much federal resources activity for more than a century, was not explicitly stated in the legislation and has not been fully realized. It was the aim of so planning and carrying out the works for river regulation and use that the region in which the basin is located would enjoy maximum practicable stimulation of its economic and social growth.[52]

The roots for such concern seem to go deep in United States public works and public land development. Gallatin's plan for waterway improvement, Powell's program for treatment of the arid lands, and Newland's early pleas for federal subsidy to western irrigation had presumed wise use of public material resources or capital to stimulate economic growth.

To take our two examples, the Hoover Dam was recognized by its proponents and opponents as likely to promote economic growth in the lower Colorado basin and in Southern California as well, although it was not specifically designed to effect such growth in a particular way. The Tennessee Valley projects were believed to be beneficial in sparking growth of what had been regarded as a backward and depressed region, but beyond this hope and the direction, given at the last hours in framing the act—[53]

> Sec. 22 To aid further the proper use, conservation, and development of the natural resources of the Tennessee River drainage basin and of such adjoining territory as may be related to or materially affected by the development consequent to the Act, and to provide for the general welfare of the citizens of said areas, the President is authorized, by such means or methods as he may deem proper within the

limits of appropriations made therefor by Congress, to make
such surveys of and general plans for said Tennessee basin
and adjoining territory as may be useful to the Congress and
to the several States in guiding and controlling the extent,
sequence, and nature of development that may be equitably
and economically advanced through the expenditure of public funds, or through the guidance or control of public authority, all for the general purpose of fostering an orderly
and proper physical, economic, and social development of
said areas. . .—

the Authority was not explicitly designed to occupy itself with those
questions. Research on the impact of navigation upon industrial life,
or on the relation of retail power rates to domestic electricity consumption, or on the efficiency of various cropping and fertilizing
methods was undertaken to supplement and make more effective
the water resources and fertilizer projects, rather than as a basis for
deciding what form of project should be undertaken, where, and
when. Repeatedly, the officers of the TVA, from its earliest years,
noted with satisfaction the effects of their activities upon the economy of the area loosely described as the "TVA Region," or the
"southeastern region." This was summed up most recently by Gordon Clapp, who pointed to increase in nonagricultural employment,
increase in variety of economic opportunities, rapid growth of high-
wage industries, a new pattern of industries processing raw materials, electrification of farms, and rural self-improvement projects as
indices of growth stimulated by the Authority.[54] The stimulation of
local citizen participation in valley improvement is counted both as
essential to and an important outcome of unified regional development activities.

In both the Hoover Dam and TVA examples, regional effects were
intimated but not planned, then enjoyed but not managed. They were
dimly perceived at the start, hailed when apparent, and the subject
of earnest study after the crucial decisions as to major river regulation works had been made. In each case, the criteria for selection
and financing of the construction work were restricted to a showing
of feasibility for the stated purposes of water control. Such gauges
of economic well-being as per capita income, diversification of industry and agriculture, and stability of employment did not figure
in decisive ways. These entered the discussion of the wisdom of the
projects more as rationalizations than as prior justifications.

A similar chronology occurred with the construction of Grand
Coulee Dam on the Columbia River in Washington by the Bureau

of Reclamation. Grand Coulee was part of a basin-wide scheme in the sense that it had been selected by the Bureau of Reclamation as a major element in a program for water development in the Columbia; but it was not a part, strictly speaking, because it was authorized long before substantial agreement had been reached among the Bureau of Reclamation, the Corps of Engineers, the Fish and Wildlife Service, and the various state interests as to the full outline of such a program.[55] Once underway, however, its full effects became a subject of lively speculation, leading the Bureau of Reclamation, with the leadership of Harlan Barrows, to initiate the "Columbia basin joint investigations." The scope of those studies is suggested by the participants, including nineteen federal agencies, eleven state agencies, and thirteen local and private institutions, and by the twenty-eight problems set for investigation.[56] Those problems covered questions of farm economy, farm size and layout, control of land use, village and community centers, transport facilities, recreational needs, and public works programming and financing.

Expected answers to such questions have figured repeatedly in the justifications given before public groups for Grand Coulee and also in descriptions given of results expected from many other basin programs modeled, in part, upon the TVA experience. Thus, the arguments made in favor of the Damodar Valley Authority in India have carried statements that, "The Corporation is to execute and operate schemes for irrigation, the generation of power and flood control. Besides those three main purposes, the Corporation will promote navigation, afforestation, public health, and industrial, economic and the general well-being of the people of the Valley."[57]

The distinction which seems crucial here is between engineering works which are planned and carried out with the sole purpose of gaining the direct benefits, such as power production or flood damage reduction from the water regulation, and engineering works which are intended to promote basic changes in the quality of life of the residents of the region. Under the second view, two considerations enter which are ignored under the first. The direct benefits become a means to an end, rather than an end in themselves; and engineering becomes one of several possible instruments, including land use, which may serve the needs of regional change. In the Miami Conservancy District plan, there was earnest desire to prevent future flood losses and to stabilize the then prevailing economy, but no concern to effect any significant changes in the distribution or character of urban occupancy in the basin: the aims stopped with the achievement of flood control. As the TVA unfolded, however, the control of floods was seen as a method of advancing a new economic and social well-being of residents of the Tennessee Valley, and power and navigation were viewed similarly. The moment in an analysis of a river basin development opportunity that this first distinction is made and that the construction of physical works takes on larger implications, the second distinction arises. If engineering is a means toward an end, then other means deserve attention as possibly contributing to or alternately serving the same end.[58]

Under the broadened view, the Bureau of Reclamation becomes as much concerned with the maintenance of family-sized farms as with optimum dimensions for distribution canals; and the Tennessee Valley Authority, from the beginning of its power program, lays stress on the marketing practices of farmers' electricity cooperatives as well as on the integration of hydroelectric and thermal-electric generating stations in the maximization of firm power production. Land use improvement, the economics of potential mineral exploitation, and the relation of freight rates to traffic movement and manufacturing location, are among the lines of inquiry that vie in importance with standard hydrologic and civil engineering practices in the planning of the river basin development.

It may appear that to take this broader view of river basin development is to expand water resources planning to encompass all aspects of natural resources as related to economic growth, including the cultural conditions of the society, and it will be necessary to ask whether or not there is any viable line which may be drawn between one and the other. For if there is no viable line, the attempts to carry water resources analysis beyond the traditional concepts of multiple-purpose, basin-wide development must inevitably lead to comprehensive regional development schemes in which water, in many instances, would play a secondary role.

Without attempting to answer that question, it may be useful to observe that no clear pattern of associated regional activities has yet emerged from the experimentation that currently is in progress in river basin development. It is not even possible to find association of electricity distribution and electricity production activities where hydroelectric plants are in operation. For example, the Compagnie National du Rhône avoids any connection with marketing policies for the power which it sells in bloc to Electricité de France.[59] In general, the basins having more nearly unified administrative control of their water regulation works have a larger number of associated activities directed at steering regional growth.

There is little clarity as to whether the aim of such growth should be specialization or balance, and the indices of results are scattered and rudimentary at best. Measures of aggregate economic growth are handicapped by lack of data, as well as by incompleteness of the theory of the process of growth. Less tangible noneconomic growth is even more difficult to quantify. There have been earnest efforts to improve the measurement devices, but these are still far from keeping pace with the claims of project proponents.[60] The claims for "secondary benefits" from investment in irrigation facilities continue to be in controversy among federal agencies.[61]

In its lack of definition and in difficulty of gauging the results of its use, the idea of comprehensive regional development has less precision and form than either multiple-purpose storage or basin-wide plan.

III. Associated Ideas

In addition to the three ideas described above, two others have enjoyed some popularity and individually have found expression in one or more development programs. They are less widely applied than the three already described and less firmly grounded in either theory or demonstrated practice, although they have attracted, in some instances, greater attention. One is the concept of articulated land and water programs; the second, the concept of unified administration. The former has commanded adherence over long periods and receives lip service generally today, but is honored more in theory than in practice. The latter bloomed suddenly and vigorously after 1933 and has been kept alive by controversy and opposition rather than by acceptance.

A. Land and Water

Early in the histories of scientific agriculture and river engineering, the idea took shape that the management of land and its vegetative cover is closely linked with the proper management of the flow of water in streams, that the magnitude, variation, and quality of water moving in a drainage basin is, in some measure, influenced by the treatment of the land. Although the origins of this idea are not clear,[62] it does seem to have been widely held in western Europe by the beginning of the nineteenth century. Summing up much of the study in the middle decades, Marsh found a weight of evidence that forests affect the rise and fall of springs and normal volume of rivers and the character of floods in rivers and torrents.[63] French engineers and foresters long had been concerned with relationships between their two lines of action, principal attention being given to the restoration of denuded sections of the High Alps. Agitation for regulation of forest and grazing practices as a means to controlling mountain torrents were suggested by administrative officers in 1819, and the first timid national legislation was passed in 1859, with major legislation following in 1882.[64]

Action was much slower in the United States, where problems of denudation were then considered less acute. For example, Gilbert, on the basis of reconnaissance studies in Utah, was not concerned with threats of accelerated erosion and concluded that if man did have an effect upon stream flow in the Rocky Mountain region, it probably would be advantageous by increasing the supply.[65] In the 1890s, however, some of the foresters began agitation for federal acquisition of forested lands in the drainage areas of important streams and for the retention of federal control over such lands where they had not passed to private ownership.[66] This culminated in the Weeks Law of 1911, which established federal acquisition of such forested, cutover, or denuded lands as "may be necessary to the regulation of the flow of navigable streams."[67] The forest program was expanded in 1924 to deal more explicitly with production problems, and there was a flurry of interest in forests as a preventive for floods following the 1927 disaster.

Then, three events placed agricultural workers directly in the river basin development field. The establishment of the Soil Conservation Service,[68] under Hugh Bennett, led to vigorous studies of channel and reservoir silting, to the establishment of experimental watersheds, and to the study of erosion conditions and corrective measures on a drainage basis. In 1936, the Flood Control Act, which shouldered federal responsibility for flood damage reduction on a national basis, authorized the Secretary of Agriculture to make surveys of "measures for runoff and water flow retardation and soil erosion prevention."[69] The TVA Act also permitted the establishment of divisions of forestry and agricultural relations. Out of these surveys by agricultural experts, there came a series of reports recommending appropriate measures for land-use treatment in selected drainage areas. These were the subject of extensive and often controversial review by the Corps of Engineers and Bureau of Reclamation. Eleven of the basin plans were completed, others delayed, and, with interruptions caused by war conditions and by problems of administrative jurisdiction within the Department, the program slowed down until the Department of Agriculture's interest in river programs was sharply revived by enactment, in 1954, of a new watershed-protection and flood-prevention program, under which direct action could be taken more readily by departmental agencies in cooperation with the land owners.[70]

A notable feature of the evolution of action by the land-management agencies is that, with a few exceptions, the work recommended on the land was not of such a character or degree of reliability in its effects upon water flow as to warrant making any changes in engineering works proposed for the same drainage areas. That is, flood-control reservoirs were not rendered needless by the prospect of stopping flow upstream, and storage reservoirs were not obviated by retention of water through forest or cropping practices. Such a conclusion was strongly at variance with the beliefs of many advocates of land management as an alternative to engineering and provoked a series of running debates which, in later years, centered successively upon the work of the President's Water Flow Committee, the Trinity River study, and the cooperative investigation of the Washita Basin.[71]

While this is not the place to enter into the jousting over land management versus engineering in river basin development, it may be useful to offer a few observations on the problem by way of partial explanation for some of the difficulties. The idea that land management may have an important effect upon water flow has received wide publicity in the United States in a series of reports over half a century.[72] It has not been translated into action in more than a few instances, because the responsible engineers have been either unable or unwilling to recognize direct and reliable connections with their programs.[73] One difficulty has been that proponents of each view have tended to exaggerate their own claims and to minimize those of others, so that foresters have at times claimed that reforestation would reduce the greatest Ohio floods, and engineers have maintained that soil erosion control had no part whatsoever in the sediment movement of streams.

The lack of scientific knowledge on which to settle disputes of that character has prevented clear answers and has permitted the controversies to continue inconclusively. Much information still is lacking. For example, the mechanism by which sediment finds its way into small watercourses is understood imperfectly, and the effect of land management upon groundwater flow is uncertain in many areas. It is only in recent years that it has been practicable for scientists to suggest valid generalizations as to the relationship between timber cutting and water yield or as to the possible effects of terracing upon flood flows.[74]

Thoroughly articulated programs of land and water development in the same basin, then, must be regarded as more hope than reality. They often have been joined in statements, but rarely in action.

B. Unified Administration

Following the enactment of the TVA Act, proposals for the creation of similar government agencies followed thick and fast for other basins. The distinguishing and common idea was the creation of an administration having full authority for dealing with whatever were the water and associated resource problems in the basin involved. Prior to that time, water projects characteristically were handled by the agency responsible for the major purpose. This might be a conservancy district for a flood-control problem, a city engineer department for a municipal water supply, an irrigation district and the Bureau of Reclamation for irrigation, a city or a drainage district and the Corps of Engineers for flood control. The Ruhr probably had the most complex administrative organization for its control works, involving several cities, two regional groups, and three drainage-area agencies. Now, the thinking turned to a single agency.

In the United States, proposals ranged from a general plan for subdividing the United States into valley authorities[75] to individual agencies for the Columbia, the Missouri, and the Arkansas. None received congressional approval, although it may be argued that some had large nuisance value in forcing livelier activity and cooperation among federal agencies where the independent agency threatened.[76] Indeed, the number of federal and state agencies involved in water resource development has tended to increase.[77]

Outside of the United States, the pattern of unified administration has been adopted sparsely. The Compagnie National du Rhône enjoys much the same quality of independence of regular government agencies, functioning as a joint stock company.[78] In the United Kingdom, a system of catchment basin authorities was established in 1930 to unify efforts to deal with land drainage, pollution abatement, flood control, and associated problems.[79] These have been established in forty-six areas and function independently of the Central Electricity Board. When the North of Scotland Hydroelectric Board was set up in 1943, however, its functions were limited to power production.[80]

Other agencies having unified control over the waters of entire basins include the Carini Valley Authority in Venezuela, the Caucas Valley Corporation in Colombia, the Damodar Valley Authority in India, the Snowy Mountains Hydroelectric Authority in Australia, and the Helman Valley Authority in Afghanistan. Each has variations from the TVA pattern. From the information available, however, all include an agency with powers to design, construct, and operate water-control works and to conduct associated activities, although they differ considerably in other respects. There has been no careful review of their functioning to date. It would appear that the idea of unified administration has spread from the Ruhrverband, TVA, and the English catchment basin experience to a few other areas, but is not, thus far, closely linked with large-scale river basin development.

IV. International and Interstate Applications of the Concept

Efforts to deal with basins and regions as units have raised major problems of scale and of political organization. These are suggested in part by the location of basins with respect to political boundaries. Complexities of administrative jurisdiction clearly have played a role in the retardation of some river basin development programs but also have contributed to decisive action in others.

It will be observed from map 4.1 that a large proportion of the major drainage basins of West Europe, Africa, Southeast Asia, and South America are international in character. In the United States, the Soviet Union, and China, the opposite is true, although each has one or more major streams having international drainage. If basin-wide plans be taken as ultimately desirable, then it is clear that the Columbia, Colorado, St. Lawrence, Rio Grande, Orinoco, Amazon, Parana, Uraguay, Rhine, Rhône, Danube, Tagus, Nile, Tigris-Euphrates, Congo, Niger, Zambesi, Indus, Ganges-Brahmaputra, Mekong, and Amur require international collaboration for their wise planning. Lesser streams could be mentioned, but this suggests the magnitude of the problem. Only the Mississippi, Don,

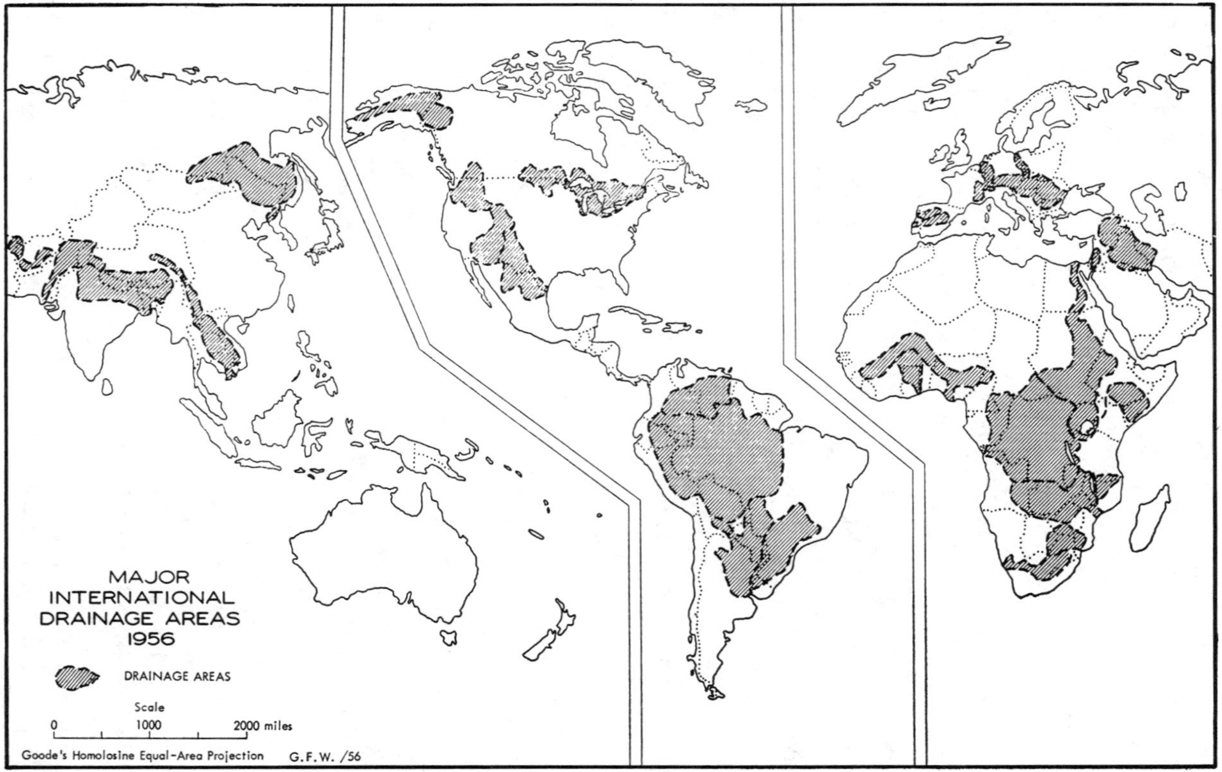

MAJOR
INTERNATIONAL
DRAINAGE AREAS
1956

DRAINAGE AREAS

Scale
0   1000   2000 miles

Goode's Homolosine Equal-Area Projection   G.F.W. /56

Map 4.1

Dneiper, Volga, Ob, Yenisey, Lena, Yangste, and Yellow, among
the great rivers, are free from the complication of crossing man-
made borders.[81]

Within individual countries, the difficulties do not stop. The complexity of political administration imposed by state boundaries is
apparent in the United States where, as shown in map 4.2, more
than three-quarters of the land area is drained by streams crossing
one or more state boundaries. The proportion would be even larger
if the streams draining into the Great Lakes were counted as interstate, since, strictly speaking, all of them drain into interstate
lake waters and through the St. Lawrence River. They are treated
on the map as intrastate units, however, although one major diversion is made down the Illinois River and although the Great
Lakes states have, during 1956, organized to begin the planning of
work, such as port development, on an interstate basis. The major
basins which are intrastate are the Central Valley of California, the
Brazos, Colorado, and Trinity in Texas, and the Altamaha, Cape
Fear, and James along the eastern seaboard.

Forty percent of the area of the United States is drained by the
Mississippi, a stream so great and complex that its planning has
been largely in terms of subbasins. Even when subdivided, the
interstate basins loom large, leaving only a few major streams such
as the Wisconsin, Kaskaskia, Green, Miami, Scioto, Muskingum,
and Yazoo lying wholly within one state. The picture for the western tributaries of the Mississippi is complicated further by the present or proposed transmountain diversions from the Colorado basin.
In view of the large diversion from the lower Colorado to the Los
Angeles basin, full consideration of basin water strategy in the
western Mississippi basin would require attention to Colorado and
southern California needs, and those now are linked with the Columbia in tentative schemes for later diversions.

Looking back over the growth of the basin-wide idea in the
United States, it is notable that the earlier examples of action were
intrastate—the Miami, Muskingum, Wisconsin, and Central Valley
of California. On the other hand, the large and dramatic examples
of both multiple-purpose and basin-wide planning came in interstate
situations, where strong leadership and participation by the federal
government was necessary to organization and financing.

The discussion, thus far, has dealt chiefly with the evolution and
demonstration of individual ideas comprising the concept of integrated river basin development. Some of the demonstrations were
not associated with other parts of the concept: thus, the Miami

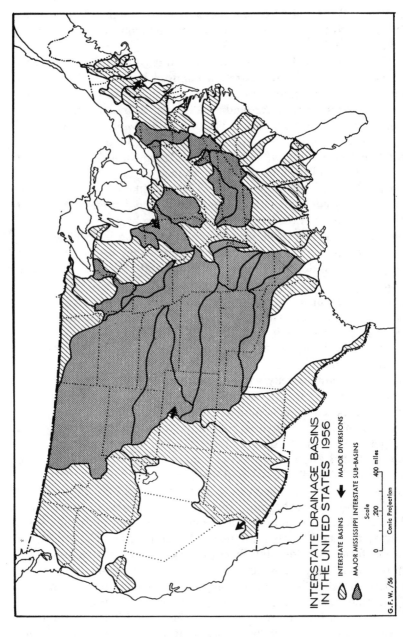

INTERSTATE DRAINAGE BASINS
IN THE UNITED STATES 1956

⬮ INTERSTATE BASINS   ➤ MAJOR DIVERSIONS

⬮ MAJOR MISSISSIPPI INTERSTATE SUB-BASINS

Scale

0   200   400 miles

Conic Projection

G. F. W. /56

Map 4.2

was strictly single-purpose; and Hoover Dam, while multiple-purpose, was not, in its first years, linked with a comprehensive basin plan. The degree to which the complete concept has been translated into action is revealed, in part, by table 4.1 and map 4.3. These show some major river basin development schemes constructed or under construction, as of 1957. Where a basin-wide plan has been projected but work is presently limited to only one project, as in the case of the Zambesi and the Volta, the entire area is shown. Where one project has been constructed without being linked with a basin plan, it is omitted. The Niger Office, responsible for the inland delta, is a borderline case. Systems of single-purpose projects also are omitted.

It is apparent from map 4.3 that the chief areas in which systems of multiple-purpose projects are under way lie either wholly within one country or are restricted to that part of the drainage located within one country. In terms of size of drainage area, the larger areas are located in Brazil (the San Francisco), China (the Huai and Yellow), India (the Damodar and Mahanadi), the Soviet Union (the Amu-Darya, Dneiper, Don, and Volga), and the United States (the Central Valley of California, Arkansas-White-Red, Columbia, Colorado, Missouri, Rio Grande, Savannah, St. Lawrence, Tennessee, and upper Ohio). Smaller complete drainages or subbasins include the Snowy Mountains—Murray scheme in Australia, the Ruhr-Emscher-Lippe in Germany, the Kitakami in Japan, the Papaloapan in Mexico, the Oum er Rbia in Morocco, and lesser streams in Puerto Rico and the Philippines.

The international streams which are being developed within only one country are the Rhône in France, the Zambesi in the Central African Federation, the Volta in the Gold Coast, and the Tigris-Euphrates in Iraq. No major international stream is receiving completely integrated treatment across frontiers. Perhaps the nearest approaches are on the Columbia River, through the International Joint Commission, and on the Rio Grande, through the International Boundary Commission. Parts of the Indus system are receiving attention on both sides of the India-Pakistan border, without agreement as to the precise terms of regulation and use, and there is a loose agreement affecting Nile waters.

In mapping the more important schemes, two criteria were used: they must involve systems of multiple-purpose projects, and they must involve the entire drainage basin or subbasin within the country of operations.

Table 4.1 Representative integrated river basin development programs

Basin	Area (Sq. Mi.)	Major purposes A Fertilizer manufacture F Flood control I Irrigation M Manufacturing N Navigation P Electric power S Soil conservation T Forestry	Brief description of works (Dams are storage unless otherwise designated. Installed hydroelectric capacity is shown in kilowatts.)
Columbia (United States)	partial 219,500	F–I–N–P	16 dams (4,500,000 KW). 41 nonfederal hydroelectric plants. Irrigation diversion dams and canals. At least 15 other dams planned. Navigation works.
Damodar (India)	8,500	A–F–I–M–N–P–S–T	7 dams underway (200,000 KW). Navigation channel. Land treatment. Industrial plant.
Huai (China)	67,200	F–I–N–P	15 detention dams. 7 storage dams. Levees and drainage works. Irrigation works. Locks. 2 hydroelectric plants (35,000 KW).
Kitakami (Japan)	3,950	F–I–P	7 dams (172,000 KW). Irrigation works.
Oum er Rbia (Morocco)	13,500	I–P	6 dams and hydroelectric plants (158,000 KW). Irrigation works and improvement. Diversion dams.
Rhône (France)	partial	I–N–P	2 dams and one other hydroelectric plant (700,000 KW). Locks and canals. 11 other dams and hydroelectric plants planned.
Tennessee (United States)	40,670	F–N–P–A–S–T	27 dams (2,600,000 KW). Locks. Fertilizer plant. Fuel electric plants. Agricultural demonstration.
Tigris-Euphrates (Iraq)	partial 192,193	F–I–P–N	4 dams (700,000 KW). 7 diversion dams. 7 others planned. Land drainage and irrigation. Locks.
Volga-Don (U.S.S.R.)	695,700	I–N–P	3 major dams (3,860,000 KW). Locks. Navigation canals. 3 other hydroelectric plants. 2 others planned. Irrigation works.

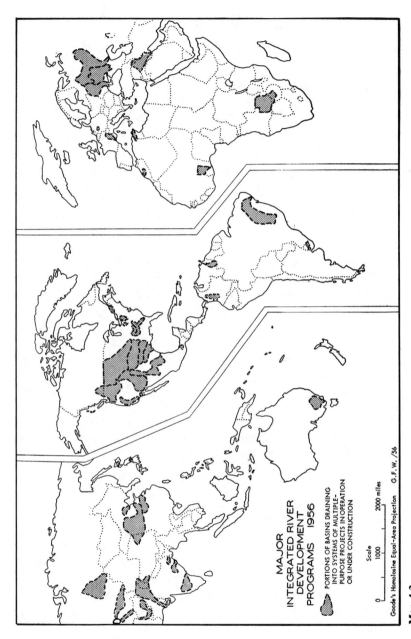

MAJOR
INTEGRATED RIVER
DEVELOPMENT
PROGRAMS   1956

PORTIONS OF BASINS DRAINING
INTO SYSTEMS OF MULTIPLE-
PURPOSE PROJECTS IN OPERATION
OR UNDER CONSTRUCTION

Scale

0   1000   2000 miles

Goode's Homolosine Equal-Area Projection   G.F.W. /56

Map 4.3

V. Aspirations and Stirrings

Spread of the essential ideas in unified river basin development has been outlined in terms of action taken across the continents. It might also be perceived, though less precisely, in the aspirations which reflect themselves in citizen, national, and international plans commanding public attention, but not yet channeled into construction works. These would include some basins, such as the Connecticut, where plans have been made, but in which unified action has not yet been launched, and other basins where discussion is far ahead of any detailed planning, as in the upper Amazon basin.

The Nile basin is still in the preliminary stage: notwithstanding the waters agreement, plans for the Aswan High Dam do not take account of possible future needs and regulatory action in the Sudan, Ethiopia, and Tanganyika sections of the basin. Likewise, the Jordan River has been studied repeatedly, without arriving at a satisfactory agreement. The Rhine River, though the subject of Dutch-German and German- French-Swiss negotiations, has not been subjected to a single, unified plan.[82] Both official and citizens groups now are working to perfect more precise arrangements for handling the Rhine waters in flood and drought. Danubian cooperation embraces only a portion of that basin. Congo planning for integrated land and water transportation facilities is well advanced without comprising a system of water regulatory works for other purposes. There are large new storage projects in the upper Ganges and river channel improvement plans for the Mekong that may, in time, evolve into unified systems of projects. An agreement concerning the Amur has been negotiated during the past year.

VI. Significance of the Concept in Practice

Reviewing the evolving ideas of the past sixty years as they now find expression in landscape and livelihood and aspiration, a few aspects stand out:

The idea of multiple-purpose water storage, while once considered of doubtful practical value, now is firmly established in present construction technology.

The idea of unified basin plans has moved slowly from theoretical acceptance to practical application: in the United States, first slowly in intrastate drainages, then quickly in the more challenging interstate drainages, but, thus far, across no international boundary to cover a complete basin.

The idea of comprehensive regional development has gained gradually in application, the vague vision of regional growth being more persistent than efforts to define or measure it.

Associated with the second of these ideas, the idea of land improvement integrated with water use and control has remained in popularity, but has been applied only rarely in genuine plans and programs.

Spreading rapidly in the realm of public discussion, the idea of unified administration has skipped from the TVA and the Rhône to a few other places, but has inspired more controversy than imitation.

Throughout the period, there has been conspicuous lack of careful appraisal of the work accomplished. A tool capturing imaginative support, as this one does, deserves penetrating assessment, and such examination has been largely absent. For every hundred studies of what might or should be done with a river system, there is hardly one that deals with the results. In so far as the results are physical, some may be measured readily—kilowatts of power produced, cubic feet of water delivered, inches of reduction in peak flow, acres irrigated. Some physical results have been touched only lightly, as for example, effects of water storage and mixing on downstream erosion and water quality.

The social results are perceived dimly; there are only the roughest gauges of the effects of river works upon economic growth and community stability or change. Little basis exists for comparing the effectiveness of one multiple-purpose plan with another for the same basin or with an alternate method of fostering social change. Thus far, there has been no genuinely searching assessment of the full impact of the TVA operations, and it is needed.[83] Yet, if it were to be undertaken, it would require as a first step a sharpening of devices for measuring construction efficiency, economic growth, and cultural advancement. Strict adherence to the ideas of multiple-use regulation for entire basins leads inevitably to confrontation of regional aims and social processes.

Evaluating the Consequences of Water Management Projects

Next, we ask what we know about the impact of these management schemes—dams, channel improvements, flood control

works, large irrigation projects, small water structures, large hydroelectric storage and diversion structures—over the last forty or
fifty years. One must immediately note that the literature and scientific investigations in respect to water management have been
very largely on the normative side. We could fill a large room with
documents drawing up what are considered the best plans for and
analyses of problems in river basins around the world. United
States reports and comprehensive plans alone would fill at least
half of such a room. The United States and Canada together have
developed a very substantial volume of analysis on what might be
done on both the St. Lawrence and the Columbia. Most of these
are normative, in the sense that they propose what should be
done and what would be desirable to do in terms of rather restrictive economic and engineering analysis.

On the other hand, the literature about what has happened after
any of the projects have been carried out can be assembled on
one end of a small table. There is no tradition for making retrospective or evaluative studies of the consequences. With a few
exceptions, some of which are quite notable, we do not have systematic, discerning knowledge about what have been the consequences of these large projects. For example, no evaluation of the
Tennessee Valley Authority has been undertaken. Although we
know a great deal about what's happened to the fish in the reservoirs, to transport on the stream, we have no satisfactory explanation of why, thirty years after the TVA was started in order to
develop the economy of the region, the major part of the Tennessee Valley is still considered an underprivileged area of Appala-
chia, deserving special subsidy contributions from the federal
government for further improvement. Nor, in a more elementary
way, do we have any very accurate kinds of measurements of
what is happening in downstream channels in irrigation project
areas and the like.

Unpublished paper, Columbia University, 21 March 1971, 2.

Notes

1. For a major attempt at such an audit in various parts of the
world, see below, vol. II, chap. 5 (Day et al.).

2. A basin may have a large stream flow and still suffer a deficiency of water for crop or processing needs. The amount and duration of this deficiency in humid areas is greater than is commonly recognized. See C. W. THORNTHWAITE and J. R. MATHER, THE WATER BALANCE (1956).

3. See Pruss, *Wasserversorgung und Abwasserbeseitigung im Ruhrgebiet auf genossenschaftlicher Grundlage*, in ORDNUNG UND PLANUNG IM RUHR, RAUM 58 (1951); Goepner, *Die Wasserversorgung des rheinischwestfaelischen Industriegebiets*, in *Geographisches Taschenbuch 306* (1956).

4. Various definitions of the same general complex of ideas are current. The Bureau of Flood Control of the Economic Commission for Asia and the Far East uses "multiple-purpose river basin development" to include multiple-purpose use, unified development of entire basins, social benefits and costs applied to a region, comprehensive development of all resources, and unified control. ECONOMIC COMM'N FOR ASIA AND THE FAR, EAST, MULTIPLE-PURPOSE RIVER BASIN DEVELOPMENT pt. I, at 1–8 (U.N. Pub. Sales No. 1955.II.F.I). It draws, in part, from the report of the President's Water Resources Policy Commission, whose legal staff defined comprehensive development of water resources as "basin-wide development for optimum beneficial uses of a river system and its watershed." PRESIDENT'S WATER RESOURCES POLICY COMM'N, WATER RESOURCES LAW 383 (1950). A common variation is illustrated in ROY E. HUFFMAN, IRRIGATION DEVELOPMENT AND PUBLIC WATER POLICY 153 (1953): "A comprehensive development program for a river valley involves working with three resources of equal importance—water, land, and people. In the past, water has tended to receive major emphasis, while the other two factors played a minor role or were disregarded altogether in project and program formulation."

5. A. A. HUMPHREYS AND H. L. ABBOT, REPORT UPON THE PHYSICS AND HYDRAULICS OF THE MISSISSIPPI RIVER 406–11 (2d ed. 1876). They considered, but dismissed as inapplicable, the plans of French and Italian engineers and of the Americans, Ellet and Morris, for reservoir systems.

6. WILLIAM LUMSDEN STRANGE, INDIAN STORAGE RESERVOIRS WITH EARTHEN DAMS iv (1904); and G. W. MACGEORGE, WAYS AND WORKS IN INDIA 107–215, 431–42 (1894).

7. See PAUL L. KLEINSORGE, THE BOULDER CANYON PROJECT (1941).

8. 45 STAT. 1057, 43 U.S.C. §617 (1952).

9. Signed at Santa Fe, New Mexico, Nov. 24, 1922, pursuant to Act of Aug. 19, 1921, c. 72, 42 STAT. 171, approved by Congress, Act of Dec. 21, 1928, c. 42, 45 STAT. 1064, 43 U.S.C. §617l (1952).

10. See INTZE, ENTWICKELUNG DES THALSPERRENBAUES IN RHEINLAND UND WESTFALEN VON 1889 BIS 1903 (n.d.).

11. See LEFFEL'S CONSTRUCTION OF MILL DAMS AND BOOKWALTER'S MILLWRIGHT AND MECHANIC (1881).

12. Bureau of Reclamation projects are described in its *Dams and Control Works*, various editions.

13. For a recent review of the controversy, see *Hearings before the Committee on Public Lands of the House on H. R. 5964*, 77th Cong., 2d Sess. (1941) (amending the Raker Act).

14. W. WILLCOCKS, THE NILE RESERVOIR DAM AT ASWAN AND AFTER 13–26 (1901).

15. E. MATTERN, DER THALSPERRENBAU UND DIE DEUTSCHE WASSERWIRTSCHAFT 100 (1902).

16. Inland Waterways Comm'n. *Preliminary Report*, S. Doc. No. 325, 60th Cong., 1st Sess. 18–25 (1908). M. O. Leighton's "Relation of Water Conservation to Flood Prevention and Navigation in Ohio River" is an appendix, *id.* at 451–90, critically reviewing the opposition to the Ellet plan for reservoirs. Ellet's plan for a reservoir system on the Ohio was hailed by his publisher in 1853 as "the foundation of a new branch of engineering, which, in the progress of the country in wealth and population, is destined to acquire increasing interest from year to year." CHARLES ELLET, JR., THE MISSISSIPPI AND OHIO RIVERS vii (1853).

17. GEORGE F. SWAIN, CONSERVATION OF WATER BY STORAGE 24–25 (1915).

18. U.S. BUREAU OF RECLAMATION, DEP'T OF INTERIOR, DAMS AND CONTROL WORKS 7–10, 49–51 (1938). Willcocks, in his design of Aswan, saw it as marking a new "epoch in dam building," especially as the use of concrete might come to be perfected, but he thought of the typical combination of purposes as being flood control and irrigation. WILLCOCKS, *op, cit. supra note* 14, at 4.

19. The Miami Conservancy District pioneered with draglines for building hydraulic fill dams with clay cores. The pioneering worker on the physical properties of soils as applied to engineering was Karl Terzaghi. Smaller earth dams had been constructed over at least 1,600 years in India and the Middle East.

20. This view of electric power as the integrating factor in river basin development was widely preached by Morris L. Cooke. See Cooke, *Multiple-purpose Rivers*, 237 J. FRANKLIN INST. 251 (1944). See also Lepawsky, *Dams and Democracy*, 29 VA. Q. Rev. 533 (1953).

21. 34 STAT. 386.

22. 41 STAT. 1063, as amended, 49 STAT. 838 (1935), 16 U.S.C. §§791a–825r (1952). This history is described in PRESIDENT'S WATER RESOURCES POLICY COMM'N, *op. cit. supra* note 4, at 391–408 (1950).

23. STATE WATER SUPPLY COMM'N OF NEW YORK, STUDIES OF WATER STORAGE FOR FLOOD PREVENTION AND POWER DEVELOPMENT IN NEW YORK STATE UNDER PUBLIC OWNERSHIP AND CONTROL 14–15, 18 (1908).

24. *Hearings before the Committee of the House on Flood Control on Mississippi River Floods,* 64th Cong., 1st Sess. 295–96 (1916). One of the members, Morris Knowles, dissented, calling this statement "an entirely unnecessary and unfair discussion purporting to show that reservoirs cannot be made useful for flood prevention, together with other purposes; whereas we know, on the contrary, that, notwithstanding all these statements, it is possible to operate reservoirs with several purposes in view, and that is actually done in some of the great reservoir systems of Europe and America without conflict of interests." *Id.* at 301.

25. "Storage capacity below spillway level cannot be devoted to both power and storage unless at all times some portion at the upper level is reserved for flood use and kept empty, and, therefore, always ready for the flood emergency. In this country such joint use of reservoirs has not been attempted, their purpose always being water storage for power, municipal water supply, or navigation, so that any decrease of flood tendencies has been merely incidental." H. K. BARROWS, WATER POWER ENGINEERING 180 (1927).

26. Act of Feb. 17, 1914, OHIO REV. CODE ANN. c. 6101 (Page 1954).

27. ARTHUR E. MORGAN, THE MIAMI CONSERVANCY DISTRICT 473 (1951). In passing, it may be noted that one of the paradoxes in river basin development history is that the head of the Miami Conservancy District at that time was Arthur E. Morgan, who later became a member of the first board of the Tennessee Valley Authority and a proponent of a multiple-purpose reservoir system for the entire Mississippi; while a leader in the negotiations culminating in the dam on the lower Colorado was Herbert Hoover, then Secretary of Commerce, who later came to oppose expansion of federal power activities in favor of private management of these resources. Hoover never shifted his basic position on the desirability of linking private power with public management of other resources, but he did find himself, as late as 1955, trying to stem a tide of public power activity which had taken strong impetus from the apparent success of Hoover Dam. See I COMM'N ON ORGANIZATION OF THE EXECUTIVE BRANCH, WATER RESOURCES AND POWER 119–22 (1955).

28. Act of April 16, 1906, 34 STAT. 116–17, 43 U.S.C. §567 (1952). This permission to make provision for water supply and power did not take the form of a directive until Act of Aug. 4, 1939, 53 STAT. 1194, 43 U.S.C. §§485 (1952).

29. Act of March 3, 1909, 35 STAT. 822, 33 U.S.C. §604 (1952); Act of March 1, 1917, 39 STAT. 950, 33 U.S.C. §701 (1952).

30. Act of June 10, 1920, 41 STAT. 1068, 16 U.S.C. §801 (1952).

31. Specific instructions to consider wildlife were contained in Act of Dec. 22, 1944, 58 STAT. 887, and Act of Aug. 14, 1946, 60 STAT. 1080, 16 U.S.C. §§661–66c (1952).

32. MISSISSIPPI VALLEY COMMITTEE OF P.W.A., REPORT 25–29 (1934); NATIONAL RESOURCES BOARD, REPORT 271–75, 286 (1934).

33. John Dixon, *Planning an Irrigation Project Today,* in CENTENNIAL TRANS. AM. SOC'Y CIVIL ENG. 357, 362 (1953).

34. WILLCOCKS, *op. cit. supra* note 14, at 13–26; W. WILLCOCKS, IRRIGATION OF MESOPOTAMIA (2nd ed. 1917).

35. J. W. POWELL, REPORT ON THE LANDS OF THE ARID REGION 10–14 (1879).

36. Inland Waterways Comm'n. *supra* note 16, at iv.

37. See BOARD OF WATER SUPPLY OF THE CITY OF NEW YORK, CATSKILL WATER SUPPLY (1928).

38. Mill owners did, however, maintain organizations to deal with their interests in an entire stream, and engineers saw the possibilities of large-scale unified development. See, *e.g.,* JOSEPH FRIZELL, WATER-POWER 588–605 (1903).

39. See ROBERT W. HARRISON, LEVEE DISTRICTS AND LEVEE BUILDING IN MISSISSIPPI: A STUDY OF STATE AND LOCAL EFFORTS TO CONTROL MISSISSIPPI RIVER FLOODS (1951).

40. Act of Aug. 8, 1917, 40 STAT. 250 (repealed in 1920).

41. Law of Oct. 16, 1919. See Arbelot and Dupin, *L'evolution des idées en matière de régularisation de l'énergie hydroélectrique,* in 2 TRANS. FIRST WORLD POWER CONFERENCE 150–54 (1924).

42. U.S. WAR DEP'T, REPORT OF THE CHIEF OF ENGINEERS, U.S. ARMY, 1926, pt. I, at 1793, 1794 (1928).

43. 44 STAT. 1015. The list was prepared by the Federal Power Commission and Corps of Engineers, in accordance with Act of March 3, 1925, §3, 43 STAT. 1190.

44. H. R. Doc. No. 308, 69th Cong., 1st Sess. (1927).

45. 48 STAT. 58 (1933), 16 U.S.C. §831 (1952).

46. H. R. Doc. No. 328, 71st Cong., 2d Sess. (1930). It is doubtful whether or not Congress was aware of the possible implications of the Corps of Engineers' low-dam plan for later high-dam construction at the time that the act was passed. It might have been thought that only one or two new structures (Cove Creek and Wheeler locks) would be built. Later legislation was required to define clearly the authority for river basin planning. See C. HERMAN PRITCHETT, THE TENNESSEE VALLEY AUTHORITY, A STUDY IN PUBLIC ADMINISTRATION 3–47 (1943).

47. See NATIONAL RESOURCES COMMITTEE, DRAINAGE BASIN PROBLEMS AND PROGRAMS (1936). See also revision of committee reports in 1937.

48. See Gilbert Tournier, Rhône: Dieu Conquis (1952). Planning had been authorized by an act of May 27, 1921, and an administrative order of Jan. 13, 1931.

49. Plan Nacional de obras hidráulicas (1933).

50. See Hugh G. Hansen, Central Valley Project: Federal or State? 21–37 (Cal. Assembly Interim Comm. Rep., Vol. 13, No. 6, 1955).

51. See TVA, TVA as a Symbol of Resource Development in Many Countries (1955).

52. The term "region" is used here in its broadest sense to include any area designated for study or action. For discussions of more precise uses, see National Resources Committee, Regional Factors in National Planning and Development (1935); and Platt. *Discussion: Nature and Scope of Regional Science,* in 2 Papers and Proc. Regional Science Ass'n 46 (1956).

53. 48 Stat. 69 (1933), 16 U.S.C. §831u (1952). The broad, regional studies were not undertaken at once, but they followed soon, and the legislative authority was sufficiently general to permit their being undertaken. Ackerman, *Tennessee Valley Authority Planning: Methods and Results,* in Papers of Int'l Conference on Regional Planning and Development (1955), gives a careful review of the development of those activities.

54. Gordon R. Clapp, The TVA: An Approach to the Development of a Region 54–56, 65 (1955).

55. See Charles McKinley, Uncle Sam in the Pacific Northwest 138–45 (1952).

56. U.S. Bureau of Reclamation, Dep't of Interior, Columbia Basin Joint Investigations: Character and Scope (1941).

57. Economic Comm'n for Asia and the Far East, *op. cit. supra* note 4, at 78.

58. A succinct statement of the idea of comprehensive regional development, as it took shape in the 1930's and 1940's, is Alvin H. Hansen & Harvey S. Perloff, Regional Resource Development (1942).

59. See Tournier, *op. cit. supra* note 48, at 304–05.

60. Problems of economic indices are outlined in Krutilla, *Criteria for Evaluating Regional Development Programs,* 45 Am. Econ. Rev. 120 (1955). Moore, *Regional Economic Reaction Paths, id.* at 133; and the discussion that follows. *Id.* at 149.

61. Federal Inter-agency River Basin Committee, Proposed Practices for Economic Analysis of River Basin Projects (1950). See also Technical Assistance Administration, Formulation and Economic Appraisal of Development Projects (U.N. Pub. Sales No. 1951.II.B.4.); and U.S. Bureau of Recla-

MATION, DEP'T OF INTERIOR, REPORT OF PANEL OF CONSULTANTS ON SECONDARY OR INDIRECT BENEFITS OF WATER-USE PROJECTS (1952).

62. No attempt is here made to trace the earlier evolution of this idea. It occurs in much Greek, Hebrew, and Roman literature. For a brief review of evolving views of the unity of nature, see Glacken, *The Origins of the Conservation Philosophy,* 11 J. SOIL & WATER CONSERVATION 63 (1956).

63. GEORGE PERKINS MARSH, THE EARTH AS MODIFIED BY HUMAN ACTION 225, 227 (1885). Marsh did not feel warranted in asserting, however, that forests, by their presence or absence, increase or lessen the total volume of water discharged by rivers or torrents.

64. See M. P. MOUGIN, LA RESTAURATION DES ALPES 145–60 (1931).

65. Gilbert, *Water Supply,* in POWELL, *op. cit. supra* note 35, at 57. Interest then centered on prospects for climatic fluctuations.

66. See GIFFORD PINCHOT, BREAKING NEW GROUND 238–40 (1947).

67. Act of March 1, 1911, 36 STAT. 961, as amended, 16 U.S.C. §§515–16 (1952).

68. Act of June 16, 1933, 48 STAT. 195, and Act of April 27, 1935, 49 STAT. l63, 16 U.S.C. §590 (1952).

69. 49 STAT. 1570 (1936), 33 U.S.C. §701(a) (1952). This was extended by Act of Aug. 28, 1937, to cover all drainage areas previously authorized for survey by the Corps. 50 STAT. 877, 33 U.S.C. §701(g) (1952).

70. Act of Aug. 4, 1954, 68 STAT. 666, 16 U.S.C.A. §§1001–06b (Supp. 1956).

71. See President's Committee on Water Flow, *Report,* H. R. Doc. No. 395, 73d Cong., 2d Sess. (1934).

72. Some of the notable statements are: W. W. Ashe, *Special Relations of Forests to Rivers in the United States,* S. Doc. No. 325, 60th Cong., 1st Sess. 514–34 (1908); U.S. Dep't of Agriculture, *A National Plan for American Forestry,* S. Doc. No. 12, 73d Cong., 1st Sess. 299–461, 1509–36 (1933). UPSTREAM ENGINEERING CONFERENCE, HEADWATERS CONTROL AND USE: A SUMMARY OF FUNDAMENTAL PRINCIPLES AND THEIR APPLICATION IN THE CONSERVATION AND UTILIZATION OF WATERS AND SOILS THROUGHOUT HEADWATER AREAS (1937).

73. Thus, the ASCE Special Committee reported in 1916: "The effects of forest growth in preventing erosion on hillsides are sufficient to justify reforestation for that purpose, but there has been no quantitative determination of its influence on stream flow which would justify its employment as a method of flood prevention." *Hearings, supra* note 24, at 298.

An example of the contrary view is: "The forest tends to equalize the flow throughout the year by making the low stages higher and the high stages lower.

"Floods which are produced by exceptional meteorological conditions cannot be prevented by forests, but without their mitigating influence the floods are more severe and destructive." Raphael Zon, "Forests and Waters in the Light of Scientific Investigation," in National Waterways Comm'n, *Final Report,* S. Doc. No. 469, 62d Cong., 2d Sess. app. V, at 205, 273 (1912).

A more reconciling view is this: "Forest rehabilitation is not urged as an alternative to engineering works for flood control. It is supplementary to the engineering program, but it is a supplement of such importance that no complete plan of flood control can omit it." Sherman, "Protection Forests of the Mississippi River Watershed and Their Part in Flood Prevention," in U.S. Dep't of Agriculture, *Relation of Forestry to the Control of Floods in the Mississippi River,* H.R. Doc. No. 573, 70th Cong., 2d Sess. 51 (1929).

74. See Wilm, *Timber Cutting and Water Yields,* in YEARBOOK OF AGRICULTURE 593 (1949); LUNA LEOPOLD AND THOMAS MADDOCK, THE FLOOD CONTROL CONTROVERSY (1955).

75. See McKinley, *The Valley Authority and Its Alternatives,* 44 AM. POL. SCI. REV. 607 (1950).

76. It also may be argued that the results of these forced unions were more legitimate than in the public interest. See Hart, *Governing the Missouri,* 41 IOWA L. REV. 198 (1956).

77. See White, *National Executive Organization for Water Resources,* 44 AM. POL. SCI. REV. 593 (1950).

78. See GILBERT TOURNIER, L'AMENAGEMENT DU RHÔNE (1953).

79. Under the Land Drainage Act of 1930, 20 & 21 GEO. 5, c. 44, catchment boards were established in England and Wales. They are composed of representatives of central and local government agencies concerned.

80. See MacColl, *Hydro-electric Developments in the Scottish Highlands,* in 4 TRANS. FOURTH POWER CONFERENCE 2158 (1950).

81. And even the Mississippi is not entirely free from international complications. A tributary of the Missouri, the Milk River, drains both Canadian and United States territory and is the subject of an international agreement.

82. See, Wehle, *International Administration of European Inland Waterways,* 40 AM. J. INT'L. L. 100 (1946).

83. McKinley, *TVA Management in the Perspective of Two Decades,* 16 PUB. ADMIN. REV. 109 (1955); Fisher, *Resource Problems and the Social Sciences,* in RESOURCES FOR THE FUTURE, ANN. REP. 15 (1956).

5 Introductory Graduate Work for Geographers

During his presidency at Haverford College from 1947 through 1955, Gilbert White observed the turmoil of an extraordinarily able undergraduate body in its encounters with graduate school. Students frequently complained of their initial experience, telling of elementary and uninspiring introductory lectures, lack of early orientation to the field, and an absence of diagnostic review.

Moving from Haverford to Chicago in 1956, White prevailed upon his colleagues to develop a student-centered program with an introductory semester that permitted students to learn from each other, the faculty, and their own strengths and weaknesses.[1] As students in that program, we recall it with pleasure and an excitement not reflected in White's brief and cautious report.

At least three problems arise in arranging programs of study for entering graduate students in geography. One is that students come with widely different formal preparation in the elementary disciplines of geography and related fields. A second is the wide spread in command of skills among those who have had the same formal courses. A third is the uncertainty of new students as to the areas in which they wish to specialize.

The ordinary transcripts and application papers do not reveal the student's true strengths and weaknesses. The title of a course is often a poor guide to its content or what the student learned from it. And the printed catalog fails to provide the kind of understanding of research fields and faculty interests that is essential to wise choices of formal courses and of more advanced studies.

In an effort to deal with these problems somewhat more helpfully, the Department of Geography at the University of Chicago has developed during the past two years a different kind of introduction to graduate work. It is a course which claims the full academic

Reprinted from *The Professional Geographer* 10, no. 2 (March 1958): 6–8.

program of all new graduate students during the autumn quarter. To meet the Registrar's requirements, it carries two prosaic titles— "Geography 300: Pro-Seminar, 1 credit," and "Geography 301: Field and Library Methods , 2 credits"—but there its resemblance to the usual courses stops.

It seeks to serve two aims at the same time: 1) to assess the full strengths and weaknesses of each entering student in dealing with geographic ideas and methods, and 2) to acquaint the new students with what the faculty consider to be major concepts and skills for professional work and with the faculty's own interests and research programs. All of the members of the Department take part, so that by the end of the quarter each has the experience of intensive work for a week or more with the entering group.

This year the quarter began with a series of diagnostic tests covering diverse facts, concepts, and techniques of geography. Norton Ginsburg and Agnes Whitmarsh followed with basic skills of map intelligence, including map interpretation and the use of the map library. Wesley Calef took the group into the field for work which ranged from simple mapping to urban land classification and functional studies. Chauncy Harris managed to condense into two and a half weeks the essential parts of the full course which he formerly gave on library and bibliographic methods. Ginsburg carried out intensive appraisal of representative research aims and methods, drawing on selected journal articles. Harold Mayer covered several concepts in urban geography, particularly economic base, spatial interaction, and gravity models. Philip Wagner examined central ideas in cultural geography. C. W. Thornthwaite introduced the group to climatic classification and the water balance. This was followed by work on concepts of resources management. Ann Larimore and Edwin Munger undertook a sample regional analysis, using an African area. A combination workroom and classroom was set aside for the group. During the quarter, students were expected to submit such evidence as field maps, appraisals of published maps, field reports, finished maps and charts, bibliographic analyses, short papers on numerous topics, oral reports, and outlines for more intensive work.

At the end of the quarter the faculty make independent and group assessments of student performance and needs. An individual, custom-made program for each student is then drawn up by two faculty members in consultation with the student. These programs range from advice that the student withdraw from graduate work to advice that he go directly into research with an agreed faculty member. In between there may be as many different programs as there are students. A few examples may indicate the kinds of solutions that are worked out.

Some students are encouraged to skip over most of the formal courses in geography and to go directly into reading courses or directed research. For some of these it may be necessary, however, to work intensively on foreign languages where these are involved in specialization. Thus, a student interested in the Middle East is steered into a year of Arabic and anthropology before returning to geography. A student concerned with economic geography may be weak in statistics or economics and be advised to take formal work in those fields. If climatology is in view, meteorology and botany will be required.

On the other hand, a student who has been teaching human geography but has been found weak in physical geography may be required to take the basic undergraduate work before going on to the graduate courses. A student who is unable to write lucid English is asked to work on it intensively and to defer any advanced work until he can meet the writing requirements.

Often two needs can be combined: a man who has a smattering of French and German but has never used it professionally, and who isn't clear as to whether he wants to specialize in urban geography, may be assigned a quarter of reading and writing from the French and German literature in the field.

In these and other ways the emphasis is placed upon identifying individual assets and liabilities as soon as practicable. The faculty feel that individual attention given early in a student's career may be more fruitful than that given later: the time to catch the difficulties and also to nourish the growing edges is at the outset.

In most instances we have found that the student's own appraisal of his needs and skills corresponds closely with that of the faculty. This is the case even where the judgment may argue strongly against moving ahead to an advanced degree. All too often in graduate work a student spends a year working for a master's degree before it is clear that he can't write. The doctor's preliminary examination can be too often an embarassing time of appraisal for all concerned. We are trying to identify and anticipate possible difficulties.

So far we are encouraged by the results. The system seems more promising than requiring a prescribed set of courses. We feel enthusiastic about its usefulness in our particular situation and would like to report it to our colleagues in the Association and to invite their comments.

Geography in Liberal Education

To restate this more explicitly, and at the risk of gross oversimplification, we might say that the final and only significant test that
need be applied to the adequacy of instruction would consist in
asking a student to put his finger in a random fashion on the
globe. Once he had located a spot on the globe, he would be
asked to predict what association of landscape features he might
expect to find at that place, indicating the scale of generalization
which he would give for that area and the degree of diversity
which he might expect among those characteristics. He also
would be asked what he would expect to be the consequences of
altering an element in the ecological complex.

Geography in Liberal Education (1965b), 18

Notes

1. For the context of this effort and earlier ones at Haverford, see
below, vol. II, chap. 7 (Feldman).

6   Strategic Aspects of Urban Floodplain Occupance

Returning to academic teaching and research as professor and chairman of the Department of Geography at the University of Chicago in 1956, White sought a research topic that would advance his long-term goal of broadening societal response to floods and at the same time involve colleagues and graduate students in a common undertaking.[1] After almost twenty-five years of professional life, this would be the first field study under his direction, the cheapest (it cost $8,000), and the most influential. With funds from the fledgling research institution Resources for the Future, the assistance of colleagues Wes Calef and Harold Mayer, and the summer service of three graduate students, White attempted to assay what had happened to the urban floodplains of the United States in the two decades since the passage of the Flood Control Act of 1936. By looking into the land-use history of seventeen sample urban areas, he hoped to resolve the paradox of why, after twenty years and $4 billion of expenditure, urban flood losses were not declining and were, perhaps, even rising.

The major research finding, that growth in the development of the floodplain had taken place everywhere, even in places with declining overall population, served as empirical grounding for a renewed effort to change the direction of public policy. Today we view this finding as a classic case of positive feedback, in which engineered reduction of the magnitude and frequency of flooding can paradoxically amplify people's exposure to floods by encouraging new floodplain occupance or retaining waters behind overtopped levees. It was the concrete demonstration of the paradox of seeming to make things better while perhaps actually making things worse that gave these findings such influential force.

But besides the attraction of paradox in its findings, there are three other characteristic White qualities in this study. The study is strengthened by the force of history—it is a 28-year retrospective. It examines only seventeen sites, but this is much more than the

Condensed from *Journal of the Hydraulics Division, Proceedings of the American Society of Civil Engineers* 86, no. HY2 (February 1960): 89–102.

usual two or three case examples of academic research, and they are part of a larger sample of 1,020 urban places identified as having "well-defined" flood problems. Finally, the study has been well-disseminated to different target audiences: a book-length research paper for serious students and practitioners (1958c), and in articles for conservationists (1958d), state officials (1959a), urban planners (1961a), and most important, for the civil engineers who dominate the flood control field. It is this last paper we reproduce here (abridged by excising the two case examples of Boulder and Denver). Its ending, a checklist of next steps in policy and research, is characteristic as well.

Introduction

A striking feature of national efforts at flood control in the United States is that in the more than two decades since a national policy was launched in 1936, the mean annual toll of flood losses has continued to rise. While at least $4 billion have been spent for engineering works to reduce and control floods, the economic losses from occasional flood disaster have mounted. To understand this paradox it is necessary to examine what has been happening in the human occupance of floodplains. To find out the full import of changes in floodplain occupance it is essential to identify those factors which enter into both public and private decisions as to floodplain use.

This paper outlines briefly the broad problem of flood-loss reduction in the United States, reviews the findings from a recent study of changes in urban occupance of floodplains, and then suggests three strategic aspects of such occupance that seem likely to affect the results of any further engineering efforts at flood control.

The broad problem of flood-loss reduction is that the rate at which flood losses are being eliminated by construction of engineering or land-treatment works is of about the same magnitude as the rate at which new property is being subjected to damage. Even though the data on which estimates of national flood loss are based seem too inaccurate to warrant any precise comparison of the two rates, it is clear, both from the aggregated damage statistics and from the record of selected floodplains, that the heavy investments in flood protection have effectively curbed the losses in many areas but that new damage potential is being built up at the same time.

Studies of seventeen selected urban areas having flood problems reveal a general and persistent encroachment of urban structures upon the floodplains during the period 1936–57, even in areas in which there was net decrease in total population. They show certain distinctive patterns of encroachment, and indicate highway construction and flood-control works as two major stimulants to growth. . . .

In the light of that experience it is argued that at least three aspects of urban occupance not ordinarily considered in flood-control plans in the past must be taken into account in the future if the tide of rising flood losses is to be turned. First, it must be recognized that engineering works are only one of the possible human adjustments to flood hazard. Second, it must be seen that the complex of elements entering into decisions as to future occupance of floodplains includes many considerations in addition to the traditional one of cost-benefit evaluation. Third, the range of choice now permitted property managers in dealing with the flood hazard is so restricted that radical changes must be made in public policy to broaden the range of choice among possible adjustments and to assure a full appraisal of each choice.

Trends in Flood Losses

There are two major sources for national estimates of flood losses. The Corps of Engineers issues, from time to time, an estimate of the total potential flood losses and of the losses that have been prevented or will be prevented by works constructed or planned by the Corps. . . . In 1954 the Corps found a potential mean annual loss of $964 million.[2] (All estimates are adjusted to 1957 price levels.) After all works then authorized had been completed, there would remain a balance of [losses of] approximately $210 million that would not be prevented. . .

The Weather Bureau issues the other series of flood-loss statistics, having published each year since 1903, an estimate of reported losses for each of its districts. The national totals for 1902–55 are shown in the lower part of the bars in the graph in figure 6.1. These totals have been adjusted to 1957 price levels, and the full height of the bar shows the totals on a comparable basis.

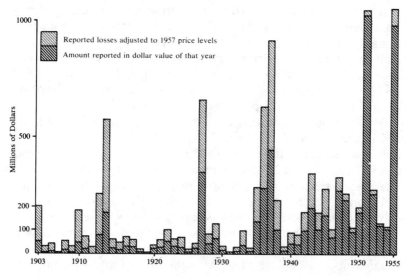

Figure 6.1 Estimated annual flood losses in the United States, 1903–55.

No attempt will be made here to appraise the validity and the discrepancies among these two sets of estimates and other less comprehensive series. This has been done in a separate publication.[3] Aggregate data alone are unsatisfactory and need to be checked in more detailed fashion. It is sufficient here to point out one characteristic which the two national estimates have in common. Both suggest a pronounced upward trend in the size of annual flood losses since 1936, when the first national flood control legislation was enacted. The estimated mean annual losses in 1936 were about $212 million. The Corps estimate of remaining annual losses after allowing for works completed was $444 million in 1954 and $700 million in 1959 (at 1959 prices). The Weather Bureau estimates of mean annual losses for 1924–53 were 25 percent lower than those for 1944–55.

Reasons for Mounting Flood Losses

One partial explanation for the apparently rising toll of recorded losses is that the inflation in dollar values has exaggerated the size of recent estimates. However, as shown in figure 6.1, even when values are adjusted to a common year there still remains a clear and sizeable increase.

A second reason advanced for the apparently mounting losses is improvement in enumeration and estimating methods. This unquestionably has played a role as Weather Bureau procedures have been made more uniform and as Corps of Engineers studies have reached areas not previously covered. Some who have been studying this problem believe that changes in enumeration methods may account for as much as 10 to 15 percent of the increase.

Perhaps of greater importance is the difference in number of large and infrequent floods. William G. Hoyt, F., ASCE and Walter B. Langbein, F., ASCE have shown a 35 percent increase in their "flood index" from the first 25 years of the century to the second 25 years.[4] There was a remarkable bunching of rare flows in the lower Missouri and Northeastern basins during the 1950s. This hydrologic record may account for as much as 25 percent of the increase in losses.

Probably the most important reason for the rising trend in flood losses is to be found in the continuing encroachment of human occupance upon floodplains. This takes the form of new structures, of changes in the intensity of existing structures, and of structures which so reduce the hydraulic efficiency of valley sections as to increase the hazard in affected reaches of the stream. Although there have been numerous studies of flood-control projects, there has been no comprehensive investigation of the changes that have been occurring in the areas subject to flood. One recent study, conducted by University of Chicago geographers, dealt with urban occupance because it has a high and increasing proportion of flood losses, is compact, and is more susceptible to change. It shows that there are at least 1,020 urban places with a population of more than 1,000 that have well-defined flood problems. That study also gives a precise picture of the changes in seventeen urban areas selected for their diversity in valley section, flood frequency and height, size, population growth, and types of land use.

Growth Patterns

The most evident and widespread trend that is to be observed in the urban floodplains is one of growth. In every place studied, including several in which the total population declined during the 21-year period, the number of structural units in the floodplain increased.[5] The growth rate ranged from less than 2 percent in a stable city like Wheeling, West Virginia, to more than 600 percent in a rapidly expanding place such as Dallas, Texas.

In most places the commercial and industrial structures grew at a more rapid rate than residential structures, although in a few of the places such as in sectors of Los Angeles, California, the residential expansion was substantial. Public structures in most places showed substantial growth rates.

Certain areal patterns of change seem to repeat themselves in urban floodplains (fig. 6.2). One typical pattern is found where residential areas in or bordering on the floodplain enlarge by moving down nearer the channel. A second is that in which industries already established in a flood-hazard zone expand laterally along the stream bank and at right angles away from it. A third occurs where a central business district spreads out into the floodplain, following the major traffic arteries.

There is no readily discernible relation between rate of population growth and rate of change in structures on the floodplain. Too many other factors such as urban function and availability of land play a part. It is clear, however, that two forms of public action have had a powerful stimulating effect upon invasion of floodplains. These are the highway program and the flood-control program. New highway construction in urban areas tends to follow the low gradients of stream valleys and the less densely settled sectors of some floodplains. It has caused substantial removal of low-quality residential

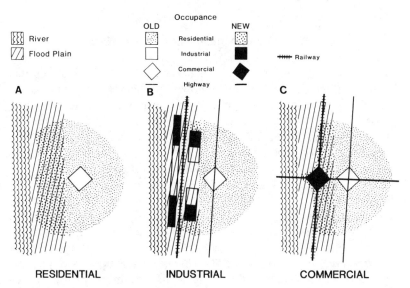

Figure 6.2 Schematic maps of three strategic land-use associations, showing patterns of growth in the floodplain.

structures but it also has been an inducement for commercial and industrial establishments to follow along into the flood-hazard zones.

As might be expected, the construction of new flood-protection works frequently has been the signal for accelerated movement into the floodplain. Thus, the completion of reliable works along the Trinity River in Dallas, Texas assured development of a large-scale commercial district behind the levees. But there are certain implications of flood-control works that are not as obvious. The lack of protection is not necessarily a deterrent to urban invasion of floodplains. In cities which have had no serious flood in 50 years, as well as in cities that had unprecedented flooding only a few years before, there has been extensive building without any immediate prospect of protection. Moreover, in certain valleys, such as in the Tennessee at Chattanooga, the completion of river-control works upstream has been followed by further movement into the floodplain where frequency has been reduced but where high flows still are possible. Although most federal flood-control works are built to protect against a project flood and conceivably will one day, however infrequently, be exceeded by a larger flow, there is a universal disposition to believe that the rare flow will never come. This means that the number of situations in which a catastrophic disaster may follow one of those rare flows is increasing as new levees are completed. . . .

Range of Choice in Adjustment to Flood Hazard

The traditional approach to flood losses in the United States has been to either control the flood flows by engineering works or to bear the losses, letting the public come to the aid of the less prosperous flood sufferers. This is shown by the heavy public emphasis upon investment in federal flood-control works and by the Red Cross and other emergency aid programs.

There are, of course, other possible adjustments which individual property managers can make. The whole range of choice may be listed as follows:

Bear the Losses.—A good many households and firms bear the losses when they occur, and a few actually make financial plans to do so.

Emergency Evacuation and Rescheduling.—Property may be removed from the reach of floods, and property movement and production operations may be rescheduled so as to avoid losses through interruption. This requires a relatively accurate system of flood fore-casting and a plan for emergency action when the critical forecast is received.

Prevent Flows.—In some areas it is practicable to prevent certain flows, particularly the more frequent flows, by land treatment and associated measures. The experience with such measures is only beginning to build up.

Elevate Land.—By landfill it is practicable in some situations to raise property above the level of flood waters. The effect upon chan-nel capacity varies according to location with respect to flowage and pondage areas on the floodplain.

Control Flows.—This is the conventional engineering solution, involving channel improvement, levees, cut-offs, and storage or de-tention dams.

Change Structures.—In advance of a flood warning, structures may be altered so as to prevent or reduce flood losses when the water rises. These changes may include such alterations as packing of machinery, bricking in of low openings, cut-off valves on sewers, and rearrangement of electrical circuits.

Change Land Use.—The use of floodplain land may be changed so as to introduce a use that is less susceptible to flood loss. This may range from the transfer of an entire town from a riverine to an upland site, to public acquisition of flood-hazard areas for recrea-tional or parking purposes. It may be guided by public regulation, including zoning, building ordinances, subdivision regulation, and land acquisition.

Insure.—Although insurance against flood losses is not generally available in the United States, there is an inactive program for fed-eral-state subsidized insurance under the Flood Insurance Act of 1956, and there are a few instances of coverage by private companies where special structural adjustments have been made.

Public Relief.—Through the medium of the Red Cross or by direct grants and loans under Public Roads, Civil Defense, Corps of En-gineers and small business programs, public aid is given to individ-uals and local governments that suffer heavily from floods.

There are concrete examples of each type of adjustment, and there are strong arguments that can be made for and against each type of adjustment, depending upon the local situation. To understand how and why particular adjustments are made in a given situation it is necessary to look to the various elements that enter into decisions as to management of urban floodplain properties.

Elements in Decisions as to Floodplain Use

As with most other resource-management situations, at least seven considerations seem to enter into decisions as to floodplain adjustments. It may be helpful to examine how each one figures in the urban situations which have been studied.

Estimating the Resource.—There is widespread ignorance of the flood hazard and a tendency to minimize it. Many people building or buying in floodplains are unaware of the precise hazard they are running or grossly misinterpret the technical estimates. This applies even in places where there have been public plans for flood protection. A man says he need not worry about floods because a 200-year flood occurred the year before. A federal housing insurance office says it does not insure mortgages for new buildings in floodplains but has no map showing where the flood-hazard areas are located.

Discounting Future Benefits and Costs.—It is common in federal flood prevention and flood-control studies to discount future benefits and costs from the proposed works. These cost-benefit ratios have significance in Congressional decisions so long as they are below unity. There is a tendency to group all programs with a ratio of more than unity together and to pay little attention to the means by which the ratios were calculated. So far as private planning of floodplain adjustments are concerned there is little evidence of such discounting procedure. A few managers of large industrial units make cost-benefit calculations but most managers do not.

Harmonizing Two or More Uses.—In the federal plans there is a disposition to harmonize flood-control plans with other water uses such as irrigation and navigation. At the level of local governments and private managers such multiple uses tend to be ignored. Few attempts are made to prevent harmful encroachments, and there is little attention to combining levees with highways, or to harmonizing a park development with a floodway improvement. One of the hopeful moves in this direction is under planning supported by the Urban Renewal Administration.

Projecting Future Demand.—Typically, the federal agencies assume that there will be little change in the demand for floodplain land while private managers tend to overestimate the effect of protection and to rush in at the hint or prospect of some control work.

Projecting Technological Change.—Private managers tend to overestimate the physical benefits from any kind of prevention or protection work. The completion of a single-purpose power dam upstream, (and even downstream), the beginning of a watershed improvement, or the clearing of an upstream channel is taken as assurance that floods will be abated in the future. Thus, a little protection work may encourage a great deal of channel encroachment.

Integrating Regional Uses.—While the federal agencies try to assess the consequences of a flood-control program upon other water uses in the same or adjoining basins, there is little attention to this on the part of local and private managers, and metropolitan area plans for storm water disposal are rare. Even less common are efforts to assess the possible use of floodplains in serving regional requirements for land and for riverine facilities.

Setting Social Guides.—We come now to one element which affects all the rest. Society, through its public agencies, definitely restricts some kinds of decisions by property managers and clearly encourages some other kinds of action. Federal agencies tend to guide the occupance of floodplains by providing prevention and protection plans, by giving public relief, by issuing storm and flood warnings, and by offering information on flood occurrence. Seven states attempt to regulate channel encroachment. A few cities and counties exercise some regulation over floodplain use. One federal agency cooperates with state planning agencies in assisting with local plans for floodplain use.[6]

Broadening the Range of Choice

In the present circumstances the social guides in the form of information, regulation, and investment affecting floodplain occupance tend to encourage further encroachment upon the floodplains at the same time that they lead to heavier federal expenditures for flood control. It is not entirely whimsical to say that the national situation is somewhat like a local situation where investigation showed forty new houses in a flooded area could not be economically protected by levee works. It was noted, however, that if twenty additional houses were to be constructed protection then would be feasible. Given the existing policies one may expect that the additional twenty houses will in time be built and that federal largess will follow in due course.

We know that the managers of floodplain properties often underestimate the hazard, fail to discount future benefits and costs, overestimate the demand for their land, overestimate the effects of engineering and land-treatment works, and ignore possible combinations of flood-loss reduction with other local improvements or with regional plans for land and water use. We also know that as a practical matter most managers who face a flood threat have a choice between only two alternatives. They can bear the losses, or they can press for a federal project of some sort. They do not receive technical advice as to the possibilities of emergency evacuation, land elevation, structural changes, or land-use changes. They do not have insurance readily available. If they suffer unduly heavy losses they have the prospect of public aid. If they are energetic enough they may receive public protection. With few exceptions they do not know the precise character of the flood hazard, and they are not subject to public curbs against any further encroachment upon the stream channel.

If we are to break out of the present situation in which flood losses promise to keep pace with flood protection, for at least several decades it seems essential to broaden the range of choice. Each manager—public or private—should be given the opportunity to choose among the whole range of possible adjustments, with the public agencies strictly limiting those choices which would cause damage to others or to the public safety. This calls for a new and fresh approach to the problem of flood loss reduction in the United States. An important beginning in that direction was made at the conference called by the Council of State Governments in December, 1958.[7] The new approach does not require abandonment of existing public programs. It requires that they be supplemented by new or expanded efforts to broaden the choice open to all who face flood hazard.

The chief directions in which it now seems necessary to move are these:

1. Publication and wide distribution of flood hazard maps and reports for all important flood hazard areas, to be made available to both property managers and to public lending and construction agencies.

2. Technical advice to property managers as to the means and costs of reducing flood losses by emergency measures, by structural changes, and by land-use changes.

3. A cooperative federal–state program of flood insurance in which premiums are in proportion to hazard.

4. State regulation of any further channel encroachments which would cause damage to others or to the public safety, and, federal requirement that additional federal expenditures for flood control be contingent upon such regulation.

5. An improved system for issuing flood warnings and for getting them into the hands of the property manager concerned.

6. Technical assistance to local governments in drawing up plans for reducing flood losses by whatever combination of means may seem most feasible.

Role of Geography in Water Resources Management

Support for geographic research on water has been relatively minor compared to support for other lines of geographic study. Typically, if the work is to be pertinent, it requires collaboration with investigators in other disciplines, and thus does not lend itself to the devotion of a single investigator. When geographers have sought funding for cooperative studies they generally have been successful, but they have had to take the initiative. In the long run, the demands made upon researchers from all the social sciences will be linked with the kinds of questions asked by water planning agencies, and these, in turn, will be influenced by interest groups involved in making the final decision. If these groups are satisfied to regard water management as a discrete, conventional field serving ambiguous ends, there will be little need for research on alternatives and their consequences and little disposition to strive for holistic assessments of them. To the extent that alternative aims and means are studied with conviction and without preconceived selection, the appetite for guidance from the findings of social research will enlarge.

The promise far exceeds the record. Yet, the record is substantial enough to warrant arguing that a program of investigation of water resources management alternatives which does not take into account the analytical techniques that geographers have developed is bound to be inadequate. To put the conclusion succinctly, any group establishing a research program in water management or carrying out a comprehensive river basin planning investigation is likely to gain from the viewpoints and skills of geographers. Geographers may keep the group from overlooking major aspects of relationships among natural and social systems in the area to be manipulated by water management. They can aid in conscientiously examining the full range of choice open to society in dealing with water, including engineering works as well as readjustments in patterns of land use, water use, or in social constraints affecting such use. They may bring new understandings of the ways in which managers of resources perceive the resource and therefore respond to new information or social incentives and constraints in use of the resource. Geographers will not assure cool and comprehensive consideration of all these points. However, they have demonstrated that they have a significant and unique contribution to make. They also have shown a capacity to work with other disciplines in probing for new answers to the puzzle of man's interrelationship with water and land.

(1974c), 118–19

Notes

1. See the discussion on the "harvest of research" below, vol. II, chap. 2 (Platt).
2. U.S. Congress, House Committee on Public Works, "Costs and Benefits of Flood Control Programs," House Committee print 1, 85th Cong., 1st sess., 1957, 2.
3. Gilbert F. White, Wesley C. Calef, James W. Hudson, Harold M. Mayer, John R. Sheaffer, and Donald J. Volk, "Changes in Urban Occupance of Flood Plains in the United States," 1–19. University of Chicago Department of Geography Research Paper, no. 57 (Chicago, 1958).
4. William G. Hoyt and Walter B. Langbein, *Floods* (Princeton: Princeton University Press, 1955).
5. Changes in occupance were measured in "structural units." A unit was defined for different occupance classes as: Residential, A—a single- or double-family dwelling, B—a multi-family dwelling for 3–6 families, and C—a multi-family dwelling for more than 7 families; Commercial, Industrial and Transport, separate structures in multiples of 10,000 sq. ft. and open working space of 25,000 sq. ft.; Public, each separate building.
6. Francis C. Murphy, "Regulating Flood-Plain Development." University of Chicago Department of Geography Research Paper no. 56 (Chicago, 1958).
7. Gilbert F. White, "A New Attack on Flood Losses," *State Government* 32, no. 2 (Spring 1959): 121–26.

7 Industrial Water Use: A Review

In 1953 White wrote, "It is common knowledge that there now are large sections of the United States in which certain types of industrial development are impracticable because of the inadequacy of existing water systems" (see selection 3 in this volume). Seven years later, in this review of industrial water use, he critically queries the "common knowledge."[1]

This review embodies a combination of research and presentation techniques White has used repeatedly throughout his career. He begins with a specific resource and carefully reviews what is known empirically about its use, supply, demand, and technological change. Multiple estimates of each characteristic are sought, contrasted, and held up to the mirror of international practice. The range of choice, central to his floodplain paradigm, is enumerated, and examples of experience are brought together. Along the way at least one existing myth is challenged; one bit of conventional wisdom is overturned.

In this case two such views are challenged. The excessive "demands" for industrial water use embodied in planning projections are deflated. The view that water is a major limit on industrial location is debunked. So persuasive is the quiet combination of reason, fact, and alternative views that twenty years later, the common knowledge changed. The nonlinear extrapolation of water requirements is now common practice, and industrial water reuse and recirculation are common features of plant design.

A lively interest in the industrial use of water has arisen in recent years in response to a growing concern with water resources and with multiple-purpose development of rivers in areas of expanding industrial plant. This interest is reflected in a series of recent inventories and assessments of such use, of which the United Nations report on "Water for Industrial Use"[2] is perhaps the most succinct

Reprinted from *Geographical Review* 50, no. 3 (1960): 412–30.

and comprehensive and the United States 1954 Census of Manufactures[3] the most detailed. A more precise picture of the industrial role of water is beginning to take shape—and in some respects a radically different one from that presented in American geographical literature.

While the facts about the quantities of water consumed by industry in the United States have been gathered and published for the first time on a national scale, a pioneer though fragmentary attempt has been made to compare experience on a world scale. Geographers in Belgium, Germany, and the United Kingdom have called fresh attention to industrial water needs. Rapid expansion of manufacturing capacity in some industrialized countries, introduction of industry into new areas, and heavier competition for water sources from growing cities and supplemental irrigation enterprises have combined to point the increasing importance of industrial use as an aspect of water economy. Much more precise data have been assembled on the quality requirements of industry, and this research, too, has been spurred by greater complexity of processing techniques and by more exacting public standards of water purity for such uses as wildlife propagation and recreation.

The new material collected under these incentives enables refinements in the description of the areal distribution of industrial water use, but it also raises several special problems as to the role of water in industry, and it throws into question some of the alarming predictions of recent years that parts of the United States, and other countries as well, may soon run short of water for industrial purposes.

In 1952 the Paley Commission's monumental report, *Resources for Freedom*,[4] predicted that "by 1975, access to good water may become the most important factor in deciding where to locate industries." Since then the popular and technical press has carried many references to this allegedly impending change in the water situation.[5] For example, in 1957 the *Wall Street Journal* reported: "The nation's water use has increased six-fold since the turn of the century. It now stands at 250 billion gallons daily. Economic growth in some areas already is stunted by water shortages, and demand is expected to double by 1980."[6] The Paley Commission projections have been accepted without much critical examination and are widely used to support statements that water shortages may soon become acute. The same line of thinking is reflected in the United Nations report:[7] "the selection of industries requiring a minimum amount of water will avoid to a large extent the difficulties" in areas of "water shortages."

The data currently at hand suggest that growth in industrial water use need not be as large as predicted and may well be smaller, and that there may not be large areas whose growth is or need be stunted by shortage of potential water supplies. The data also reveal that there are still critical gaps in knowledge about the conditions in which industry uses water, and that in the absence of precise knowledge, it is difficult to venture final judgments on the part water has had, or may be expected to have, in industrial location.

These recent data and studies are reviewed here to indicate their content and to call attention to elements in decisions with respect to industrial water management that they either establish or leave in doubt.

Total Quantity of Industrial Use

Until the 1954 Census of Manufactures there had been no comprehensive enumeration of industrial water uses in the United States. A beginning had been made in the 1953 *Annual Survey of Manufactures,* but the first complete coverage—reporting the water intake by 159,264 establishments with six workers or more—came in the following year. Several attempts have been made to estimate the quantities of water withdrawn from surface and ground sources. In 1950 the National Association of Manufacturers and the Conservation Foundation made a questionnaire canvass of some 3300 industrial users, from which they estimated the uses, percentage increase, treatment, and disposal of water by industries and by regions.[8] MacKichan of the United States Geological Survey, working from some earlier efforts, estimated industrial use by sources and by major hydrographic regions as part of an estimate of all water uses in the United States for 1955.[9] In 1956, Picton of the Department of Commerce estimated water withdrawal in the United States for 1900–75, and in the following year Woodward, then working at the Industrial College of the Armed Forces, estimated industrial needs for 1980.[10] And most recently projections of industrial water use for 1959, 1980, and 2000 prepared by Resources for the Future and the Department of Commerce were published by the Senate Select Committee on National Water Resources.[11]

In these efforts (except the recent projections for the Senate Committee, received too late to assess) there were significant differences in enumeration definitions, which, together with the differences in years, may help explain estimated total industrial withdrawals exclusive of fuel-electric plants of 38.7 billion gallons daily by MacKichan and 33.1 billion gallons by the Census of Manufactures.[12] Some estimates of industrial water use take into account the water used for cooling and boiler feed at plants generating steam-electric power. The Census of Manufactures does not count this. Mac-Kichan, using data from the Federal Power Commission, found the fuel-electric water withdrawals to be about twice the total withdrawals of self-supplied industrial plants. A separate estimate of the use of water in thermal-electric plants was made for 1954 by the Federal Power Commission.[13] Because the fuel-electric withdrawals are nonconsumptive and change only the temperature of the water, and because they are susceptible of great economy in shortages, they properly should be separated from the main body of industrial uses.

If we take the 1954 census as more nearly accurate, several striking facts regarding current industrial uses are apparent. Of the total number of establishments reporting, about 3.7 percent (10,226) account for 97 percent of the total water used. These are the establishments with a gross intake of 20 million gallons or more a year.[14] Among the large users about 62 percent of the intake is used for cooling and air conditioning, 26 percent for processing, and 12 percent for boiler feed and other purposes. Unfortunately, such aggregated data are not at hand for most other parts of the world. Only a few countries have any kind of estimate of national water use; the most comprehensive data come from Germany and the United States.

It is possible, however, to use the United Nations report and selected fragmentary data to show how industrial use compares with other water uses in several countries or national subdivisions that have widely different proportions of manufacturing activity to total employment. These are presented in table 7.1. The situation in areas that have not yet been highly industrialized is suggested by India and the Tigris-Euphrates region; the case of a highly industrialized economy is revealed by the United Kingdom, the United States, and the Nordrhein-Westfalen area. After the United States, the largest per capita industrial use is reported in Finland, where one large water-using industry, paper and pulp making, dominates the manufacturing structure. Per capita use in the West German area is slightly more than one-half of that in the United States; per capita use in the United Kingdom is less than one-quarter of that in the United States. These indicate that dense industrialization need not result in the voracious appetite for water which now prevails in the United States.

Table 7.1   Estimated water withdrawals in selected areas

Water Withdrawals	Finland	India	Iraq (Tigris-Euphrates)	Union of South Africa (Vaal)	United States	U.K. (England and Wales)	West Germany (Nordrhein-Westfalen)
	(In millions of U.S. gallons per day)						
	1950	1955	1957	1951	1955	1955	1951
Municipal	52	6,411	34	96	17,000	2,349	327
Industrial	716	5,531	21	70	33,088	2,040	619
Steam-electric power	?	?	—	22	72,200	18,000	?
Rural, including irrigation	40	139,932	3,193	223	112,400	?	?
	(In gallons per capita daily)						
Total population (*in millions*) and date	4 (1950)	372 (1953)	4 (1952)	2 (1951)	162 (1954)	44 (1951)	13 (1950)
Municipal	12	17	4	43	104	53	68
Industrial	177	14	4	30	204	46	130
Steam-electric power	?	?	—	10	420	408	?
Rural, including irrigation	10	376	655	101	700	?	?

Sources: Finland: "Water for Industrial Use," *United Nations Publication E/3058, ST/ECA 50,* Department of Economic and Social Affairs, New York, 1958 (Sales no: 58.II.B.1). *India:* R. C. Seipp, "Ecomomic Growth and the Organization of Scientific Research" (Thesis, Ph.D., University of Chicago, 1958), 120. These are rough estimates but are believed to indicate the magnitude of the different uses. *Iraq:* W. H. Al-Khashab, "The Water Budget of the Tigris and Euphrates Basin" (*University of Chicago Department of Geography Research Paper* no. 54, 1958), 61 and 68. Residential and industrial uses are estimates based on samples selected from Baghdad and Basrah. *Union of South Africa:* "The Vaal River: Report on the Water Supplies of the Vaal River in Relation to Its Future Development" (Union of South Africa Natural Resources Development Council, Pretoria, 1953), 10 and 37. Estimates are for the Vaal below Vaaldam. *United States:* Industrial use from *1954 Census of Manufactures: [Subject] Bulletin MC–209,* U.S. Bureau of the Census, 1957; other uses from K. A. MacKichan, "Estimated Use of Water in the United States, 1955," *U.S. Geological Survey Circular 398,* 1957, 11 and 13. *United Kingdom:* Municipal and industrial estimates from "First Report . . . Sub-Committee on the Growing Demand for Water" (Ministry of Housing and Local Government, London, 1959), 4–5, 21, and 25. Estimates for industry based on questionnaire returns from six industries and on net consumptive use of nationalized industries. Fuel-electric power estimate from W. G. V. Balchin, "Britain's Water Supply Problem," *Water and Water Engineering* 61 (1957), 290. *West Germany:* Personal communication from Reiner Keller. About one-third of the municipal water is delivered to industry.

Population Estimates: Demographic Yearbook 1954, Statistical Office of the United Nations, New York, 1954, Tables 1 and 2. Population estimate for the Vaal below Vaaldam in "The Vaal River" [see citation above]. Population estimate for Nordrhein-Westfalen from Reiner Keller, *Geogr. Taschenbuch 1956–1957,* Wiesbaden, 1956, 306.

Total Use: Estimates and Projections

One question raised by the census data is why there is such a marked discrepancy between the reported industrial uses and the estimates made by the Paley Commission, the Department of Commerce, MacKichan, and Woodward for the same or the following year. The several estimates are compared in table 7.2, by interpolation from the four reports. It will be seen that the estimates for 1954 substantially exceeded the amounts reported by the census. If projection and fact continue to differ by the same proportion over a few years, the earlier estimates for 1975 and 1980 will be far out of line.

One obvious cause of the discrepancy is the difference in enumeration techniques. The Department of Commerce estimate is restricted to self-supplied water but includes commercial, motel, military, and miscellaneous establishments not covered by the census. The census estimate is based on forms returned by the manufacturers; the other estimates relied on observation. The census

Table 7.2   Estimates of industrial water withdrawal in the United States, 1950–75
(*In average billion gallons daily*)

Source of Estimate	1950	1954	1955	1975
President's Materials Policy Commission[a]	45.0	52.2[d]	54.0[d]	90.0
U.S. Department of Commerce[b]	46.0	57.2	60.0	115.4
Woodward[c] (including thermal-electric power)	69.1	93.1[d]	99.1	350.0[d]
MacKichan[e]	—	—	38.7	—
U.S. Census of Manufactures[f]	—	33.1	—	—

Sources: (a) Resources for Freedom: A Report . . . by the President's Materials Policy Commission (5 vols., Washington, D.C., 1952), vol. 5, 94. Includes some saline water; excludes thermal-electric uses. (b) W. L. Picton, ''Summary of Information on Water Use in the United States, 1900–1975,'' *Business Service Bulletin 136,* U.S. Dept. of Commerce, Business and Defense Services Administration, 1956, 4. Only self-supplied industry, including mineral, commercial, resorts, and military establishments. (c) D. R. Woodward, ''Availability of Water in the United States with Special Reference to Industrial Needs by 1980'' (Thesis no. 143, Industrial College of the Armed Forces, Washington, D.C., 1957), 33 and 39. Gross use of fresh water, including thermal-electric. Woodward's data for 1950 and 1955 are based on MacKichan, and the 1975 figure is interpolated from Woodward's 1980 estimate. (d) Interpolated by the writer, assuming linear trends. (e) K. A. MacKichan, ''Estimated Use of Water in the United States, 1955,'' *U.S. Geological Survey Circular 398,* 1957, 11. This excludes an estimated 5.7 billion from public supplies. He also published (''Estimated Use of Water in the United States—1950,'' *U.S. Geological Survey Circular 115,* 1951, 5–7) an estimate of 77,216 million gallons daily for 1950, but this did not separate thermal-electric plants and did not include water taken from public water systems. (f) ''Industrial Water Use,'' *1954 Census of Manufactures: [Subject] Bulletin MC– 209,* U.S. Bureau of the Census, 1957, 32. Includes some brackish water.

excludes water use by commercial and mining interests and by Atomic Energy Commission establishments. This latter use is large in certain areas: in Georgia, for example, a statewide study in 1955 showed all industrial withdrawals of water exclusive of those by atomic-energy plants and public supplies to be 1830 million gallons daily (1450 million gallons for steam-generating plants and 380 million for all other industries), but water withdrawn by the Savannah plant of the Atomic Energy Commission amounted to an average of 720 million gallons daily.[15] Another, possibly more significant, cause of discrepancy is the difficulty of taking into account the wide differences in amount of water used per unit of product. This will be considered later. But there is not yet any thorough regional examination that uncovers the full explanation for differences in estimates of industrial use.

Types of Industrial Use

Whatever the gross quantity of industrial water use in the United States, the 1954 census did not greatly change the picture of types of use as it had been drawn in a 1953 symposium of the American Association for the Advancement of Science[16] or in reports of American Water Works Association committees. The symposium papers were the first to attempt an assessment of the whole range of industrial water problems and to place geographical analysis alongside engineering and hydrological.

Amounts used in each major industrial group are shown in table 7.3. Three facts stand out clearly from this summary and related census data. The first is that four industries—primary metals, chemicals, pulp and paper, and petroleum and coal—use about four-fifths of the total. The second is that these heavy users are the industries withdrawing the largest volumes of water per dollar value added by manufacture; for example, the petroleum group, with 13 percent of the total intake, uses 579 gallons for each dollar value added in production, the largest of all the groups. Third, the large users are also the industries in which the water use per unit value of product is high. With the major consumers cooling is the largest single use. In most industrial use, and particularly in the water-hungry industries, water's major role is that of a transportation agent, carrying heat and materials in the course of the process.

Somewhat similar findings appear in the other countries for which data are available. In Finland, 91 percent of the use is by one in-

Table 7.3 Water withdrawn per dollar value of product in the United States, 1954

Industry Group	Total Intake (*billions of gallons*)	% of Total	Value Added by Manufacture (*millions of dollars*)	Water Withdrawn per Dollar Value Added by Manufacture (*gallons*)	% of Intake by Establishments Using Less Than 20 Million Gallons
Petroleum and coal	1,519	13	2,622	579	0.2
Primary metals	3,651	31	9,390	389	0.3
Pulp, paper	1,613	14	4,630	348	0.4
Chemicals	2,827	24	9,611	294	0.6
Stone, clay and glass	282	2	3,866	73	5.3
Textile mill	287	2	4,709	61	4.9
Rubber	115	1	1,954	59	1.7
Lumber and wood	166	1	3,242	51	19.9
Food and related	677	5	13,769	49	12.9
Fabricated metal	220	2	7,653	29	10.9
Transportation equipment	259	2	13,776	19	2.3
Electrical machinery	126	1	7,300	17	6.3
Miscellaneous	74	*	4,393	17	20.3
Leather and leather goods	25	*	1,641	15	20.0
Machinery, except electrical	143	1	12,183	12	16.1
Instruments and related	22	*	2,132	10	13.6
Furniture and fixtures	16	*	1,998	8	56.3
Printing and publishing	36	*	6,412	6	63.9
Tobacco	3	*	1,004	3	33.3
Apparel and related	16	*	5,166	3	100.00
U.S.	12,077	100	117,451	103	2.6

Sources: "Industrial Water Use," *1954 Census of Manufactures: [Subject] Bulletin MC–209*, U.S. Bureau of the Census, 1957, 52; and vol. 2, part 1, 24.
*Less than 1 percent, totalling 1 percent.

dustry; in Norway, according to a sample survey, paper and paper products, chemicals and chemical products, and primary metals account for 78 percent of total industrial use.[17] Reports from West Germany, the only other country for which detailed data are at hand, indicate that iron and steel, chemicals, coal mining, oil refining, paper and pulp, textiles, and food industries claim 86 percent of industry's total use.[18] The evidence from diverse industrialized areas shows that the overwhelming amount of water use is found in large operating units in a few industries, and that its predominant application is for cooling.

The United States census also gives new evidence on two other quantitative aspects of industrial use that are known to vary widely among manufacturing plants—the consumptive use of water, and the total withdrawal of water for different manufacturing products. The census did not attempt to collect data on consumptive use for all establishments, but for those with an annual withdrawal of 20 million gallons or more it found that about one-tenth of the total intake is reported not discharged and thus is presumed to be consumed in the product or through evaporation and system losses.

Comprehensive data are still lacking on the precise amounts of water used for manufacturing the principal products. Three general methods of expressing water use in industry have been employed. In Australia and South Africa and, more recently, in the United States, industrial water use has been expressed in terms of the monetary value of the product. The Australian data, for example, show water cost as ranging from 0.03 to 0.34 percent of the value of output.[19] In some recent forecasts, such as those for the Muskingum Basin and for Oakland County, Michigan,[20] water use is expressed in volume of water used per worker employed. More common is the practice of relating water use to total industrial production, and the most incisive comparisons of value, employment, water intake, and water depletion have been made by Kneese.[21] In all these methods, however, the essential step is the measure of water used per physical unit of product.

It is possible to compute from the 1954 census the average amounts of water for the value of product of each industry given in table 7.3. What is much more important to know is the economic range of water reuse and recirculation; technically, almost all water not consumed in processing could be reused if the manufacturer were willing and able to pay the cost of pumping and treating the waste water so that it would be restored in temperature and chemical qualities to those permissible for the process. Net water loss may be very small, for example, in a petroleum refinery where water is recirculated twenty-five times before being discharged. Even greater economy can be attained in a closed system. A completely closed system is rare, yet information is meager on precisely how far conservation has been practiced and on the conditions in which it is favored. One line of analysis followed in the United Nations report[22] is to record the range of water use in operating situations, a method that indicates a possible spread in application of technical devices. A similar compilation from United States experience appears in a University of Illinois study for the Air Force Cambridge Research

Center in which a wealth of technical reports on industrial water usage are abstracted and compiled.[23]

The 1954 census, by reporting for the large plants the number "recirculating or reusing water" and an estimate of the amount of "water required if no water was recirculated or reused," gives another kind of check on the degree to which conservation is practiced in particular industries and hydrographic regions. Ackerman and Löf, in their comprehensive and thoughtful study of technology in American water use,[24] offer an analysis of water-conservation measures in industry, examining the extent, types, conditions, and effects of water reuse, and citing many examples. They include a map showing the distribution by major drainage areas of water withdrawal, brackish-water use, and water-recirculation measures as reported in the census data. It is known, however, that within a small area there may be extreme range in use in the same industry. For example, in the Western Coal Field of Kentucky in 1951–52, the water used in washing a ton of coal was 25 gallons in one large producing mine, 1065 gallons in another, and an average of 110 gallons in the two dozen mines studied.[25] The range of water-use efficiency as revealed by a number of published studies is summarized in table 7.4 for one segment of each of the four major water-using industries. It is striking that in each of the industries the maximum use reported is at least nine times greater than the minimum reported.

Quality and Source of Water

Although it is known that the physical, chemical, and biological properties of water play radically different roles in different pro-

Table 7.4 Range of reported water use in four industries
(*In U.S. gallons per unit of product*)

Industry	Unit of Product	Minimum	Maximum
Primary metals[a]	Ton of finished steel	1,400	65,000
Chemicals[b]	Ton of sulphuric acid	780	7,270
Pulp and paper[b]	Ton of dry groundwood pulp	4,860	60,820
Petroleum[a]	Gallon of crude oil	0.8	44.5

Sources: (*a*) H. E. Hudson, Jr. and Janet Abu-Lughod, "Water Requirements," in *Water for Industry, American Association for the Advancement of Science Publication no. 45*, J. B. Graham and M. F. Burrill, eds. Washington, D.C., 1956, 19; (*b*) "Water for Industrial Use," *United Nations Publication E/3058, ST/ECA/50*, Department of Economic and Social Affairs, New York, 1958 (Sales no.: 58.II.B.1), 24–25.

cessing activities, the previously available data on water quality refer largely to particular plants. The United Nations report draws heavily on Nordell's compilation in indicating ways in which water is used and treated in particular industries, and along with this study should be noted the comprehensive listing of industrial water uses in Germany.[26]

A major contribution to an understanding of both use and treatment as reflecting quality requirements and source is to be found in three series of reports published during the past nine years by the United States Geological Survey. One series reports the quality and quantity requirements of the aluminum, carbon-black, paper, and rayon-fiber industries.[27] These studies indicate a downward trend in unit water use in processing for some industries, such as paper, but a rising trend for some others, such as carbon black. Unit water requirements for sample products are summarized in part in table 7.5.

A second series reports the chemical quality of public water supplies viewed with respect to their utility for industrial use, bring-

Table 7.5   Unit water requirements in selected industries
      (*In gallons per unit of product*)

Industry and Process	Unit of Product	No. of Plants Studied	Minimum	Maximum	Average
Paper—fine and book	Ton	9	20,000	184,000[a]	25,000[b]
Paper—kraft	"	7	17,000	82,000[a]	30,000[b]
Sulphite pulp	"	4	38,000	80,000	75,000[b]
Carbon black (contact process)	Pound	27	0.02	0.90	0.14
Carbon black (furnace process)	"	22	0.83	20.52	3.34
Alumina (after plant changes)	"	6	0.28	26.00	3.48
Aluminum reduction (after plant changes)	"	14	1.24	36.33	16.05
Rayon—viscose and cuprammonium	"	27	30	240	110[c]
Acetate	"	6	41	494	170[c]

Source: O. D. Mussey, "Water Requirements of the Pulp and Paper Industry," *U.S. Geological Survey Water-Supply Paper 1330-A,* 1955; H. L. Conklin, "Water Requirements of the Carbon-Black Industry," *ibid.,* 1330-B, 1956; *idem,* "Water Requirements of the Aluminum Industry," *ibid.,* 1330-C, 1956; O. D. Mussey, "Water Requirements of the Rayon- and Acetate-Fiber Industry," *ibid.,* 1330-D, 1957.

[a]Plants using pulpwood rather than pulp.

[b]Weighted average, reflecting author's judgment based on experience of other plants as well as of those surveyed.

[c]Median.

ing up to date a report on the situation in the larger cities in 1932.[28] Each earlier report begins with the statement: "The location of industrial plants is dependent on an ample water supply of suitable quality. Information relating to the chemical characteristics of the water supplies is not only essential to the location of many plants but also is an aid in the manufacture and distribution of many commodities."

The third series consists of detailed studies of supply and use in some twenty metropolitan areas.[29] Each report canvasses supplies, present demand, and potentialities for both ground and surface water. It is remarkable that in all these regional assessments no serious curtailment of industrial growth as a result of water shortage is predicted, though it is noted that in most areas some engineering and control measures will be needed in order to assure expanded supplies. In two areas—the Mahoning Basin and the San Francisco Bay Area—further importation of water is regarded as likely, to assure adequate quality and volume of water.

Quality, of course, bears a close relation to source of supply. Where fresh water is relatively expensive, as in the Chesapeake Bay and Texas Gulf areas, brackish water is often substituted. It also appears that water-treatment processes have become more elaborate to the degree that available sources are short of meeting quality standards.

A whole new literature has grown up on the problem of desalinization. Cost is the principal limit, and municipal and industrial uses the chief ones for which plants have so far been designed. A general review of the cost limits for industrial purposes, drawing from experiments over a wide front and including vapor distillation, ion exchange, solar distillation, and ice processes, is given in the annual *Saline Water Conversion Reports* of the Office of Saline Water of the Department of the Interior.[30]

For the large water users in the United States in 1954, about 13 percent of the withdrawal was brackish water, 12 percent came from company groundwater systems, 59 percent from company surface systems, and the remaining 16 percent from public water supplies, both surface and ground. A rather similar situation appears to prevail in other industrialized countries. This is especially significant because it indicates that well over 84 percent of the water used in industry is withdrawn by the companies themselves and that each industrial manager has been faced with an independent decision as to the source, reliability of supply, quality of water, and amount of conservation he would provide.

Problems of Growth and Location

If industry in the wide range of areas studied has the ability to vary its use of water enormously according to local conditions without any apparent serious economic consequences, it may be asked whether industrial use will necessarily continue to grow in linear relation to the level of productive activity, as is assumed in the Paley Commission and Commerce Department projections. For as water supply became shorter, the amount used per unit of product might be expected to decrease. It also may be asked for the same reasons whether water is likely to become "the most important factor" in the location of new industries. The answers to these two questions are intertwined and lie in a more nearly precise understanding of water as a locational factor than has been developed so far.

It may be argued from the evidence now available that physical supply of water has not in fact been a major factor in the location of industry in the United States and several other industrialized countries during recent years: that water has been plentiful in many areas, and where it is not readily available in suitable amount and quality, industry has made the necessary investment in supply, treatment, and recirculation facilities to correct deficiencies. There is now strong ground on which to challenge the assertion that water's role in industrial location will become more important.

Drawing on reasoning as to the growing role of water as a limiting factor, the United Nations report[31] suggests: "It is necessary therefore that demand for water by one type of industry as against different types, and the quantities and qualities of water available at reasonable cost be considered, among other things, as a guide in selecting industries to be developed in any country and area." The report hypothesizes that industries requiring large amounts of water will be at a disadvantage in areas short on water.

A more likely hypothesis is that although some water is essential to all industrial development and although, other factors being equal, the availability of water of suitable quality may play a major part in decisions on industrial location, the greater number of industrial establishments adjust their processes and facilities to available water instead of locating where water conditions are optimum. Is it possible that in giving due weight to the necessity for water in industrial processes we have overemphasized its strategic importance in industrial location? To put the case more concretely, are there, in fact, more than a few large industrial plants that would be likely to give up a proposed location solely because water supply seemed inadequate? Most manufacturing establishments are small water users. Many of those that do use large quantities are able to reduce usage by water-conserving devices and to obtain new ground or surface supplies at relatively low cost.

Several pieces of information suggest that this position may have merit. When the water actually used in industry is expressed as a ratio of the additional water that would be used were it not for reuse and recirculation, it appears that there is a tendency for the highest conservation ratio to occur where mean total supplies are most highly used for all purposes. A thorough geographical analysis of water-recirculation practices has not yet been made, and much remains to be done by way of recognizing areal differences in industrial conservation measures in relation to potential supplies. A preliminary analysis of the census data shows many anomalies.

Interviews with representatives of several national factory-location consultants and with some state development agencies have revealed so far only a few industries that have rejected a possible location solely or even largely because water supply was inadequate. Balchin[32] indicates that, although industry is of growing importance in Britain, critical problems have only begun to emerge. Likewise, a number of the intensive studies of industrial location have failed to identify water as a major locational factor.

The materials marshaled by C. Langdon White in his study of the iron and steel industry[33] might be interpreted as supporting the same argument. He shows that the Fontana plant in Southern California, in an area where water is relatively expensive, uses 1400 gallons per ton of finished steel at a cost of 2 cents per 1000 gallons. But the Fairless plant on the Delaware River, in an area where water is relatively cheap and available, uses about 37,000 gallons per ton[34] at a cost of 0.9 cents per 1000 gallons. White[35] concludes that "when once the problems of assembling coking coal, iron ore, and limestone at the blast furnaces and of supplying markets with steel have been solved, the availability of a huge supply of water of good quality and at reasonable cost becomes a major location factor." On the other hand, a group of experts under the sponsorship of the Organization for European Economic Cooperation, who in 1955–56 investigated water use in at least seventeen iron and steel works in Austria, France, the German Federal Republic, and Italy, arrived at a different judgment. They found that none of the works enjoyed sufficient water of satisfactory quality. Limitations on natural supplies, decreases in groundwater reserves, public controls on waste-water quality, efforts to reduce the cost of pumping water, and the high costs of treating natural water have led to widespread recycling of water even where there are fairly ample reserves.[36] Water use per United States ton of crude steel was found to range from 871 to 66,834 gallons. The experts observed that the new closed-circuit systems had marked advantages in reducing some production costs and that with them "iron and steel works can be more independent than hitherto in choosing a suitable site."[37] A recent report on one of the new British plants located in a water-short area makes a similar point:[38]

> . . . if sufficient water resources are not available to permit of installing a once-through system of cooling-water supply for a proposed steelworks project, the ability to economize on water requirements by installing a recirculation system could represent ability to install a new steelworks in a locality which is convenient for reasons other than availability of unlimited cooling-water supply. When the capital and running costs associated with water treatment are regarded as elements of the overall costs per ton of steel produced, the apparently high expenditure on water treatment assumes its true significance to the steelworks engineer.

The technical literature on European steel, chemical, petroleum, and paper manufacturing[39] and Ackerman and Löf's summary of American experience seem to support the same point. Advantages of quality control and of reduced pumping costs are noted for recirculation systems quite independent of problems of source of water. Gibson's study of the paper industry of northwestern England notes a range in intake of water per ton of product from 3150 gallons to more than 400,000.[40] He shows the relation of location of both water supply and waste disposal to plant location and roughly assesses the effect on water intake of plant size, raw materials used, quality of paper product, and amount of recirculation.

Sporck,[41] from his studies of industrial location in Belgium, concludes that water is a factor of importance but that changes in the technology of water and energy use are permitting industry greater latitude in the choice of new locations within large areas. Keller,[42] on the basis of his broader studies, reports somewhat similar views, giving Euskirchen, Germany, a water-intensive industrial complex in a water-poor area, as an example. The area of most intensive industrial water use in the world, the *Land* Nordrhein-Westfalen, is reported to have increased its steel output during 1952–55 while lowering its water consumption.[43] The water resources and problems of that area are the subject of a careful description by Göpner[44] that details the intricate system of public-supply enterprises and their surface and ground sources and points out lines along which further development might proceed. Hessing[45] also has summed up the water problems of the area as presented at a 1958 gathering of German water technicians, with special attention to interconnections of public and private supplies, price differentials, protection of watersheds, land for waste-disposal facilities, and distribution of social costs. Water supply is not seen as short in that area so long as known technology and land-management measures are applied with intelligence.

All this material strongly challenges the more conventional concept of water as a determining factor in industrial location. Yet some industrial analysts maintain that "technology often can overcome significant portions of the basic needs for water, but costs seem likely to rise as treatment and conservation practices are expanded. It seems almost certain that water supply is destined to become an increasingly significant factor in plant location."[46] And a mechanical engineer asserts the common belief that "availability of water is one of the primary factors in the selection of a site for a new plant."[47] Whitaker[48] takes a longer and broader view than most others and raises a series of questions as to both immediate shortages and the lines along which constructive action might be expected.

It does not seem practicable at this time to offer a clear-cut solution to these and other questions that have been posed, but it is possible to point out the places at which information as a basis for judgment is either available or lacking. These points will be noted as they relate to elements of analysis that enter into decision making regarding industrial water supply.

Elements in Decision Making

Estimating the supply. Evidence points to a general tendency among industrial managers to overestimate the amount and reliability of groundwater supplies. It is this tendency that seems to encourage some "mining" operations whose results come as a surprise to the operators. Here it is important to distinguish between "potential" supply, which is physically available but not necessarily ready for use, and "effective" supply, such as a city water supply or a readily available well or aquifer, which can be tapped with little delay or expense.

Most reported "shortages" turn out to be deficiencies in providing effective supply rather than deficiencies of potential supply.[49] Certainly this has been shown for urban supplies.[50] There are a good many reports on insufficiency of surface sources, but they relate chiefly to failure to invest in new facilities or to unexpected lowering of quality rather than to deficiency of total potential supply. Except for one investigation of steel-plant managers, which showed that they placed water supply as a definitely tertiary factor in their thinking about plant location,[51] the only broad-scale canvasses of industrial views toward water are those made by the National Association of Manufacturers and the Conservation Foundation in 1949 and by *Fortune* magazine in 1954. Almost 40 percent of the industrialists responded "I don't know" in 1949 concerning the potentialities for expansion of their supply. Among the others the largest single group in all parts of the country believed "a few additional plants" could be supplied. In all areas, but especially in the Rocky Mountain states, a few respondents believed local sources were being used to capacity.[52] The *Fortune* survey showed a relatively optimistic view about possible shortage. Of twenty-six major industrial users queried regarding adequacy of water for a 50 percent expansion, ten saw no difficulty, fifteen were confident that the situation, though tight, could be solved at reasonable cost, and only one did not see a possible solution.[53]

As noted by Gibson,[54] by the Ohio River Interstate Commission, and by others, industry tends to respond to deterioration of water supply by changing its treatment and, sometimes, its product. If the source proves less satisfactory than estimated, the treatment or processing can be changed. Thus the papermakers adopted new brown-paper products as the stream water deteriorated, and some American paper manufacturers expect that "after available water supplies are developed to their practical utmost, waste water will be reused to a larger extent, varying with the intensity of the local problem."[55] But the role that estimates of water supply play in decisions on plant location is not well documented. Only recently have systematic efforts been made to find out their importance.[56]

Harmonizing multiple uses. Until a few years ago industry had shown little widespread interest in combining industrial use of water with other uses. The Nordrhein–Westfalen area, with its comprehensive unified development for both power and navigation, presents one of the notable exceptions.[57] The Beaver–Mahoning area has long received multiple-purpose treatment,[58] and the Saginaw Basin may also be noted. Other cases of such planning are more recent. For example, there is discussion of a comprehensive water-supply scheme to meet municipal and industrial needs in the German area of Magdeburg, Halle, and Leipzig.[59]

The rapid spread of supplemental irrigation, with its high consumption of water in the driest growing periods, has perhaps inspired more concern among industrial groups than any other single development.[60] It is revealing that when the Tennessee Valley Authority Act was drawn up in 1933, it did not refer to three water uses which increasingly claim attention in the valley—industrial, recreational, and irrigation. The whole program for water control in the Ohio Basin may well be reoriented in the next few years to reconcile low flow with flood control, and efforts may be made there, as well as in areas such as the Kansas Basin, to transfer storage capacity previously allocated to flood control to the maintenance of supplies for urban areas. Only as such emerging uses are seen in balanced relation to the more conventional demands on water resources will it be possible for industrial operators to make informed decisions regarding possible combinations of their needs with the needs of others.

Discounting future gains and costs. From the few major water-supply projects studied, it seems that there has been a tendency in the semipopular discussions of industrial water to underestimate the costs and to place high intangible values on supply. A study by the Rand Corporation, while noting the local view that water is essential to further growth, shows that the Metropolitan Water District of Southern California has provided water far ahead of active needs at relatively high cost.[61] Blake,[62] in his careful historical study of the evolution of major city-water-supply programs in the United States, indicates that public health aims figured heavily in the planning of new supplies. However, he gives no evidence that industry was a major claimant for additional water or that industrial growth ever did suffer from the chronic delays in construction of new facilities. These aspects deserve more thorough study looking to a better delineation of demand functions.

Projecting demand. The typical method of estimating future demand for industrial water has been adjustment of total withdrawal to estimated growth in gross national product and total population on a fixed-ratio basis.[63] This was the general procedure used by the Paley Commission and the Department of Commerce in the projections given here in table 7.2, and it has been used in the United Kingdom[64] and in West Germany.[65] In California and Texas an effort has been made in statewide studies to improve such projection techniques in estimating future industrial demand for water. One refinement has been demonstrated by Thomas and Teeters for the Muskingum Basin, using different assumptions as to labor productivity and use per employee.[66] A study of the San Francisco Bay Area[67] estimates industrial use on the basis of water used per acre occupied by various types of industries. All these projections commonly assume that conservation devices will not be used in greater number in the future than in the past.

Yet there is still remarkably little evidence on the elasticity of demand for water. We know that water is much more valuable for industrial use than for most other bulk uses.[68] We do not know how responsive a particular industry is to a change in the price of water. No adequate demand curves have been constructed, and most judgments are speculative. As has been suggested in several studies, a principal explanation for the abundant or excessive use of water both in industry and in municipal systems is the common practice of making water available at little or no cost for the basic supply, in contrast with land and mineral resources, where a charge is conventional.[69]

The engineering journals offer a scattering of case histories of reduction in water use in specific plants. For example, a Philadelphia electrical-equipment plant using city water for heat treating, welding, and air conditioning was faced with an increase in rates that jumped its annual bill from $24,000 to $33,000.[70] Under this incentive the plant installed a recirculation system with evaporative cooling and reduced its bill to less than $15,000 by an investment that was expected to bring a return in two years. On the other hand, the Economies Group of the British Sub-Committee on the Growing Demand for Water reports:[71]

> The concept of water as a valuable raw material of industry is comparatively new in this country, particularly where supplies are drawn from private sources and are not subject to the occasional restrictions imposed by public water undertakers in times of drought. Moreover, there is little or no financial inducement to economy because, as far as the Group could discover, in no industry does the cost of water amount to a significant proportion of the value of the product.

Forecasting technological change. As a general rule, industry has been slow to expect marked improvements in the technology of

obtaining and treating water. There is little evidence of plans for
new sources of supply based on expectations of new devices for
conserving water. Although air cooling is proving more economical
than water cooling in some manufacturing processes,[72] the tendency,
as reflected in the Department of Commerce projections, is to plan
new installations so as to use cheap water in large quantities as long
as it is at hand. Indeed, the cost incentive apparently has been small:
when water supply accounts for only 0.07 percent of the total value
of a product as in the case of industrial metals or 0.15 percent as in
the chemical products,[73] the tendency appears to be to adopt or seek
improved technology only when supplies are short or unusually expensive. A clear demonstration of this occurred in Illinois during
the war period 1940–45 when a shortage of materials for well equipment forced numerous industries to adopt for that period alone water-
saving measures that were available both before and after the materials shortage. Here, materials supply rather than water cost seems
to have been determinative. Ackerman and Löf[74] give the most
searching examination of this whole problem yet available.

Public guides. In a pricing system that prevails in most parts of
the world in which water is a free commodity, the chief public guides
to industrial use have been the administration of water rights and
the requirement of standards of quality for effluents. Neither riparian
nor prior-appropriation water-rights systems appear to have exercised a strong guiding effect on industrial water use. The Vaal River
case is one of the few reported in which a water-supply situation
has led to explicit public discouragement of new industrial location.[75]
It seems that pollution-abatement regulations as they affect the cost
of water disposal both in the United States and in the United Kingdom have provoked a greater tendency of industry to discriminate
in location than have any public guides affecting the supply.[76] In
much of the literature cited, waste-disposal facilities as required by
public regulation emerge as a more frequent control of industrial
location than water supply. It seems likely that future water-quality
requirements of the type enacted in the United Kingdom in 1951,[77]
rather than water supply, will turn the course of new manufacturing
investment.

Even if water were to play a less significant role in industrial
location than has been commonly assumed, and even if it were less
likely to limit industrial growth than the Paley Commission suggested, its enlarged use for industry would nevertheless be the occasion for critical reexamination of basin water-management plans.
As is already evident in the Ruhr, Vaal, Ohio, and Kansas Basins,

new and sharper views of integrated river development grow out of the enlarged industrial demands. In areas such as the Gulf drainages of western Texas, it may be that increasing industrial demands will be used to justify the construction of regional water transport and storage systems. Yet it should be recognized that industry—meaning essentially a few large plants in four industries—has the capacity to reduce use within the plant and to develop its own supplies. A Bensenville, Illinois, manufacturer recently constructed his own "internal" water supply by storing in an artificial lagoon rain water from five acres of roof and five acres of parking area to meet all plant needs (except drinking water from a well) at a cost of about half a cent per thousand gallons.[78] Nevertheless, a group of plants may find that a joint supply from a distant source permits them to continue heavy water use less expensively or more conveniently than individual adjustment. In Great Britain one of the products of increasing attention to integration of multiple purposes and of private and public supplies has been intensified discussion of water policy[79] and geographical research to indicate the lines along which rationalization of the water industry might proceed. Gregory's work on water sources and the pattern of urban development has been especially pertinent for planning decisions.[80]

In several countries awakened interest in industrial water has led to discussion of the need for improved networks to collect data relating both to hydrological phenomena and to industrial use. In the United Kingdom, Balchin has been one of the more vigorous advocates of improved hydrological research and data-collection facilities,[81] and a special government committee is now dealing with the problem.[82] The United Nations report considers the collection of basic water data an urgent and fundamental need wherever industrial growth is to be anticipated.

The foregoing generalizations are based on examination of scattered studies of particular industrial locations, the more important of which have been noted. Together, the studies do not afford a full picture of how industrial managers take water supply into account as a factor in location. Nor do they elaborate the areal distribution of either demand or production functions for water. They do give some hints of the character of management decisions, and taken with the newly available national and international statistics they suggest that our concepts of the role of water supply in industrial location are undergoing a thorough reevaluation. Water supply clearly is of growing importance in the economic life of market economies, but it is not necessarily, or even often, a decisive factor in industrial location. Its part in limiting present or future economic growth in most areas is doubtful.

A Time to Choose: America's Energy Future

First, in the preparation of the three scenarios it was necessary to make a series of assumptions as to how technology will unfold, how people will act, and how society will respond to new information and conditions. In the circumstances these are useful for purposes of projection. However, experience would suggest that some of the assumptions will turn out to be wrong and that at least a few may be terribly wrong. The assumption that consumer preferences reflected in demand curves and life styles will remain the same may grossly underestimate the capacity of Americans to change: how many reports in the early 1960s predicted the shift in values that marked the environmental movement at the end of the decade? Projections of the year 2000 population may be wide of reality. And so on.

To make such assumptions is necessary to the analysis. To challenge them and suggest disturbing alternatives calls for imagination and a stubborn willingness to confuse the calculations with doubts and possibly fanciful observations. While the scenarios are a good start, people should be encouraged to stretch their minds to explore the effects of still different views. Just as the scenarios break out of the rather slavish linear extrapolations which characterized electricity growth projections for so long, the new figures deserve fresh and continuing appraisal.

(1974j), 410–11

Notes

1. For a review of this and other "myths" of water supply, see below, vol. II, chap. 1 (Wolman and Wolman).
2. "Water for Industrial Use," *United Nations Publ. E/3058, ST/ECA/50,* Department of Economic and Social Affairs, New York, 1958. (Sales no.: 58.II.B.1.)

3. "Industrial Water Use," *1954 Census of Manufactures:* [*Subject*] *Bull. MC–209,* U.S. Bureau of the Census, 1957.

4. *Resources for Freedom: A Report . . . by the President's Materials Policy Commission,* 5 vols. (Washington, 1952), vol. 1, p. 50.

5. *America's Water Needs: Their Investment Significance* (Dominick and Dominick, New York, 1957). An example of many from the technical press is "Needed, An Extra 250 Billion Gallons of Water a Day by 1975," *Chemical and Engineering News* 36, no. 12 (1958), pp. 50–53 and 80.

6. J. A. Reynolds, "Scientists Ready New, Inexpensive Methods to Desalt Sea Water," *Wall Street Journ.,* Apr. 17, 1957.

7. *Op. cit.* (see note 2 above), p. 17.

8. *Water in Industry: A Survey of Water Use in Industry* (National Association of Manufacturers and the Conservation Foundation, New York, 1950).

9. K. A. MacKichan, "Estimated Use of Water in the United States, 1955."*U.S. Geol. Survey Circular 398,* 1957.

10. W. L. Picton, "Summary of Information on Water Use in the United States, 1900–1975," *B*[*usiness*] *S*[*ervice*] *B*[ull.] *136,* U.S. Dept. of Commerce, Business and Defense Services Administration, 1956; D. R. Woodward, "Availability of Water in the United States with Special Reference to Industrial Needs by 1980" (Thesis no. 143, Industrial College of the Armed Forces. Washington, 1957).

11. "Industrial Water Use," Select Committee on National Water Resources, United States Senate, 86th Cong., 2d sess., *Committee Print No. 8,* 1960.

12. The term "water use" is employed here as synonymous with "water intake," or water withdrawn from surface or ground sources by either private or public suppliers. It represents the total withdrawal and does not distinguish between consumptive and nonconsumptive use. The term "water requirements" is used in some places, but its definition is loose and may mean potential demand as well as actual intake.

13. "Water Requirements of Utility Steam-Electric Power Plants, 1954" (U.S. Federal Power Commission, 1957).

14. Although the census data are given on an annual basis, the common measure for water use is millions of gallons as a daily average. Where there is seasonal variation in production or in cooling requirements, the actual daily use may greatly exceed the average.

15. M. T. Thomson, S. M. Herrick, Eugene Brown, and others, "The Availability and Use of Water in Georgia." *Georgia Geol. Survey Bull. no. 65,* 1956, p. 36.

16. J. B. Graham and M. F. Burrill, eds., *Water for Industry, American Association for the Advancement of Science Publication no. 45,* Washington, D.C., 1956.

17. "Water for Industrial Use" (see note 2 above), pp. 21 and 42–43.

18. *Statistisches Jahrbuch für die Bundesrepublik Deutschland, 1958,* Statistisches Bundesamt, Wiesbaden, 1958, p. 211. See also "Die Wasserversorgung der Industrie im Bundesgebiet 1952" (Bundesministerium für Wirtschaft, Bonn, 1954).

19. "Water for Industrial Use" (See note 2 above), p. 44.

20. "Water Requirements in the Muskingum River Basin in 1975: A Report to the Muskingum Watershed Conservancy District" (Dept. of Conservation, School of Natural Resources, University of Michigan, Ann Arbor, 1957); N. R. Heiden, "Industrial Water Needs, Oakland County, Michigan" (Oakland County Planning Commission, 1957).

21. A. V. Kneese, "Water Resources: Development and Use" (Federal Reserve Bank of Kansas City, 1959).

22. *Op. cit.* (see note 2 above), pp. 22–28.

23. *Climatic Criteria Defining Efficiency Limits for Certain Industrial Activities*—Section 42, "Water Supply," U.S. Dept. of Commerce, Office of Technical Services, PB 111454, n. d. This material was summarized in considerable measure by H. E. Hudson, Jr., and Janet Abu-Lughod in their paper, "Water Requirements," for the A.A.A.S. symposium (Graham and Burrill, *op. cit.* [see note 16 above], pp. 12–22; reference on pp. 19–21).

24. E. A. Ackerman and G. O. G. Löf, *Technology in American Water Development* (Baltimore, 1959), pp. 407–41.

25. B. W. Maxwell, "Public and Industrial Water Supplies of the Western Coal Region, Kentucky," *U.S. Geol. Survey Circular 339,* 1954, p. 11.

26. Eskel Nordell, *Water Treatment for Industrial and Other Uses* (New York and London, 1951); W. Pürschel, *Der Wasserverbrauch von Industrie, Handwerk und Kleingewerbe: Literaturauswertung, Ermittlungsmethodik und technische Grundlagen zu seiner Bestimmung* (Deutscher Verein von Gas and Wasserfachmännern, Frankfurt am Main, 1958).

27. O. D. Mussey, "Water Requirements of the Pulp and Paper Industry," *U.S. Geol. Survey Water-Supply Paper 1330–A,* 1955; H. L. Conklin, "Water Requirements of the Carbon-Black Industry," *ibid., 1330–B,* 1956; *idem,* "Water Requirements of the Aluminum Industry," *ibid., 1330–C,* 1956; O. D. Mussey, "Water Requirements of the Rayon- and Acetate-Fiber Industry," *ibid., 1330–D,* 1957.

28. The final report in the series of nine is E. W. Lohr and W. F. White, "The Industrial Utility of Public Water Supplies in the New England States, 1952," *U.S. Geol. Survey Circular 288,* 1953. This series of Geological Survey *Circulars* (197, 203, 206, 221, 232, 253,

269, 283, and 288) and the report on cities, *Water-Supply Paper 658,* all now out of print, have been superseded by *Water-Supply Papers 1299 and 1300,* 1954.

29. These are published as Geological Survey *Circulars.* The numbers, metropolitan areas, and dates of the principal reports are as follows: 148, Atlanta, 1951; 173, Buffalo–Niagara Falls, 1952; 174, Lake Erie shore, Pennsylvania, 1952; 177, Mahoning Basin, 1952; 183, Detroit, 1952; 216, St. Louis, 1952; 246, Rochester, 1953; 247, Milwaukee, 1953; 254, Birmingham, 1953; 273, Kansas City, 1953; 274, Minneapolis–St. Paul, 1953; 276, Louisville, 1953; 315, Pittsburgh, 1954; 323, Grand Rapids, 1954; 339, Western Kentucky coal region, 1954; 340, Wheeling–Steubenville, 1955; 366, Indianapolis, 1955; 369, Eastern Kentucky coal region, 1956; 372, Portland and Vancouver, 1956; 373, Mobile, 1956; 378, San Francisco Bay, 1957.

30. Issued annually since 1953. For discussion of individual projects and local situations see "Saline Water Conversion" (Hearings, March 20 and 21, 1958, on S. J. Res. 135 and S. 3370, Subcommittee on Irrigation and Reclamation, Committee on Interior and Insular Affairs, U.S. Senate, 85th Cong., 2nd sess.), 1958.

31. *Op. cit.* (see note 2 above), p. 17.

32. W. G. V. Balchin, "A Water Use Survey," *Geogr. Journ.,* Vol. 124, 1958, pp. 470–82 (discussion, pp. 483–93). This paper and the accompanying discussion give a broad picture of attitudes toward water development in Great Britain.

33. C. L. White, "Water—A Neglected Factor in the Geographical Literature of Iron and Steel." *Geogr. Rev.,* Vol. 47, 1957, pp. 463–89.

34. R. L. Leffler, "Water and Steel: Fairless Works Water Supply," in *Water for Industry* (see note 16 above), pp. 35–42; reference on p. 36.

35. *Op. cit.,* (see note 33 above), p. 468.

36. *Water Economy in Iron and Steel Works, Project No. 298* (European Productivity Agency of the Organization for European Economic Cooperation, Paris, 1958), pp. 1–71.

37. *Ibid.,* p. 51.

38. I. M. E. Aitken, E. L. Streatfield, and H. C. White, "Water Recirculation in Steelworks," *Journ. Instn. of Water Engineers,* Vol. 13, 1959, pp. 253–84 (discussion, pp. 284–303).

39. For example, D. J. Tow, "Cooling Water for Industry," *Petroleum,* Vol. 19, 1956, pp. 233–36; E. L. Streatfield, "Some Recent Developments in the Economic Utilization and Purification of Water," *Chemistry and Industry,* Vol. 27, 1958, pp. 841–46.

40. J. R. Gibson, "The Paper Industry of North-West England, Part 2—Influence of Water Supply and Effluent Disposal upon Location," *Paper-Maker and British Paper Trade Journ.,* Vol. 136, no. 4, 1958, pp. 64–66.

41. J. A. Sporck, "L'eau et la géographie de la localisation de l'industrie," *Cercle des Géographes Liégeois Travaux No. 98,* 1956 (from "Livre de l'eau," Vol. 3, 1956, pp. 119–52).

42. Reiner Keller, "Wasserhaushalt und Wasserwirtschaft der Stadt Euskirchen, ihre Bedeutung für Industrie, Gewerbe und Bevölkerung," in *650 Jahr Stadt Euskirchen,* Vol. 2 (Euskirchen, 1955), pp. 345–67.

43. *Water Economy in Iron and Steel Works* (see note 36 above), p. 17.

44. Werner Göpner, "Die Wasserversorgung des rheinisch-westfälischen Industriegebiets," *Geogr. Taschenbuch 1950–1957,* Wiesbaden, 1956, pp. 306–11.

45. F. J. Hessing, "Wasserversorgung in dichtbesiedelten Industriegebieten," *Raumforschung und Raumordnung,* Vol. 16, 1958, pp. 173–77.

46. T. D. Best and R. C. Smith, "Water in Area Industrial Development," *Battelle Tech. Rev.,* Vol. 6, no. 11, 1957, p. 8.

47. J. Nachbar, "Water Supply for Industrial Plants," *Air Conditioning, Heating and Ventilating,* Vol. 54, no. 12, 1957, p. 53.

48. J. R. Whitaker, "Water in the Future," in *Water for Industry* (see note 16 above), pp. 105–20.

49. "Is Water Short?" *Economist,* Vol. 182, 1957, pp. 305–6.

50. K. A. MacKichan and J. B. Graham, "Public Water-Supply Shortages, 1953," *U.S. Geol. Survey Water Resources Rev. Suppl. No. 3,* 1954, p. 2.

51. A. H. Doerr, "Factors Influencing the Location of Nonintegrated and Integrated Iron and Steel Centers in Anglo-America." *Southwestern Social Sci. Quart.,* Vol. 34, no. 4, 1954, pp. 39–44.

52. *Water in Industry* (see note 8 above), pp. 26–28.

53. Francis Bello, "How Are We Fixed for Water?" *Fortune,* Vol. 51, March, 1954, pp. 120–25, 140, and 148.

54. *Op. cit.* (see note 40 above), p. 64.

55. R. E. Fuhrman, "Water Resources and Industrial Expansion," *Tappi,* Vol. 41, no. 4, 1958, pp. 214A–216A.

56. K. S. Watson, "Need for Water Management Program in Industry," *Journal American Water Works Association,* Vol. 47, 1955, pp. 973–81.

57. Göpner, *op. cit.* (see note 44 above).

58. W. P. Cross, M. E. Schroeder, and S. E. Norris,"Water Resources of the Mahoning River Basin, Ohio," *U.S. Geol. Survey Circular 177,* 1952.

59. Helmut Hübner, "Die wasserwirtschaftlichen Verhältnisse Mitteldeutschlands als Grundlage der künftigen Wasserversorgung," *Geogr. Berichte,* Vol. 1, 1956, pp. 2–12.

60. For discussion of some of the questions raised by expanding supplemental irrigation see M. C. Boyer, "Water Supply versus Irrigation in Humid Areas," *Proc. Amer. Soc. of Civil Engineers*, Vol. 84, 1958, pp. 1–13; and J. R. Davis, *Water Demand Potential of Irrigation in Humid Areas: Report to Indiana Water Resources Study Committee* (Purdue University, 1956).

61. J. C. De Haven, L. A. Gore, and Jack Hirshleifer, *A Brief Survey of the Technology and Economics of Water Supply* (Rand Corporation, Los Angeles, 1955).

62. N. M. Blake, *Water for the Cities* (Syracuse, 1956).

63. For examples of recent projections of water demand, assuming different patterns of irrigation use, see H. J. Thiele, "The Water Problem in the West," *Western Business Rev.*, Vol. 2, 1958, pp. 163–70; and "Water Availability: A District Problem?" *Monthly Rev. Federal Reserve Bank of Kansas City*, February, 1959, pp. 3–8.

64. "First Report," Central Advisory Water Committee [Ministry of Housing and Local Government], Sub-Committee on the Growing Demand for Water (London, 1959).

65. Hans Plett, "Die Entwicklung der öffentlichen Wasserversorgung in der Bundesrepublik Deutschland," *Geogr. Taschenbuch 1956–1957*, pp. 301–5.

66. "Water Requirements in the Muskingum River Basin in 1975" (see note 20 above), pp. 6–8. For a review of the types of projections of total municipal demand made by consulting engineers, waterworks engineers, and state sanitary engineers see "Present and Future Estimates of Water Consumption," *Public Works*, Vol. 87, no. 12, 1956, pp. 73–77, 152, 154, and 156. These estimates indicate a uniform belief in rising per capita use of water over the next two decades.

67. H. F. Matthai, William Back, R. P. Orth, and Robert Brennan, "Water Resources of the San Francisco Bay Area, California," *U.S. Geol. Survey Circular 378*, 1957, pp. 50–51.

68. E. F. Renshaw, "Value of an Acre-Foot of Water," *Journ. Amer. Water Works Assn.*, Vol. 50, 1958, pp. 303–9.

69. Gilbert F. White, "The Facts about Our Water Supply," *Harvard Business Rev.*, Vol. 36, 1958, pp. 87–94.

70. Walter Jacoby, "Reclaiming of Process Water Cuts Bills 50 Per Cent at I-T-E," *Plant Engineering*, Vol. 13, no. 2, 1959, pp. 92–93.

71. "First Report" (see note 64 above), p. 11.

72. Best and Smith, *op. cit.* (see note 46 above).

73. "Water for Industrial Use" (see note 2 above), p. 44.

74. *Op. cit.* (see note 24 above).

75. "Water for Industrial Use" (see note 2 above), p. 22.

76. L. B. Dworsky, "Industry's Concern in Pollution Abatement and Water Conservation Measures," *Public Health Repts.*, Vol. 69, 1954, pp. 37–47.

77. Report of the Ministry of Housing and Local Government, 1957 [*British Command Paper*], *Cmd. 419,* London, 1958, pp. 43–44. This report also carries an excellent discussion of water-supply problems and policy since the passing of the Water Act of 1945.

78. "Firm Meets Own Water Needs with Unique Supply System," *Area Development Bull.,* U.S. Dept. of Commerce, Vol. 4, Oct.–Nov., 1958, p. 5.

79. "Policy for Water," *Planning,* vol. 24, no. 418, Political and Economic Planning, London, 1958, pp. 1–15.

80. S. Gregory, "Conurbation Water Supplies in Great Britain," *Journ. Town Planning Inst.,* Vol. 44, Sept.–Oct., 1958, pp. 250–54; *idem,* "The Contribution of the Uplands to the Public Water Supply of England and Wales," *Inst. of British Geogrs. Publ. no. 25: Trans. and Papers 1953,* London, 1958, pp. 153–65.

81. Balchin, *op. cit.* (see note 32 above). See also W. G. V. Balchin, "Hydrological Research," *Water and Water Engineering,* Vol. 61, 1957, pp. 150–52.

82. [Report to the] Central Advisory Water Committee [by the] Sub-Committee on Information on Water Resources (London, 1959).

8 The Changing Role of Water in Arid Lands

Within the natural resources field there has been a continuous search for integrating concepts. Beginning in the 1950s, White took part in UNESCO's new integrated research program on arid lands; this program promoted a return to a unified perspective last seen in the United States in the writings of J. W. Powell[1] and in Europe by Brunhes in his work on North Africa.[2] Thus for White, arid lands joined river basins and metropolitan planning as regional foci where land and water management might meet and the natural and human sciences collaborate.

As dams and levees were to floodplains, canals and wells were to arid lands. The prevailing societal response in the latter case was to provide more water, as it was to control and reduce water in the former. Just as dams and levees caused increased occupance, canals and wells, by providing water, created a demand for even more water and an accompanying physical degradation of soil and water alike.

This Reicher Memorial Lecture at the University of Arizona is, as White notes in the opening line, "perhaps courageous or audacious," but not because it concerns arid lands, for that was surely fitting. Rather, it is audacious because it attacks, in 1962—long before the growth problems of the Sunbelt were recognized—favored public attitudes toward expanding irrigation, green urban oases, and excessive water development and consumption.[3]

To come to Tucson to speak of arid lands is like going to Mecca to speak of Islam: it is either courageous or audacious. It may be excused on the ground that the whole effort of international arid lands research for which the University of Arizona is a unique center is in itself courageous and audacious.

Arid land research is one of the deeply exciting intellectual adventures of our generation. Cutting across established scientific dis-

Reprinted from *University of Arizona Bulletin Series* 32, no. 2 (November 1960).

ciplines, it seeks to relate the findings of pure science to technological development in its social applications: radiation physics in Tucson is seen in relation to energy collection devices in New Delhi and the political structure and daily rhythm of a remote Mexican village. More than two dozen disciplines are involved, from pure physics and chemistry to biology and the complex realm of human ecology. The efforts at correlation span fifty nations, and embrace not only halting steps to fashion research institutes in the young educational systems of Cairo and Pakistan but sophisticated ventures in transferring the established research machinery of Rothamstead or Wageningen to new situations. To try to work across these many boundaries of outlook, scientific disciplines, and nations is at best courageous and possibly foolhardy.

Perhaps these attempts are the more challenging because they center their interest upon a meager environment: the arid lands in which the margin between human failure and success commonly is narrow and in which delicate changes in rainfall and land use can trigger profound shifts in soil, vegetation, sediment movement, and water flow. These are risky lands. Much of the table is bare, but where chances are taken the stakes are high.

In any arid landscape with its thin soils, sparse vegetation, and exposed erosion forms, the green-fringed evidence of water long has been taken to mark the land's most valued asset. Lack of water defines the arid lands. The presence of water in oasis well, or ephemeral stream bed, or massive exogenous river has been regarded as a stamp of resource wealth, an essential key to economic growth. It has been a commonplace to say with Jean Brunhes, "Water is pre-eminently the economic wealth; it is, for men, more truly wealth than either coal or gold." And a great state can wage a bitter battle at the polling booth on the simple argument, "Water for people, for progress, for prosperity."

We now are moving into a time when the more traditional formulations of the role of water are challenged and altered. New pressures upon the arid lands inspire this challenge and set the scene for social change on an unprecedented scale. In the western fringe of the Australian Desert, the dry wastes of Turkestan, the sweeping plains of the north-central Sahara and the southwestern United States, both society and landscape are undergoing drastic transformation. Against this shifting backdrop of denuded hills and sprawling cities, water continues to play an important part but one that is beginning to differ in radical degree from its part in the past.

Unique Aspects of Arid Lands

A geographer must at once register the reservation that no two points of the arid zone are alike in all respects: each has its unique combination of terrain, soil, vegetation, and moisture. For this reason, generalities about arid lands are to be approached with the caution of a hydrologist assessing a drainage basin, knowing that one steep sector may be perennially dry while a nearby valley fill is usually saturated. Potential water supply in a stretch of the Egyptian desert, where fresh water accumulates chiefly as a thin layer on a brackish underground supply, is not to be regarded as the same as in an area of the central Sahara crossed by seasonal streams rushing from a rocky massif to dissipate themselves in salty playas. A plan or policy for resource development which fails to take explicit account of this tremendous diversity in local environment is likely to go astray.

Nevertheless, from the spotty experience of the arid one-third of the earth's land surface, a few trends may be noted. To describe those trends is not to venture a prediction. The current vogue of projecting our resource future by extrapolating past trends in the use and supply of natural wealth has heuristic value in trying to pierce misty uncertainties of the future, but it carries the danger of these trends being accepted as possible guides for that growth. Sometimes they obscure realization that a present view of resources in itself shapes future use of those resources.

This is my excuse for reviewing familiar experiences with water management. Today's vision of the role of water in the Santa Cruz Valley, in the Colorado Basin and in the arid zone as a whole, seems likely to affect the complex of individual decisions determining how remaining waters of those areas will be allocated or reallocated for human use in the decades ahead. According to the framework of public attitudes in which both familiar and unfamiliar hydrologic facts are arranged, water may be cast in roles which will either inhibit or advance social aims. These public attitudes may be expected to reflect in some measure the sweeping changes in progress and the forces shaping them.

Changes in Arid Lands

Arid lands now are a place of remarkably rapid social change. In the great reach of dry mountains, plateaus, and plains stretching from Central Asia and Rajasthan to the Atlantic coast of Morocco, the cultures of centuries of nomadic life are being battered and warped. Sedentarization has been a watchword of recent social policy in many sectors. On a massive scale the nomads of the Soviet Union have been attached to permanent bases and prescribed ranges, their flocks limited, their institutions altered to meet the needs of sedentary agriculture. A population of something more than ten million people, divided among at least one hundred ethnic and language groups, has been taken from its tents and seasonal migrations and settled on farming plots by government plan and force. Iran and Iraq carry out calculated programs to impede the nomad, to settle him on cultivated lands, or to run him into the labor market of the city: the wandering Kurd becomes the taxi driver of Baghdad. Along its Mediterranean coast, Egypt has sought to make stable homes for the Bedouins who have roamed the Western Desert.

Even where government policy has not sought to eliminate nomadism, the turn of political and economic events combines to do so. While agricultural agents in disputed Saharan territory work to improve the flocks, water points, and pastures, other forces render nomadic life less tenable or desirable. Air and truck transport eliminates caravan trade and cripples the caravanseries. Law enforcement and administrative regulations curb the movements of pastoral peoples. Markets for animals become less encouraging. Jobs in petroleum and other mineral exploitation and in growing urban centers attract the nomadic men, offering them income and amenities and a different kind of insecurity. Throughout the Old World deserts nomadism is dead or ill. Possibly the only growing edge of nomadism in a fresh context is the restless invasion of North American deserts by trailers and pickup trucks.

In the dry lands of the New World as well as in nomadic territory, at least two other aspects of social change have special significance for water management. Perhaps as evident in the Santa Cruz Valley as in any arid area, they need only to be enumerated to suggest the scope of their impact. One is the rapid advance of technology for finding, lifting, storing, transporting, and treating water. Experts have pointed out the implications of new methods for locating and pumping underground water, storing and moving water on a large scale, improving water-application techniques and dry-land plants, treating raw water and waste water, transmitting electrical energy, and scheduling water use. Within six or seven decades the concept of where and how water could be stored, moved, and treated has changed dramatically. The technical means are at hand to go deep and far for water at low cost, to increase the returns from the volume withdrawn and to change its quality.

An important feature of the technology of water management is its rate of advance. Opportunities to regulate and treat water are unfolding much more speedily than they are being used. While there are conspicuous examples of applications of the new knowledge—nets of tube wells where gas engines are novelties, massive concrete arches where baskets are the common means of moving earth, and intricate hydrologic models for educational use where most of the citizens are illiterate—much water use lags far behind. Irrigators in a Peruvian valley are less open to innovation than the young engineer who designs a new dam. Fifty years ago a farmer or manufacturer could be less profligate with water in field or plant than today because he could not have been aware of opportunities to line ditches or use brackish waters or recirculate industrial coolants. Tragically, a number of water developments in Southwest Asia are running on to rocks of inept management as soon as they are launched; the first division of a project begins going out of cultivation before the others are complete. Thus, technological advance enlarges the possibilities of wise water management at the same time as it widens the margin between knowledge and performance.

Another significant aspect of social change is the increased demand upon arid lands for recreational and industrial use. Throughout the arid zone, cities are growing more rapidly than total population. In Egypt, where an increase of 1.5 percent per year took place in total population during 1927–47, urban population increased at 3 percent per year.

Pressures of Urbanization

Cities are becoming the dominant features. In countries of Southwest Asia a principal element in the strategy of economic development is industrialization; manufacturing is seen as the great expanding sector of the economy, but an essential aid to such expansion is strengthening agricultural production, and so irrigation development is a step toward intense urbanization. Rather than representing competition between city and country for water, as in the southwestern United States, irrigation expansion is seen as an aid to ultimately preponderant manufacturing activity.

With urbanization, water use changes in several ways. Industrial uses assert themselves and claim larger volumes of water than the entire urban population. In industrial centers of the United States the manufacturing intake may amount to one-third to fifty times the municipal intake. Total water needs therefore may increase more rapidly than total population. Manufacturing commonly leads to a rise in per capita income, which in turn is reflected in increased per capita water use for household purposes. Industry is more demanding upon water for waste disposal than for direct use: dilution requirements mount tremendously with manufacturing aggregations and may quickly attain troublesome proportions where streams are thin or ephemeral. Growing urban populations reach out for recreation amenities, either extending living quarters to the dry lands or going to them for holiday use. This in turn raises the standards of water quality for the few available stream courses. Workers create more rigid demands for clean water in picnic areas, residential areas, and fishing grounds at the very time their plants are requiring enlarged volumes of water to dilute industrial wastes.

These phenomena of urban growth are familiar signs in the southwestern United States, but they are proliferating in other sectors of the arid zone and they give every indication of becoming more intense in the decades ahead.

The Changing Land

Hand in hand with social transformations have gone changes in the land itself. The gashes of new highways have exposed fresh terrain to grazing and recreational use. Increased population has intensified the destruction of natural vegetation with concomitant alterations in fauna and in soil. Just as agricultural stages of occupance of the older arid lands had cleaned out fuel-wood supplies in brush and gallery woodland for miles around, the new cities are invading scenic landscapes, appropriating them for building and recreational sites and stretching along improved roads beyond former limits of exploitation. The net effects of these physical changes are difficult to assess. If the ideal drainage area for catching water is a paved surface, some of the growing edges of the arid zone are approaching the ideal. Broader influences upon infiltration rates, sedimentation rates, and evapotranspiration are less certain. Much of the arid zone— and particularly its semiarid fringes—is in a delicate balance that, once triggered by modified vegetation or drainage, may be severely dislocated. Knowledge of fundamental processes in the hydrologic cycle is growing but still does not permit refined estimates of effects of changed land use and technology upon extreme flows, water quality, and total water output.

Use vs. Supply

In this setting of social flux and physical development and degradation, a dominant fact is that for the first time in many areas the volume of consumptive water use has approached the potential supply. Because it has been common to describe arid lands as places that are deficient in precipitation, there is a tendency to think of them as being short of water. If potential water supply is taken as the volume of water available to an area from surface or underground sources, regardless of the amount effectively available through engineering works, then there are only a few drainage areas in the arid lands where use yet approaches potential supply. Shortage stems from political ineptitude in developing supplies or from the heavy cost of doing so rather than from limits of potential supply. Paradoxical as it may seem, in the arid zone large volumes of water annually evaporate from playas or flow into the sea without having served human needs.

To be sure, basins like the Ili and the upper Salt already are used virtually to full capacity: no flood spates go uncontrolled, no water stands until evaporated purely by chance. On a larger scale, the entire southwestern United States is approaching that condition. Consumptive use has not come close to potential supply, but withdrawals are nearing the mean runoff and, with reuse of irrigation waters, are exceeding it in some areas. On the other hand, great basins such as the Nile and the Tigris-Euphrates still discharge 40 percent or more of their mean flows without economic return. Indeed, the prevailing condition in much of the arid zone is one of large-scale losses by evaporation and transpiration of water that has served no human use. This margin generally is dwindling. Under the impact of new technology and new uses, the physical limits are in sight.

Always before in arid areas the threat of water shortage has been accompanied by knowledge of unused potential supplies awaiting exploitation. Now, as limits loom ahead, only the most heroic of schemes for long-distance transport or for new technology can stand in the way of compulsory choices among alternative uses for a given supply. Engineers and ranchers and civic leaders have known physical supply has clear limits, yet much of their action, as in both

Arizona and California, presupposes that somehow, somewhere it can be augmented.

A sobering aspect of any attempt to assess arid zone resources and potential is our state of knowledge as to basic natural processes. At many critical points it leaves us in doubt as to the grounds upon which new technologies can be developed. Ignorance of some of the vital mechanisms still is great. The mechanism by which rain droplets form in a cloud, the precise fashion in which evaporation takes place from land surfaces, and the means by which mineral nutrients in the soil become available to plants are among the baffling unknowns. Descriptions of hydrologic relations are largely empirical. To the extent that scientific understanding of such processes can be deepened, the way may be paved for more efficient water use plans, and perhaps, for drastic innovations in such directions as weather modification and irrigation agriculture.

Public Attitudes

Important as it is to fill these gaps in scientific knowledge, it seems likely that even more urgent is a drastic reorientation of public attitudes toward water. Unless water can be seen in a different role, much of the new research and investment may yield meager and frustrating returns. The prevailing view of water in the social and economic growth of arid lands faces revision in at least three regards.

Most obvious and probably the most difficult is recognition that using more water may not be the panacea for water ills. The typical solution for water problems in the past has been to go deeper or farther for water or to irrigate more land or to build a bigger dam. Many an area has been like the proverbial drunkard who thinks the best corrective for excessive use is another little drink or a large one early the next morning.

Shifts in Irrigation

One of the more startling dilemmas in the arid lands is the rate at which old irrigated lands are lost while new ones are wrested from the sagebrush and mesquite. Salting and waterlogging are the two most active thieves in the ancient lands of southwestern Asia and northern Africa.

While ambitious programs for harnessing the Indus, the Tigris-Euphrates, and the Nile are commanding both public attention and public funds for the enlargement of irrigated acreage under new storage dams and canal systems, equivalent acreage is quietly going out of cultivation. Without fanfare and with only limited police measures to stop the robbery, salt accumulation is taking as much as 40,000 acres a year in West Pakistan. More than 60 percent of the irrigated lands of Iraq are seriously affected by salting, the product of overapplication of water and unsuitable soils. In the fruitful delta of the Nile, as much as one-quarter of the land is threatened by a high water table resulting from inadequate farm drainage. Declining yields of cotton testify to the curtailed productivity. There now is doubt that the investment in massive new works is any more than offsetting the inconspicuous dissipation of old investment in farms that are abandoned to salt-crusted fields and soggy soils.

Cities are also beginning to steal irrigated land, as in Southern California where the demands of urbanized areas are accompanied by a retreat of irrigation and an intensification of agriculture on the urban fringe.

Needed: Water Budgeting

In the western United States, according to a recent report by the Agricultural Research Service, it would be practicable to enlarge the irrigated acreage from 29,600,000 to 55,500,000 acres by increasing the total intake of water by 307 million acre-feet if no improvement in efficiency were to be made, and by less than 200 million acre-feet if there were acceptance of known and tested methods of conserving water as it flows in streams, canals, and distribution ditches so as to cut the losses from seepage, phreatophytes, and excessive irrigation applications. Irrigated land could be increased greatly at the cost of making better use of the water already in effective supply. Intelligent use of water budgeting, for example, might achieve tremendous savings in water without altering the usual methods. It would be sanguine, however, to think this would come soon. Much easier in a political sense than altering the irrigation practices and canal system in the Gila or the Sind is constructing a new project to apply more water to another area by the same inefficient methods.

Much the same situation applies to industrial use of water in arid lands. Although there are successful instances of low water intake per manufacturing product, as in the case of steel production in the

Great Basin or petroleum refining along the Texas coast where intake per unit of product may be as little as one-thirteenth of intake in some other plants, the prevailing tendency in national as well as international projections of water use is to assume new industry will continue its present habits of lavish water use. Projections of water need, assuming currently wasteful practices will persist, may become self-fulfilling prophesies. They predict shortage and encourage the water user to think his expanded activities inevitably lead to shortage. The common solution where potential supplies seem limited is to seek new supplies rather than actively to encourage widespread economies in water use. Moreover, much water law permits the careless user to remain entrenched in his indolence.

In disposing of waste a similar attitude prevails. During recent decades relatively little attention has been given to improved methods of treating waste: much more concern is expressed for regulating streams so as to provide adequate dilution for the growing volume of untreated or partially treated waste.

Several conditions help explain this widening gap between knowledge and application in water use, and the tendency to rely upon more water rather than better management. Illiteracy, poor adult education facilities, water law and administration, pricing systems, inflexible social organization, lack of capital, marketing and production methods, and the high risks of transition are among the reasons.

To put the argument more sharply, the cure for threatening water shortage is not necessarily more water. Indeed, some predictions of shortage may hasten or assure the shortage by discouraging remedial action which may be taken with present supplies. The farmer who is running short may find the needed supplies in seepage from canals, in excessive applications, in evaporation from reservoirs. A manufacturer may decide to move into a water-short area, knowing that by waste reduction, recycling and reuse, he can operate as effectively and even as profitably as at a site where water is abundant. As technology for water use advances and as the possible gap between knowledge and application enlarges, the opportunities to benefit from water-saving measures will increase. The cure for water shortage may be using available knowledge of how to deal with water.

Needed: A Fair Price on Water

A second public attitude facing revision has to do with value of water. It is shifting, but so slowly as to promise little effect before the pattern of water development in many basins has been frozen into dams, canals, and court orders that sturdily defy revision.

Like air, water is accepted as a God-given commodity outside the ordinary pricing system. Throughout large sectors of the arid zone initial water rights are acquired at nominal charge, and no tax is levied on water holdings or water use. There is a widespread assumption that water will be provided by public agencies at cost. Market mechanisms rarely are used to determine a fair price for water, and when a substitute for cost is sought it generally is in the direction of estimates of what the user can bear. Implicit in much of the public policy toward irrigation development in arid lands is the belief that water regulation and the cultivation of new lands are inherently good, and that water, while an item of inestimable value for national growth, should be treated as an item of lowest possible value for computing costs and returns.

This is illustrated by the rate schedules for municipal water supplies. In the days when the germ theory of disease was still not widely accepted, and the backyard well and the outdoor privy seemed good enough, public and private water companies sought to promote new connections and the installation of water-using equipment by rates favoring large use at low cost. The more a customer used the less the unit cost. These promotional rates still prevail even though leaders in the water industry decry the undervaluation of their commodity with its resultant restrictions upon service and plant expansion. A city such as Denver, while compelled to ration water for lawn sprinkling during dry periods, encourages maximum intake among its customers, reaps the financial benefits of enlarged revenues and earnestly seeks additional supplies. Tel Aviv and Karachi, rapidly expanding cities in arid situations, until recently based their water charges upon the assessed valuation of property. Under this scheme the owner of a large establishment is impelled to use very large quantities of water since he is paying for it anyway. In Chicago, with annual rainfall of thirty-seven inches and a tremendous source within a mile of the shore line, a middle-class family pays as much for its water as does a similar family in Boulder, Colorado, with annual rainfall of seventeen inches and a glacier source twenty miles away.

Recent pioneering studies sponsored by Resources for the Future give fresh perspective on the economic productivity of water. The San Juan investigation suggests that in terms of returns in employment and volume of production an acre-foot of water used in industry may be a hundred times as valuable as if used in agriculture. Recreational use of water flowing in a fishing stream may yield returns ten times larger than those from the same volume of water in an irrigated farm downstream.

Obviously, these values presume a demand for the industrial and recreational use, although it seems ridiculous to envisage the deserts blanketed with factories, homes, and playgrounds. To the extent, however, that national economies may favor such development, the systems of planning and regulating new water development may require heavy revision, for irrigation will have to give way to other claimants, and pricing systems will have to take account of other demands if wise allocations are made. It does not seem too much to expect rate schedules that charge least for the first gallons used and increase rapidly as usage exceeds a reasonably efficient level for the particular type of use. Nor is it unlikely that means will be found to expedite the transfer of water rights for new uses with adequate compensation for the present owners. The full social consequences of this type of transfer are not yet apparent.

The people of areas such as the Santa Cruz Basin or the lower Indus must recognize sooner or later that water may be more useful in the service of industry or recreation than of agriculture. Their cities may grow literally at the expense of fields, and farming operations of crops which can be grown elsewhere at similar levels of cost may be transferred out of water-short areas, letting the transport of the product substitute for transport of water into the area.

Current pricing systems, water rights law, and public attitudes are largely opposed to readjustments in use priorities. They can be expected to change slowly at best and then only in response to severe cases of misallocation or to persistent public education that paves the ground for revisions in public attitudes and institutions.

Needed: A New Look at Water Programs

A third shift is needed in attitudes toward investment in water facilities. They have come to enjoy a kind of sanctified status in arid lands that sometimes belies the facts of their effects. From the Helmand to the Tigris-Euphrates and the Zambezi there exists a sort of magic about investment in water storage and transportation, a spell of solid confidence that to be sure does not necessarily fall over the international banker or the taxpayer of more humid climes, but that inspires enthusiastic public action within the arid lands.

A simple example is in public investment in flood protection, a program claiming heavy federal outlays in cities of the arid United States. Since the national flood control policy was initiated in 1936, more than $4 billion have been expended on reservoirs, levees, and channel improvements to curb the annual flood losses. Twenty-four years later the toll of flood losses remained high. The explanation lies in part in methods of counting losses and in the incidence of floods, but in large measure it rests in continuing encroachment upon natural floodplains. Even in basins with elaborate reservoir systems, new building in hazard zones has tended to offset greater protection so that while the Corps of Engineers valiantly beats back the enemy on one front, their position is threatened from the rear.

This experience has lessons for other water regulation programs. By assuming that only engineering works were needed to curb the cost of unruly streams, other possibly effective means were neglected. Little or no attention was paid to such alternatives as land-use regulation or flood-proofing of buildings. By assuming the engineering works would do what the benefit-cost calculations had solemnly estimated they would do, without attempting to verify the practical results in land use, the public reaped quite different effects. A single-purpose channel improvement may invite further encroachment upon natural flow ways; a single-purpose levee may set a confident scene for later catastrophe; a single-purpose reservoir may appropriate a unique dam site without assuring complete reduction of flood losses.

Another case is public investment in irrigation as a means of stabilizing grazing. For a long time new federal reclamation schemes in the western United States have been justified in part as measures to protect the economies of nearby grazing areas from fluctuations in forage supplies. When one of the older projects in this category, the Uncompahgre in western Colorado, is examined, it is found that the grazing industry in fact was dislocated by the investment: the combination of reclamation and forest land policies attending the project made the ranching economy less stable than before and probably supported further deterioration of the range.

A prevailing attitude toward water in arid lands is that public investment, usually framed to stabilize the economy and promote growth, will yield the expected returns. Water, having been vital to past development, is taken as the easy key to the next stage. But water, although necessary, may not be sufficient, and the other requirements of sufficiency may be hard to define. It is much easier to get $100,000 to investigate a new project than $1,000 to find out what happened after an earlier one was completed. Methods of assessing the effects of investment in water now are being refined, and while much remains to be learned as to the redistribution of income gains and losses, rough estimates are practicable.

When questions of this sort are raised about the impact of water development projects, the traditionally complacent view of "the program" for a given basin is shaken. To reestablish confidence that public aims are in fact being met requires more searching studies and a wider range of choice than ordinarily is available. In practice, new water projects are put forward as the best judgment of an engineering staff which has investigated hydrology, geology of dam sites, soil conditions, structural requirements, and expected demand for water. A single plan judged most feasible is recommended for legislative or administrative approval. The choice is between "yes" and "no," or "yes" with minor modifications. Only passing reference is made to the hundreds of other water control plans that might have been designed as alternatives, or to the possible strategies of investment, promotion, or regulation in other sectors of the economy which might achieve similar results.

If the public welfare is taken as the broad, albeit vague, aim for an arid basin and if industrial and recreational production are seen as principal features in sound economic growth, it may well be that water development may be a necessary though secondary aspect of investment promoting cheap transport of raw materials, efficient marketing facilities and amenities conducive to a stable labor force. We can only speculate at this stage as to what might be the effects of investing in such facilities in lieu of certain of the great dams under construction in the upper Colorado Basin. It must be speculation because there have not yet been full plans which provide the executive and legislative agencies with a range of choice. The nearest approach has been in some of the International Bank studies of investment needs in underdeveloped countries.

Pitfalls of New Technology

At this point my argument may be challenged on the ground that having begun with the tremendous repercussions of new technology in water development I do not properly credit the possibilities of future advances in that direction. It may be asserted, for example, that saltwater conversion, or evaporation repression, or weather modification may so enlarge the potential of effective supplies of water as to render invalid much of what has been said about shortage of supplies. There is some support for this but it is shaky.

Take the case of saltwater conversion. A few years of international experimentation has spawned what is almost a modern mythology about brackish water. Important improvements have been made in methods of removing minerals, and experimentation with both pilot and laboratory plants is proceeding on a wide front with fruitful collaboration among the technologists of a score of nations. The costs thus far do not promise saltwater as a source for more than a few situations where urban needs can support costs of conversion that exceed the total cost of delivering fresh water supplies. They are still far above common irrigation water costs. Demineralization for agricultural use seems more likely to develop where a slight reduction would bring brackish water below toxic limits for irrigation or stock use. Undoubtedly brackish water will be cheaper to convert and more widely used in future, but the evidence does not yet warrant expectations that its availability will work miracles in the use of most arid lands. Possibly more promising would be a scientific development which would permit a reduction in transpiration, a process that accounts for more than two-thirds of water use in some areas.

Even if the more optimistic assessment of new means of removing salt or increasing rainfall were to be proven sound, their proper application would be hampered by the very conditions that hamper the use of technologies already in hand. Until they have been developed so as to be practicable alternatives to present water sources, the public emphasis upon them diverts attention from basic scientific research and from applications of other water knowledge. The prospect of going to saltwater for additional supplies induces complacency toward current shortages and toward the exploration of other solutions. We must find ways to prepare the patient and the dosage so that the promise of new technology becomes a stimulus to intelligent water use rather than a sedative.

If these three modifications of view are to be achieved, there will be need for a much deeper understanding of social processes in the arid zone than we now enjoy. It is not enough to know the available techniques or the combinations of them which would optimize economic returns. Only as there is understanding of the elements that enter into decisions governing water use by farmers or manufacturers or public servants can there be adequate programs to cultivate new patterns of use. We must find out, for example, how the resource managers view their land and water, what uses they think they can make of it, how they estimate the demand for and return from their products. In this fashion we may move toward recognition of conditions in which the farmer will stop overirrigating his fields, the manufacturer will install recycling devices, and the state legislator will ask more about a proposed dam than how much it will cost and who will get the water from it.

Summary

By way of summary, let us examine two cases of water use on quite different scales.

First, consider the lawn of one house in one oasis town of the western United States. Generally, the owner applies much more water than a water-budget analysis would show it needs. When in doubt he applies a little more. This is because the owner either doesn't know about or care about water-budgeting. He tends to use as much water as he is entitled to, and, indeed, the local water department hopes he will, so long as he doesn't exceed restrictions at times of peak demand. He resents any increase in rates as an infringement upon his inherited rights, and he thinks the more he uses the less he should pay per gallon.

An alternative would be to have no lawn at all and to cultivate a patio and border vegetation which gave him cooling relief with less water. He is slow to consider this or any other ways of dealing with his arid climate. The house is unsuited to such a change because it was designed as though for a Connecticut seacoast village. His midwestern tradition tells him a neat lawn is a mark of respectable status. Moreover, watering, in either the homely tug-of-war methods of yesteryear or the missile-control-board method of tomorrow, has therapeutic value after a trying day. Here is the water user who thinks more water is the sure solution for a dry environment, consistently undervalues his commodity and is reluctant to consider any alternatives to his heavy usage.

Next, consider the state water plan for California. The plan essentially provides for moving water from surplus areas to deficit areas by a system of reservoirs and aqueducts. What I find lacking is a careful appraisal of how much water might be gained for the state by applying other means of using or reusing water, or what effect differences in the price of water could have upon its use, and of alternative measures which might be taken to guide the redistribution of income, land use, and water in the state. Perhaps such appraisal would reveal the approved plan to be the best. Whether or not it did, a state facing gigantic expansion in its use of water deserves to know the best available answers to these questions before making its decision.

Because the arid lands are currently marked by widespread social change, it is not unreasonable to expect that views toward water may be revised drastically. To deal intelligently with conditions created by unfolding technology, new uses, and dwindling margins of supply will require new orientation as to the role of water where the potential supply is small relative to natural demands. The gap between technology and application must be narrowed, the standards of value reappraised and the prospective impacts reexamined. This would add to the already long bill of needs for physical and biological research, another set for social science research. On the scale of a single lawn or an entire state, these new views of the role of water constitute a basic challenge to both research and public education. We cannot expect sudden revolution in attitudes toward water, but we can expect that as they are revised, many promising opportunities for making effective use of the limited potential supplies will be realized.

Science and the Future of Arid Lands

Essentially, the task of helping make the most of the sparse and variable arid lands is one of discovery and dissemination: discovery of the fundamental facts about social behavior, water, soils, and plants on which refined technological advances may be grounded; and dissemination of the new and practicable knowledge to those who should use it. Thus, scientific research and education march shoulder to shoulder, each reinforcing the other, not only in the university but in the dry landscape. There is no magic for breeding new discoveries and no universal rhetoric for making them known and understood. Fresh insights into natural and human processes come erratically, but there are ways of fostering the climate in which they can arise, and of speeding up the whole social mechanism for dealing with new facts and ideas so that further discovery can be inspired and technical applications and scientific training can be accelerated. It is this effort to sharpen and speed up the scientific and educational processes that has special urgency for the arid zone while its waters are not yet fully appropriated, while its soils and plants still have some recuperative capacity, and while its minerals still provide the stimulation of investment capital.

(1960e), 92

Notes

1. John Wesley Powell, *Report on the Lands of the Arid Region of the United States* (Washington, D.C.: Government Printing Office, 1879).

2. Jean Brunhes, *Étude de géographie humaine: L'irrigation, ses conditions géographiques, ses modes et son organisation dans la péninsule ibérique et dans l'Afrique du Nord* (Paris: C. Naud, 1902).

3. For the current status of these attitudes and problems, see below, vol. II, chap. 4 (Johnson).

9 The Choice of Use in Resource Management

The 1950s were an exciting period for the social sciences. New crosscutting concepts were emerging in studies of systems, of structure, and of decisions. In this, the most theoretical of his papers, White attempts to conceptualize key decision making in resource management—the choice of use.

Theory does not come easily to White; it is always there in his writing, but most often it is implied. Even this theoretical paper begins and ends with pragmatic public policy issues. Most of the paper illustrates the theory of resource use choice by means of the now familiar (to his readers) empirical example of floodplains. In describing his view of how decisions are made, White makes his own choices among existing theories in the social sciences. He equates individual, group, and institutional behavior, implying that each form can have a common explanation. He accepts the cultural definition of resources with its assumptions of relativism in human behavior. He joins in the anti-utilitarian rejection of the then prevailing model of economic man. And in choosing his key elements in decision making, he presupposes an interactive process of nature, society, and technology without acknowledging that such a three-dimensional analysis stands in contrast to one-dimensional theories of determinism or two-dimensional theories of conflict or equilibrium.

Thus he limits his theory to the specifics of the resource manager, eschewing the opportunity to relate to more general notions of human behavior. This is not because his theoretical ideas are simply derivative from more general views. While he cites the anti-utilitarian choice literature of the late fifties, his own work predates it. His interest in the range of choice and modes of selection is apparent in *Human Adjustment to Floods,* as was his rejection of simplistic economic justification. Yet it is always the functional need for practical explanation that dominates, not the need to describe human behavior in general. In this case there was the need to know how people choose to use specific resources before an optimal choice can be prescribed. The issue is of pressing importance since choice

Reprinted from *Natural Resources Journal* 1 (March 1961): 23–40.

is now embodied in a new law related to multiple-purpose forest management.

His framework of key decision elements—physical, economic, technological, spatial—appears and reappears throughout this volume. Only the variable of spatial linkage appears in other guises. We see it in social guides (selection no. 14) and in its effects on other people (selection no. 20). The use of the keyhole of decision making as a way of observing resource management was to dominate resource and hazard management studies for twenty years as well as to launch an interdisciplinary field of environmental perception.[1]

The enactment into law of the "principle of multiple use" in the Multiple Purpose Forest Act of 1960[2] reminds us that much thinking concerning choice of resource use is based upon very rough understandings of the process by which choice is exercised. In the national forest operations, for example, multiple use has been a watchword of administration since the earliest days of federal activity,[3] but there has been relatively little attention to the way in which either private operators or public servants arrive at schedules for using particular pieces of forest land in particular ways.

The 1960 Act defines multiple use to mean:

> The management of all the various renewable surface resources of the national forests so that they are utilized in the combination that will best meet the needs of the American people; making the most judicious use of the land for some or all of these resources or related services over areas large enough to provide sufficient latitude for periodic adjustments in use to changing needs and conditions; that some land will be used for less than all of the resources; and harmonious and coordinated management of the various resources, each with the other, without impairment of the productivity of the land, with consideration being given to the relative values of the various resources, and not necessarily the combination of uses that will give the greatest dollar return or the greatest unit output.

This is an extremely difficult task, especially when there is sparse knowledge of the factors that are at work in setting present uses and of the effects of such use.

It is easier to specify what would seem a wise combination of uses for a forest unit in the light of physical research and national aims than to determine the process of choosing present uses, and, indeed, as McConnell[4] and McKinley[5] have pointed out, there may be administrative advantage in having a vague charter and a flexible method in applying public policy where the special interest groups are thorny and sometimes incompatible.

It has been common to advocate multiple-use of water resources since the Inland Waterways Commission and the National Conservation Commission advanced their broad aims.[6] The view held by many workers in public service has been summed up simply by the Commissioner of Reclamation as follows:

> Sound water resource development is not limited to irrigation and hydroelectric power. Other purposes deserve adequate consideration, and facilities, therefore, should be constructed when shown to be economically justified. These other purposes include flood control, navigation, municipal and industrial water supplies, pollution and salinity control, protection of public health, enhancement of recreation potentialities, and preservation and propagation of fish and wildlife.[7]

What were for many years crude attempts to establish economic justification for water projects have been refined by the highly fruitful work of Eckstein,[8] Hirshleifer,[9] Krutilla,[10] McKean,[11] Wollman,[12] and others who have advanced analysis of the efficiency criteria for public investment. This goes far beyond the thinking when the federal "green book" first attempted to formulate uniform methods of project evaluation[13] although it has not yet had a major impact upon federal policy. We now have keener tools for estimating optimal combinations of complementary water uses which permit sharper judgment of uses in terms both of contributions to national product and of income redistribution, and which also allow for difference in timing.

These improved methods of weighing various resource allocations nevertheless leave much to be desired in explaining present allocations and in indicating the conditions in which wiser allocations might be achieved. Considerations of economic efficiency may not be determining factors. One of the pervasive problems in resource management today is the widening gap between practice and knowledge.[14] Many forest lands are operated at far less than what is optimal use for production; efficiency of water use by either economic or physical criteria is low. This requires explanation in terms that both state the problem and point to possible solutions.

In an effort to find a general method of analysis which would help describe actual choices in resource use and promote systematic appraisal of possible new uses, I have experimented with several approaches. The broad formulations of Ackerman,[15] Zimmerman,[16] and others concerning the relation between levels of living and natural resources have been helpful. So also have been the decision-making schemata of Banfield,[17] Firey,[18] Lee,[19] Simon and March,[20] and the Sprouts.[21] From work with problems of floodplain occupance, water use, and recreational land use, a framework for describing resource decisions has emerged which is presented here for discussion. It will be stated briefly, and its possible application will be illustrated by detailed reference to floodplain use and by passing reference to prevailing policies of national forest and water management.

I. Managers and Establishments

In this analysis decision making affecting natural resources will be considered as centering upon managers who are responsible for establishments. Managers are defined as the individuals, including corporations and government agencies, who act as a unit in the management of an establishment. An establishment is defined as a single residence, agricultural, commercial, manufacturing, transportation, or public service organization that has a distinct usage of area. Examples of establishments would be a city residence, a ranch, a shop, a manufacturing plant, a national forest, a county and a government dam. Examples of managers for such establishments would be a city dweller who rents or owns a house, a farmer, a merchant, the board of directors of a manufacturing company, the regional office of the Forest Service, the county zoning board, and the bureaucracy of a government agency. The manager, so defined, may be either an individual or a group. There are, of course, situations in which the responsibility is divided, as between a legislative body and an administrative officer, or is shared, as between a land owner and tenant.

Ownership is not a necessary attribute of a manager. He may be a farmer who wholly owns his land, a tenant owning no interest in the land but enjoying the privilege of using it as he pleases, or a government bureau following general policies laid down by a legislative body.

The size of an establishment may vary from a fraction of an acre, as in the case of many city dwellers and small-scale farmers, to thousands of acres, as in the case of large ranchers, timber corporations, and government land and water-development agencies.

Establishments also may overlap. An irrigation farmer may be fully responsible for the use of his own land subject to a county zoning ordinance, but belong to a water district where the responsibility for distribution of water is shared by an elected district board. And in the same instance he may have sold the mineral rights to oil underlying his land and be subject in his summer grazing operations to decisions by the regional forester.

II. Resources, Adjustments, Uses, and Multiple-uses

A resource is culturally defined. It is considered as an aspect or combination of aspects of the natural environment suited to some human use. Whether resources are described by physical classes—soil, water, minerals, and the like—as in the traditional conservation literature, or by areal combinations, as in the tradition of much geographic study, they may be regarded as being delimited by human assessment of possible use. Such use may be for a single purpose or for complementary purposes, and for any given purpose the resource manager may employ some variety of adjustments of technology and managerial methods to achieving that use. Thus, given a piece of land with a homogeneous forest stand, the manager may use it for timber alone or for various combinations of uses, but he also may adjust his technology, such as logging methods, and his management system, such as timing of harvest, in several ways to reap the returns he seeks. It is helpful to think of there usually being several possible kinds of adjustments by which a given use may be realized.

Notwithstanding widespread acceptance of the concept of multiple-use in both land and water management there has been relatively little investigation of the possible limits to multiple-use combinations. The compatibility of pairs of land uses are summarized by Clawson, Held, and Stoddard,[22] and of pairs of water uses by Ackerman and Löf.[23] These are based chiefly upon observation of experience in resource management and tend to group three distinct types of limiting criteria in one classification—technological, economic, and social. For any pair of uses the degree of complementarity may be different for each of these three criteria and subject to wide variation over area.

A. *Theoretical Range of Choice.* A decision in resources management involves, theoretically, a choice among a large number of possible adjustments and uses which might be applied to the particular physical environment. The manager has the option of using the resource in any one of the ways in which such a resource has ever been used by man, and there is always the possibility that he may invent a new way of using it. Thus, a chernozem soil may conceivably be used to grow a large number of crops in farming systems in which the operating unit varies from a few acres to more than 1,000 acres and in which livestock may play no role or a dominant role. Many such variations are known to be practiced in chernozem areas. But, in addition, a farmer may devise a new type of use, for example, cultivating water fowl in poorly drained phases of land.

The theoretical range of choice may be defined as that number of adjustments and uses that have been practiced in any similar environment, plus a possible innovation. This number may reach astronomical proportions for certain types of area as in the case of alluvial valleys in middle and low latitudes. Maass and Hufschmidt have pointed out the immense number of possible combinations of water use and control in a sample drainage area.[24] It becomes necessary to examine the adjustments and uses according to their major types and distributions, and this has been done by geographers in many observed situations.[25]

B. *Practical Range of Choice.* No manager ever has open to him in practice the full theoretical range of choice. He has some choices which he may quickly reject because they seem to him unwise. There always are others that are closed to him, however wise or unwise they might appear to him or to others upon careful investigation. Usually, these blocked choices seem to derive from one or both of two aspects of social guides in the manager's culture. One aspect is awareness, the other restraint.

Although a scientist may be aware of theoretical possibilities for extracting ore, the mine manager may be entirely ignorant of them. Or even if the manager has heard of a different method of ore treatment he may not be aware of its relevance to his situation. Thus, a miner may consider that a cyanide process for obtaining gold would not work with his tungsten ore although in another place the process has been successful. Lack of awareness of the possible choices is a product of the social environment and of the kind of communication system that it affords its members. Illiteracy; poor schools; paucity of books, libraries, and journals; and lack of communication with other areas are some of the conditions effectively shutting a manager off from knowledge of ways that otherwise would be open to him. There is always a time lag in the diffusion of new techniques. Even where institutions of teaching and publication are well advanced, habits of perception may prevent certain situations or ideas from being recognized by the manager.

With or without awareness of all the possible choices, there are many social restraints which block uses in various situations. At one organizational extreme are the formal strictures of a political agency, as in the case of prohibition of opium culture. At the other extreme are the subtle but no less rigid restraints of social attitude and tradition. Dietary preferences, religious beliefs, tribal customs, social status systems, or personal values may prevent a manager from considering any substantial departure from the use he currently is practicing. In between these extremes are innumerable shades and forms of restraint, many of which have been recognized in institutional economics. Social guides combine to limit in greater or less degree the choices that a manager perceives and can feel free to consider. The practical range of choice always is narrower than and often tremendously reduced from the theoretical range of choice. A manager may not be conscious of making a choice among many possible uses. He simply may feel that he must without question go on doing in one year what he did in the preceding year and what his father did before him. This is the most elementary form of choice— the reaffirmation of the past.

III. Elements in Decision Making

Given a practical range of choice that is open to the resource manager it remains to note the kinds of considerations that enter into the process of selecting a course of action. It will help to assume that a theoretical description of elements of that process for the most complex market society under governmental restraints is capable of being used to arrange those possible elements in other societies. This implies that an element of decision making in nonmarket economies might be found in market economies, an assumption which has not been fully tested. If the possibly crucial elements in decision making can be distinguished, the next step is to ask whether or not they are recognized, and, if recognized, to examine the quality and areal implications of the way in which they are handled. The questions then are: How does the manager deal with this aspect of decision making? How does his appraisal compare with that made by others? In each case the description of the element in choice involves first identifying the resource manager's view and then assessing its weight in his decision.

To isolate such elements in decision making is not to describe that process directly in all its social aspects; it merely describes possible elements in it and thereby lays the groundwork for determining peculiar combinations of elements at a given time and place. For example, the maximization of net gains—a standard economic formulation of decision making—may be found to be completely absent in one area while in another it is dominant. And even where dominant it may be applied incorrectly.

It is suggested that resource managers whether they be owners or tenants, peasants or public agents, in arriving at a choice of use or uses to be made of a specified resource, evaluate some or all of the following elements:

1. The quantity and quality of the physical resource. This is expressed in resource estimates which may range from casual and highly biased perception to accurate scientific inventory or appraisal.

2. The present value of the gains and losses accruing from future use of the resource. This commonly is expressed in discount rates and benefit-cost ratios. It necessarily includes an estimate of future demand for the utility created by use as expressed in demand curves and projections.

3. The technological change which might affect future demand, future production, and compatible use combinations. This is expressed in projections of techniques for consumption and production.

4. The relation of any given use to other resource use in contiguous or functionally linked areas. This is expressed in regional plans or spatial linkage.

From these considerations emerges a choice among possible uses or multiple-uses. Other elements no doubt enter into the decisions, but these are taken in this preliminary statement as covering considerations in many situations studied.

A. *Rationality and Perception.* Such analysis need be neither conscious nor explicit. The manager may ignore it or may exercise it by assuming invariance. He may take a position ranging from intuitive acceptance to highly sophisticated computation, but a position which in many instances can be identified. These analyses neither need to be rational nor linked together in a rational manner. Their association may dramatically bespeak irrationality. It nevertheless identifies the character of analysis that has been in fact used in managing the resource.

Another phase of the manager's appraisal also is important. It is the relation of his appraisal to other appraisals made by presumably more objective methods. How does a farmer's estimate of the potentialities of his soils compare with the classification by a government soil scientist? How does a city council's estimate of the economic gains and losses from investment in a new water supply compare with the estimates of an econometrician? Perception of environment is a basic feature of resource management and may drastically limit the practical range of the choice. From this starting point through appraisal of possible uses, income streams, technological trends, and regional impacts, the comparison of the manger's appraisal with that of others helps identify distinctive and crucial aspects of decision making.

An examination of the analysis followed for each element deepens understanding of the full decision that is made in the social setting and reveals the facts, perception, aims, and values upon which that decision is based. And it is such understanding, especially in its areal implications, that deserves cultivation. To examine these analytical problems in resources management does not necessarily require different sorts of data than in the more conventional resource approaches, but it does ask a somewhat different set of questions. Different, and possibly more rewarding answers, then may result.

According to this view, the interaction of population, environment, and technology takes place within a framework that may be considered to have three parts. The *theoretical range of choice* open to any resource manager is set by the physical environment at a given stage of technology. The *practical range of choice* is set by the culture and institutions which permit, prohibit, or discourage a given choice. The *actual* selection within those limits depends upon the way in which a manager analyzes the different elements of the decision, and it apparently does not involve continuous functions, but rather, discontinuous ones which may be recognized as turning on sensitivity points. The process of decision might be stated in mathematical form, but for this preliminary report it will be sufficient to state it in tabular form.

IV. Decisions in Management of Floodplains

A recent geographic study of urban occupance of floodplains in the United States[26] may be used to illustrate the approach. This is selected partly because some results are at hand, but also because they provide a situation in which one characteristic of the physical environment—floods—is readily measured and is subject to concrete predictable manipulation. Within a floodplain the combination of land and water resources may be relatively uniform back from the stream channel. The following illustrations assume that flood hazard in those situations is the chief variable in the resource base. This is an oversimplification, but will serve to suggest the complexity which might be expected where the resource is less uniform and the range of use greater.

A common approach to flood problems has been to ask what areas and activities are subject to flood damage, what is the character of flood occurrence, what kinds of protection have been provided, and what is the public investment and policy in flood control. These are all interesting and pertinent points, yet their elaboration from year to year has neither greatly advanced understanding of the complex decisions which have led to occupance of floodplains nor charted grounds upon which the consequence of changes in public policy affecting floodplain use might be predicted.

If floodplain use is approached in terms of methods of analysis followed by the resource managers the following questions are asked: How do managers of floodplain land estimate the flood hazard? What is their estimate of the productivity of the floodplain in the light of frequency and magnitude of flooding? How are the gains and losses of various possible adjustments calculated? What attention is given to harmonizing flood adjustments and floodplain uses with other resource uses? What account is taken of technological change in projecting future demand and adjustments? How is the solution for one area related to management of resources in contiguous or functionally linked areas? What social guides either restrain or encourage floodplain use? Five situations will be described briefly.

A. *The Case of One Manager.* A relatively simple example is the manager of an isolated residential property subject to occasional flooding. He is well informed about the hazard, he knows what other home owners in similar situations have done about flood losses, and he is committed to maintaining his use for residential purposes. What adjustments can he make to flood hazard in continuing his use of the site?

As shown in table 9.1, there is a theoretical possibility that he could obtain insurance, but this is blocked by the general refusal of insurance companies to cover flood losses. He does not consider abatement or protection because he understands this would require work off his property. Land elevation, change in structural features,

Table 9.1 Elements in choice of adjustment to flood hazard by one home owner in a floodplain
Single use: Residential

			Elements in choice				
Theoretical range of adjustments	Practical range of choice	Estimate of hazard	Techno-logical trends	Economic efficiency	Spatial linkage	Others	Actual choice
Bear the loss	o	−	+	−	+	.	R
Insurance	x	.	.	.	.	.	R
Land elevation	o	+	−	−	−	.	R
Structural change	o	+	−	+	+	.	R
Emergency action	o	+	−	+	+	.	A
Flood abatement	x	.	.	.	.	.	R
Flood protection	x	.	.	.	.	.	R
Public relief	o	+	+	+	+	.	A

x Blocked choice   + Favorable to choice   A Accepted
o Open choice   − Unfavorable to choice   R Rejected

and preparation for emergency action would be possible adjustments, and his estimate of flood hazard would encourage some action along those lines. He is discouraged from undertaking any of these by the prevailing belief in his community that because the government is doing something about water control in the region, floods will be reduced in the future. Some rough calculations of loss to be prevented convince him that it would pay him to invest in structural changes in preparing for and carrying out emergency action, that is, shifting his storage room and electrical appliance connections. By the same means he concludes land elevation would be too expensive and that it might, moreover, land him in a lawsuit with a neighbor whose drainage system would be interrupted. Although he thinks his losses may decline as engineering works improve, he thinks he cannot afford to bear the loss indefinitely. In his terms, any adjustment he might make except land elevation would not harm him through its effects upon the community. Public relief will be offered in time of disaster, and he gradually decides to rely partly upon such help and to prepare for emergency moves when the next flood does strike.

He has arrived at this position over a period of time without ever stating a formal decision, but the apparently pertinent considerations are summarized in table 9.1. His rejection of structural change is to be traced to his sanguine feeling about technological progress and his confidence that public relief will bail him out of a serious disaster.

Simpler cases might be described. The new home owner, of which there must be thousands in the United States, who is totally unaware of flood hazard and who unwittingly bears some of the loss and is a candidate for public relief would be shown as having only those two practical choices. Many resource managers in floodplains believe they have no other choice than to bear the loss.

B. *The Case of a Single-use Neighborhood.* If the same questions are asked about a neighborhood of residential property owners in which there is uniformity of attitude toward floods and the use of the floodplain, the kind of situation described in table 9.2 might be encountered.

Again, it is assumed that they are committed to continuing residential use, and in this case a zoning ordinance limits the choice of use. They are not, however, as well aware of the theoretical choices as the single owner described above. While they know of land elevation they reject it as being too expensive, probably soon outmoded, and likely to cause trouble with the community across the river. They do not consider insurance because it is unavailable. Structural change and emergency action are ignored because such adjustments are unknown. The managers are encouraged to count on public relief but they would prefer the greater security of either a Watershed Protection project or a small reservoir project. The former, from the standpoint of the community, would be desirable in all regards except that it would require extensive organization of landowners upstream and this "other" consideration seems insurmountable in a reasonable period of time. Therefore, they push for a construction project at federal expense.

Table 9.2 Elements in choice of adjustment to flood hazard by a neighborhood of N home owners
Single use: Residential

	Elements in choice						
Theoretical range of adjustments	Practical range of choice	Estimate of hazard	Technological trends	Economic efficiency	Spatial linkage	Others	Actual choice
Bear the loss	o	−	+	−	+	.	R
Insurance	x	.	.	.	.	.	R
Land elevation	o	+	−	−	−	.	R
Structural change	x	.	.	.	.	.	R
Emergency action	x	.	.	.	.	.	R
Flood abatement	o	+	+	+	+	−	R
Flood protection	o	+	+	+	−	+	A
Public relief	o	+	+	+	+	.	A

Table 9.3 Elements in choice of adjustment to flood hazard and of land use by one manager of vacant urban property
Uses: Vacant; recreational; residential; or commercial

Theoretical range of adjustments and uses	Elements in choice						Actual choice
	Practical range of choice	Estimate of resource	Techno-logical trends	Economic efficiency	Spatial linkage	Others	
Vacant	o	−	−	−	−	.	R
Residential							
Bear the loss	o	+	+	−	+	.	A
Insurance	x	.	.	.	.	.	R
Land elevation	o	−	−	−	−	.	R
Structural change	x	.	.	.	.	.	R
Emergency action	x	.	.	.	.	.	R
Flood abatement	x	.	.	.	.	.	R
Flood protection	o	−	−	−	−	.	A
Public relief	o	+	+	+	+	.	A
Recreational							
Bear the loss	o	+	+	−	+	.	R
Insurance	x	.	.	.	.	.	R
Land elevation	o	−	−	−	−	.	R
Structural change	x	.	.	.	.	.	R
Emergency action	x	.	.	.	.	.	R
Flood abatement	x	.	.	.	.	.	R
Flood protection	o	−	−	−	−	.	R
Public relief	o	+	+	−	−	.	R
Commercial							
Bear the loss	o	−	+	−	−	.	R
Insurance	o	+	−	+	+	.	A
Land elevation	o	+	−	+	−	.	A
Structural change	o	+	−	+	+	.	A
Emergency action	x	.	.	.	.	.	R
Flood abatement	x	.	.	.	.	.	R
Flood protection	o	+	−	−	−	.	R
Public relief	o	−	+	−	−	.	R

C. *The Case of Choice for Vacant Property.* In the two preceding cases it was assumed that choice applied only to type of adjustment for a single use. The description becomes more complicated when a large vacant property in the hands of one manager is considered and three possible uses are canvassed. In this instance, as shown in table 9.3, the manager considers some adjustments for residential, recreational, and commercial use. He decides to develop part of the land for residential purposes, allowing the new residents to bear whatever loss is not covered by public relief, and to develop the other part for commercial use. In the latter decision, however, he knows from a skillful architect that by a combination of land elevation, structural change, and insurance available to large commercial users he can deal with all foreseeable losses, and he goes ahead with this without relying upon other adjustments.

D. *The Case of a Zoning Agency.* If a similar piece of property is within the jurisdiction of a zoning board it would be subject to a different kind of analysis as shown in table 9.4. Here, the same range of adjustments and uses is considered as in table 9.3, but insurance is excluded because a public agency cannot assume it to be available.

Table 9.4 Elements in choice of adjustment to flood hazard and of land use by county zoning agency for vacant urban property
Uses: Residential; recreational; or commercial

Theoretical range of adjustments and uses	Practical range of choice	Estimate of resource	Techno-logical trends	Economic efficiency	Spatial linkage	Others	Actual choice
Residential							
Bear the loss	o	−	−	−	−	.	R
Insurance	x	.	.	.	.	.	R
Land elevation	o	+	+	+	−	.	R
Structural change	o	+	+	+	−	.	R
Emergency action	o	+	+	+	+	.	R
Flood abatement	o	+	+	+	+	−	R
Flood protection	o	+	+	+	−	−	R
Public relief	o	−	−	−	−	−	R
Recreational							
Bear the loss	o	+	+	+	+	.	A
Insurance	x	.	.	.	.	.	R
Land elevation	o	−	+	−	−	.	R
Structural change	o	+	+	−	+	.	R
Emergency action	o	+	+	+	+	.	A
Flood abatement	o	+	+	−	+	.	R
Flood protection	o	+	+	−	−	.	R
Public relief	o	+	+	+	+	.	A
Commercial							
Bear the loss	o	−	+	−	−	.	R
Insurance	x	.	.	.	.	.	R
Land elevation	o	+	+	+	−	.	R
Structural change	o	+	+	+	−	.	R
Emergency action	o	+	+	+	+	.	R
Flood abatement	o	+	+	+	+	−	R
Flood protection	o	+	+	+	−	−	R
Public relief	o	−	+	−	−	−	R

The judgment is reached that land elevation and certain new structures would be deleterious because of effect upon channel efficiency, causing damage to other property. It also is concluded that occupance of the land for either residential or commercial purposes would lead to a situation where investment in flood abatement or flood protection works under prevailing federal policies might later be found justified but that the community would not like to face the cost involved in contributions to the works and in relief during the intervening time before construction could be expected to begin. There is need for setting aside recreational open space in the growing community; the riverine situation has special amenities; and there is land elsewhere in the county equally well situated for residential and commercial purposes. Thus, the land is zoned for open recreational use.

E. *The Case of a Construction Agency.* The preceding cases have described a high degree of choice on the part of resource managers. These illuminate the possible variations but they should not be taken as representative of floodplain situations. It is known, for example, that while some managers such as those in Pittsburgh's Golden Triangle make the kind of analysis shown in table 9.3, there are many who apparently consider only public relief or public protection as an alternative to bearing the loss. This situation is reinforced by federal policy which has emphasized construction works.

The decision of a federal engineering agency in appraising the flood problem in a residential neighborhood of the type described in table 9.2 is narrowly defined. The agency, as shown in table 9.5, considers only whether protection works are justified or not according to federal policy. It finds that the hazard is serious and will not be reduced by technological changes upstream; and, as another consideration, it has the encouragement of a strong Congressman. It recognizes, nevertheless, that the economic efficiency is low and that the works might cause distress elsewhere. The proposed construction project is rejected and no substitute is offered to the neighborhood.

V. Implications for Policy

This tabular scheme, it must be reiterated, does not necessarily explain why the manager arrives at his final decision. It attempts to describe how he does so, and takes a necessary step toward arriving at an explanation of his behavior. It may have succeeded only in

Table 9.5   Elements in choice of adjustment to flood hazard by a federal engineering
agency
Use: Residential

	Elements in choice						
Theoretical range of adjustments and uses	Practical range of choice	Estimate of resource	Techno-logical trends	Economic efficiency	Spatial linkage	Others	Actual choice
Residential							
Bear the loss	x	.	.	.	.	.	R
Insurance	x	.	.	.	.	.	R
Land elevation	x	.	.	.	.	.	R
Structural change	x	.	.	.	.	.	R
Emergency action	x	.	.	.	.	.	R
Flood abatement	x	.	.	.	.	.	R
Flood protection	o	+	+	−	−	+	R
Public relief	x	.	.	.	.	.	R
Other use	x	.	.	.	.	.	R

raising questions that are unanswerable for the present.[27] For example, under this scheme it may become apparent that the kind of perception which certain resource managers have of the flood hazard is related to their consideration of a wider range of adjustments, and it therefore appears that an extended program of public information about flood hazard through the publication of maps would alter management decisions by others. A further check of the effect of maps shows that the other managers do not respond to such information. Without an explanation for these different perceptions of flood hazard, it may be difficult to design information programs that are effective, and much of the effort expended on flood hazard maps may be lost.

However, when the questions stated earlier become the framework of investigation, understanding of problems of floodplain occupance sometimes deepens through a more nearly precise description of crucial elements in management decisions. Accordingly, the capacity to predict probable effects of a given change in floodplain management is strengthened.

A few sample findings may be cited. Attention to estimates of the flood hazard as one aspect of the resource leads to classification of floodplain situations into those in which managerial perception of the flood hazard corresponds to the hydrologic realities and those in which it does not. This reveals in the United States, among other findings, an extraordinary unawareness of catastrophic hazards. Investigation of estimate methods also suggests characteristics of floods which limit the kinds of decisions which are made. For example, it becomes evident that while classification according to flood origins is not especially significant to managerial decision, the flood-to-peak slope of the hydrograph does set the limits within which certain classes of emergency adjustments, such as flood warning and flood proofing, can be made. A new, much more meaningful, classification of streams on that criterion then emerges.

Appraisal of methods of discounting present value of future gains and losses from various possible adjustments to the flood hazard reveals, first of all, major classes of adjustment that commonly are ignored. It later directs attention to the spatial linkages, such as the costs of stream-channel encroachment, which are given little or no weight. Appraisal of the methods of projecting demand for floodplain land shows wide variation among the managers. Private managers tend to overestimate. The federal agencies often follow a method out of line with the realities of change in land use in observed situations, and for convenience assume no increase in demand unless required to produce a favorable benefit-cost ratio. It then becomes possible to classify floodplains according to population growth characteristics, and to suggest relationships between such growth and changes in occupance as well as to point out the effects of the public demand estimates upon further occupance. Investigation of linkage between floodplain use and uses in other areas suggests the existence of definite patterns of growth that may be triggered by developments off the plain. Considerations of relative location then demand attention.

In another vein, the preparation for a zoning agency of a description, such as that in table 9.4, may promote an intelligent canvass of the place of public regulation of land use. Regulation of floodplains is not necessarily wise, and the justification for it needs to be checked against the public goals to be served in a specific situation.[28]

VI. Sensitivity Points and Social Process

In examining elements involved in decision making in the field it may be found that there are characteristics of the area in each case which establish constraints upon the management of a resource no less effectively than the institutional constraints. For example, in the urban occupance of floodplains man is not likely to go into areas which are flooded as frequently as once every year or once every two years. Where the frequency is as great as once in every three to twenty-five years the managers are likely to occupy the floodplain but with some consideration of alternative adjustments. Where the frequency is less than once in twenty-five years there is a pronounced tendency to assume that there will never be another flood, and it makes little difference whether floods come once in twenty-five or once in two hundred years. This relationship between flood frequency and estimates of flood hazard might be graphed for selected areas. In some situations it is clear that the two-year and twenty-five year frequencies are inflexion points on the curve, and that in the management of floodplain properties they are likely to be critical. In trying to provide managers with needed data on the flood hazard these may then be taken as sensitivity points to identify and show on maps.

Another example from floodplain occupance is the relationship between the feasibility of emergency adjustments and the flood-to-peak interval. As already suggested, the time periods in which it is practical to make different forms of emergency adjustment set sensitivity points. In charting the areas in a basin in which emergency adjustments would be feasible it becomes important to map these points.

For each one of the elements of analysis in a given management situation it is, then, possible to think of a curve of relationship between the feasibility of a type of adjustment and a production factor. On the curve there may be one or more sensitivity points that are significant in the management decision. These identify critical items to map in seeking differences over area.

A. *The Test of Location*. To expand this description of choice to the situation in a larger area where many managers are involved is difficult but not impossible. It must be recognized that the aggregate decisions are not the simple algebraic sums of all the factors entering into the system of equations; that is, if the economic efficiency for one use is uniformly higher than for any other the aggregated choices will not necessarily favor that use. In each case, and possibly for somewhat different reasons in each case, the actual choice may be against the economically efficient use. The grouped decisions for an area are therefore the aggregation of the individual decisions and not the summation of the various factors in each.

This is an important distinction because it is common in discussion of resources use to make normative judgments, to say that where soils are low in certain nutrient values they will generally be used in a particular way. It might be expected that land values for a given use would vary with the severity of flood hazard. In some cases these statements of an average condition may hold as a description of resource use over a large area. In other cases they may have little relevance, such as where a soil type may be favorable for grain farming, but where grain is not raised in the area because of institutional or other blocks, or where land values do not reflect flood hazard.

Study of areal differentiation in use patterns may suggest sensitivity points in management. It also may present anomalies, and in either case the geographic analysis of use patterns goes hand in hand with analysis of decision making.

B. *Social Process.* Descriptive analysis of this character does not imply any single concept of culture, social system, or social process. The validity of such concepts may be tested in part by the descriptions, and certain of the concepts may illuminate understanding of resource management decisions, but there is no necessary link. Thus, the way in which a flood hazard is perceived may be found to be closely related to cultural conditions in some societies. The awareness of possible choices such as structural change or flood protection may be shown to be a function of the social status of the resource manager. An examination of numerous floodplain situations may suggest a sequence of occupance—such as, initial avoidance of flood hazard, later invasion as settlement continues, and final decision to protect the exposed users—that can be hypothesized as having general application. The responsiveness of a community to a reliable flood warning may be related to its social mobility and organization. All of these relationships have been examined recently in floodplain studies.

VII. Choice of Use in Public Forests and Water Projects

Against the background of this framework for examining resource management decisions, the multiple-use declarations of public forest and water programs now may be recalled. One striking aspect of both the forest and water declarations is that they are based upon only meager knowledge of the decisions currently made by resource managers in three fields. While the Multiple Purpose Forest Act is, in effect, a formalization of policies prevailing over the years in Forest Service practice, it presumes multiple-use to be desirable as a matter of public policy and also assumes that wise choice can be made among the alternatives. The evidence from detailed analysis of forest management still is slim, and the methods of comparing choices are rudimentary. In the absence of precise guides from the Act itself as to ways of estimating the optimal combination of uses from a national standpoint, there is none of that routine or prescribed study which might provoke undue local irritation.

Similarly, studies looking to selection of economically justified combinations of water uses in new federal projects often spring from hortatory instructions rather than assessment of experience with water use. The conditions in which irrigators apply different amounts of water to their fields, the demand curve for industrial water, and the way in which floodplain dwellers perceive the flood hazard are examples of gaps in knowledge as to present choices.

To expect responsible public choice without the benefit of such knowledge is to expect that policies and programs will be adopted without full appraisal of their consequences. Yet, it is wholly un-realistic to expect that forest supervisors will defer a judgment as to grazing-lumbering-recreation combinations while the necessary studies are being made, or that authorization of an irrigation project will await understanding of prevailing practices of water application in the benefiting area. Administrative action will not mark time while more studies are made. Moreover, it should be recognized that there can be heavy administrative resistance to any canvass of choices which lies beyond the prescribed authority of a public agency; the Corps of Engineers has not been quick to consider flood proofing as an alternative, however attractive, to the traditional engineering works; and the Soil Conservation Service has not taken wide initi-ative in appraising means of preventing those encroachments upon floodplains which can be protected by upstream measures. Agency commitments tend to strengthen rather than remove the obstacle to choice.

A second interesting aspect of forest and water multiple-use aims in the light of this analysis is the critical part played by spatial size and organization in shaping the expression of these aims on the ground. Although both the forest and water policies are stated in terms of national goals, their application vacillates from a primarily local view to a primarily national view and necessarily takes account of intermediate regional views. The variation in assessment of re-sources and possible uses which is displayed so clearly in the de-cisions of individuals, communities, and national agencies in managing floodplains is even more apparent in the forest and water programs. Multiple-use may apply to an acre, a farm, a forest stand, or an entire national forest unit. It may be planned for a single dam, a small watershed, or an entire drainage basin; and depending upon the size, number, and organization of establishments involved, the decision making process will vary radically.

A third aspect is the self-fulfilling prophecy of some assumptions built into public choice. If it is assumed that irrigation or industrial water users will use water in inefficient quantities in terms of both physical and economic criteria, there will be a tendency to shape policy to provide supplies to care for the increasingly heavy withdrawals. Encouragement then is given to the users to continue their present practices. If the Corps of Engineers assumes that human invasion of floodplains will continue at a rapid rate and that it must maintain a construction program to protect the new damage centers, there is a strong probability that the program will indeed be needed. Only as the full range of possible choice is considered can there be clear recognition of the effects of assuming certain adjustments and uses.

In another paper, I have tried to point out the importance to public policy in resource management of widening the range of choice among programs submitted for public action.[29] This requires the administrative machinery to make choices, and it can proceed effectively only insofar as there are reasonably handy tools for assessing the different choices in terms of their probable effects. One of the necessary tools is the method of analysis by which the nature of decision making can be recognized. The method described here is crude, with a good many weaknesses, some of which have been noted. But it offers a way of examining the choice of resource use in a framework that takes account of the physical range of choice, the practical limits to choice, and the economic, technological, and spatial considerations entering into actual decisions. As the pressure for multiple-use becomes more intense and as the assessment of economic efficiency is refined, the need increases to understand management decisions and their consequences in such a framework.

History of Fire in North America

Among the shamans or mystical men around the world and in some North American tribes, including the Zunis and Ojibways, the "mastery of fire" is one of their wonderful powers. Their capacity to play tricks with fire, to manipulate blazing coals and walk across the embers, is a symbol in Eliade's view of the attainment of a condition of ecstasy, of "access to a nonconditioned state of perfect spiritual freedom."[30] There is no early promise that contemporary researchers and practical rangers who work with fire in the forest will attain such mastery or spiritual freedom. However, the sensitive, wise management of fire does call for a state of deepened knowledge and of genuine freedom from conventional modes of thought. It will only be reached, as in human coping with most other extremes in nature, by candid, innovative assessment of the whole range of possible adjustments, now, or as a fruit of research, open to man.

(1972j), 9

Notes

1. For discussions of these developments, see below, vol. II, chap. 6 (Kunreuther and Slovic) and chap. 9 (Whyte).
2. Multiple Purpose Forest Act, 74 Stat. 215, 16 U.S.C.A. §§ 528–31 (Supp. 1960).
3. Hays, Conservation and the Gospel of Efficiency (1959).
4. McConnell, *The Multiple-Use Concept in Forest Service Policy,* Sierra Club Bulletin 14 (195—).
5. McKinley, Uncle Sam in the Pacific Northwest 317 (1952).
6. Conservation Comm. Report, vols. 1 and 2, S. Doc. no. 676, 60th Cong., 2d sess. (1908–1909).
7. Address by Floyd E. Dominy, World Power Conference, in Madrid, June 5–9, 1960.
8. Eckstein, Water Resource Development (1958).
9. Hirshleifer, DeHaven and Milliman, Water Supply (1960).
10. Krutilla and Eckstein, Multiple Purpose River Development (1958).
11. McKean, Efficiency in Government Through Systems Analysis (1958).
12. Wollman, *The Value of Water in Alternative Uses* (1961). Unpublished in hands of author at the University of New Mexico.
13. U.S. Inter-Agency Comm. on Water Resources, Report on Proposed Practices for Economic Analysis of River Basin Projects (Rev. 1958).
14. White, *Alternative Uses of Limited Water Supplies,* 10 Impact of Science on Society 243 (1960).

15. Ackerman, *Population and Natural Resources,* in THE STUDY OF POPULATION 621 (Hauser and Dudley, ed. 1959).

16. ZIMMERMAN, WORLD RESOURCES AND INDUSTRIES (1933).

17. Banfield, *The Decision-Making Schema,* 17 PUB. ADMIN. REV. 278 (1957).

18. Firey, *Patterns of Choice and the Conservation of Resources,* 22 RURAL SOCIOLOGY 113 (1957); Firey, *Coalition and Schism in a Regional Conservation Program,* 15 HUMAN ORGANIZATION 17 (1957).

19. Lee, *Optimum Water Resource Development* (Calif. Agricultural Experiment Station Mimeo. Rep. no. 206, 1958).

20. MARCH AND SIMON, ORGANIZATIONS (1958).

21. SPROUT, HAROLD AND MARGARET, MAN-MILIEU RELATIONSHIP HYPOTHESES IN THE CONTEXT OF INTERNATIONAL POLITICS (Center of International Studies, Princeton 1956).

22. CLAWSON, HELD, STODDARD, LAND FOR THE FUTURE 448 (1960).

23. ACKERMAN AND LÖF, TECHNOLOGY IN AMERICAN WATER DEVELOPMENT, RESOURCES FOR THE FUTURE 89 (1959).

24. MAASS AND HUFSCHMIDT, TOWARD BETTER RIVER SYSTEM PLANNING, RESOURCES DEVELOPMENT: FRONTIERS IN RESEARCH 1133 (1960).

25. For reference to the spread of geographic work see HARTSHORNE, PERSPECTIVE ON THE NATURE OF GEOGRAPHY (1959), and Ginsburg, *Natural Resources and Economic Development,* 47 ANNALS ASS'N AMER. GEOG. 197 (1957).

26. White, Calef, Hudson, Mayer, Sheaffer, Volk, *Changes in Urban Occupance of Flood Plains in the United States,* University of Chicago Dept. of Geography Research Paper no. 57, 1958.

27. Much of this is based upon papers to be published in 1961 by Berry, Burton, Kates, Roder, and Sheaffer in a research paper of the University of Chicago Dept. of Geography.

28. Dunham, *Flood Control Via the Police Power,* 107 U. PA. L. REV. 1098 (1959).

29. PERSPECTIVES ON CONSERVATION 205 (Jarrett ed. 1958).

30. ELIADE, MYTHS, DREAMS AND MYSTERIES 93–95 (1960).

10 Critical Issues Concerning Geography in the Public Service—Introduction

Identifying a small set of goals and setting out to achieve them is characteristic not only of White's writing but also of his administrative style. Always included in such goals is at least one that he almost surely can achieve.

Taking office as President of the Association of American Geographers in 1961, White chose three such goals and by the end of his term had made progress in two—the development of the high school geography program and the college curriculum and teaching materials that have now evolved into the college geography resource papers. The third goal, encouraging the role of geographers in relating to issues of public policy, then and now provides a paradox— significant contributions are made by individual geographers, but the potential for contributions by the field as a whole goes unrecognized.

This brief paper is both interesting and painful. It predates the call for academic relevancy by seven to ten years,[1] but as a diagnosis of geographical woe it is still current. Geography today wanes in the high schools, is tenuous in the university, and rarely influences public policy.

The contributions which geographic thought can make to the advancement of society are relatively few, simple, and powerful. They are so few and simple that a significant proportion of them can be taught to high school and beginning undergraduate students. They are so powerful that failure to recognize them jeopardizes the ability of citizens to deal intelligently with a rapidly changing and increasingly complex world.

In the United States in the early 1960s it appears that geographers are falling short of applying their diverse thought effectively to at

Reprinted from *Annals of the Association of American Geographers* 52, no. 3 (1962): 279–80.

least three major sectors of public action. Teaching of geography in the high school is near a low at a time when the nation's responsibilities on a world scale are at a high. Instruction in geography in the undergraduate curriculum is not keeping pace in either quality or numbers with liberal arts efforts as a whole. And geography's voice in policy making is modest beyond the experience and refinement of analysis which it has to offer. The issues involved in stimulating more enlightened service in these three sectors are the subject of the papers by Pattison, Aschmann, and Ackerman which are printed on the following pages and which were presented at the President's plenary session of the Association at its 58th annual meeting. Without attempting to define geography as a discipline, they pose some of the questions which arise when assessment is made of the more immediate opportunities for relating the widely ranging body of geographic study to social needs.

In the field of education it is important that geographers take a fresh and more discriminating view of the content and skills growing out of geographic research that have practical meaning at the elementary, secondary, and liberal arts levels. Possibly elementary education deals most satisfactorily with geography at the present time in that it tends to emphasize a few basic ideas, but it also is troubled by the obsession that it must describe in some fashion all major surfaces of the world before the student leaves the elementary grades, and there are stirrings of a new attack upon both materials and curriculum. High school geography is in a miserable state: materials on the whole are inadequate, and the supply of competent teachers is quite insufficient to meet current needs, let alone those that emerge when school systems attempt to improve the intellectual quality of the rather pedestrian work that goes under the name of geography or of geographic aspects of social studies. At the liberal arts level it is likely that much of the introductory work in colleges and universities would be eliminated if only adequate instruction in geography were provided in the secondary schools.

Even though geographers have difficulty in describing limits for their field, some of their unique contributions to thought are well recognized and are sufficiently developed to be widely useful. These include an understanding of the extraordinary diversity of combinations of surface features around the world, and an understanding of patterns of the major distributions such as those of population, livelihood, climate, and land forms. Closely related is an understanding of the processes which give unity to the explanation and regularities of those peculiar patterns. Among these processes are energy and water balance, the diffusion of culture, and the location of economic activity. While properly mindful of great deficiencies in knowledge of both distributions and processes, geographers have an obligation to teach and interpret the tools they have forged.

There clearly are several urgent social problems on which geographic discipline has an important bearing. One of these is the tremendous wave of urbanization which is moving swiftly in both industrialized and nonindustrialized countries. Another is the pervasive question of the capacity of constantly changing natural resources to meet future population needs at rising levels of living and technology. Obviously, geographers do not have the full answers to either these or other problems which might be stated, but they do contribute important insights.

The possible applications of geographic research to public policy and action are growing in number and significance. In transportation and city growth it is becoming evident that systematic knowledge of the way in which land use changes and in which alterations in traffic facilities affect the distribution of land use and population are vital to planning decisions. In international policy and action looking to peaceful organization of the world, geographers are illuminating those conditions in which economic growth takes place and the resource-use situations that result. Their findings and their methods of areal analysis have meaning not only to programs for particular areas but for broad strategy of bilateral and multilateral aid. In advancing the wise husbandry of resources in the United States and overseas areas, there is need for more precise recognition of the ways in which patterns of resource use are affected by technology, public regulation, and more subtle social constraints. In such directions as water resource development, the occupance of floodplains, the management of western grazing lands, and the design of area development measures, geographic knowledge is playing a critical role.

The profession now is moving into a period in which it is organizing for more responsible service. The immediate tasks are to improve the quality of teaching materials and teaching methods in the high schools and colleges and to step up the application of research to public policy formation. The basic and continuing task is less susceptible to administration and less dramatic. It is to raise the level of competency of fundamental research. Research is the life blood of the developing body of scholarly thought, and on its persistent cultivation the vigor of the profession and the nature of its service ultimately will depend.

These three papers which explore specific challenges to geography in the public service are by authors having much in common. Each practices what he preaches. Each has made independent contributions to geographic research. Each writes with felicity and clarity not only from broad experience but from current involvement in organized efforts of the Association: Pattison as director of the High School Geography Project of the Joint Committee on Education; Aschmann as a member of the new Committee on Liberal Education; and Ackerman as a member of the new Committee on Research on Foreign Aid and Policy.

Report of the Past President

The most nearly momentous event of the past seven months' history of the Association has been the decision to appoint a full-time officer. Finding the man, prospecting for funds to support him, and setting the task for him has claimed an energy and a degree of cooperative and argumentative action unprecedented in our corporate group. The stage now is set, and there is earnest activity in the wings to find the leading character in a new act in which the drama of creative work might unfold. It will not necessarily be a highly productive time. This coming period in the life of our Association could be a gentle-mannered comedy of committee-room high jinks. Or it could be a tragedy of dull-eyed searchers lost or abandoned in a blizzard of paper. We must not permit it to be either.

We all have a part in the performance. Our flights of scientific fancy, our hammering on the solid facts of the earth's skin, our negotiations as teaching impresarios, our faces to the dimly seen public audience, will strike the tone for this next act. But it will be a rather drab and sodden affair unless we can perceive our roles as advancing a public cause rather than as bolstering the creaturely comforts of a professional fraternity.

We are taking a considerable risk in this decision. We are making a financial commitment which in Philadelphia eleemosynary circles ranks only one step above the deadly sins—we are dipping into capital. If it is assumed that approximately $45,000 will be drawn from capital funds over the next three years, and that the remaining expense will come from special contributions of which

almost one-half now is pledged and all is in sight, the financing of
the venture is assured until the end of 1964–65.

(1962a), 13

Notes

1. For a detailed explanation of the dilemma between scholarly
objectivity and social relevance, see below, vol. II, chap. 7 (Feldman).

II Review of *Man, Mind, and Land: A Theory of Resource Use*

It is fitting that White devotes one of only three published book reviews to this major work in natural resources theory. There are two major theories of natural resource use. In both, resources are a nonunique case governed either by ecological theories of nature or economic theories of resources. Those who think that the combination of nature and resources requires a distinctive theory remain bereft. In *Man, Mind, and Land*,[1] sociologist Walter Firey makes a major beginning toward a unique theory of natural resources, but as White notes, it is a beginning still awaiting completion.[2]

Believing that "To give consistency to his decisions the planner needs a theory which will yield precepts telling him under what conditions one or another resource use is good or bad" (p. 165), Firey sets out a general theory to describe the ways in which man makes use of resources. He does not attempt to give details of such a theory, to test it by reference to widespread examples, or to apply it in quantified terms. Writing with clarity and with careful attention to the logical structure of his argument, he develops a view of resource use that rests upon the interrelation of three sets of resource processes in a lattice model.

A resource process is defined as a space-time coincidence of happenings in resource use which recurs in time with somewhat the same combination of human and biophysical factors, for example plowing with oxen. A natural resource is the biological or physical component of the process. A resource system is a generic grouping of any set of processes, for example the cropping system, the soils that are cultivated, or a system of using animals. Each of these is a "man-mind-land structure which imposes a special kind of constraint

Reprinted from *Economic Geography* 39, no. 4 (1963): 373–75.

or necessity upon its human agents such that there is a sufficient reason for them willingly to conform their behavior to the practices which comprise that resource system." From an ecological approach, processes may be arranged in sets of those that are possible and not possible in a given physical environment. From an ethnological approach, they may be divided into those which are adaptable and those which are not adaptable in terms of a people's system of activities. From an economic approach, all processes may be divided into those which are gainful and those which are not gainful according to a formally stated degree of productive efficiency. The Zimmerman functional view of resources is accepted. Within a set of physically possible resource processes, the relation of the culturally-likely processes to those that are economically gainful is examined; and, by relating economic and social optima, different types of resource systems are explained. The three optima are shown not necessarily to coincide. Here, the theory provides a unified ordering of social returns which may work against optimum economic returns, and vice versa. Every process can be graded in the lattice according to its physical possibility, its social adaptability, and its economic efficiency.

A basic distinction is made between a resource complex and a resource congeries. A resource complex is a resource system which shows some stability in the face of external changes and has invariance properties as a structural whole, for example a peasant farming system. A resource congeries has no such stability in response to external changes, and is an indeterminate entity, for example, transitory soil destruction.

A distinction also is drawn between resource development and conservation: development is concerned with "conversion of inert natural processes into potential capital," whereas conservation is concerned with maintaining social capital and involves the constraint of a moral obligation. This concept of conservation is then related to broader thinking as to obligatory social behavior. Firey's interest in scientific grounds for planning emerges strongly in the final chapter in which he suggests what he regards as an intellectually respectable framework based on the general theory and using plannee's consent as the criterion.

Four examples are used in developing the theory: shifting agriculture among the plateau Bemba in Northern Rhodesia, the subsistence farming of the Tiv in Nigeria, open field farming in the Medieval English Midlands, and groundwater development in the South Plains of Texas and New Mexico. These are illustrative rather than essential to the argument, and with the exception of the South Plains case which seems unduly complicated by data on verbalized attitudes toward conservation, serve their purpose neatly. The inclusion of both Western and non-Western cultures is welcome, although discussion of an agricultural society subject to rapid technological advance would have been illuminating.

The whole work commends itself as a consistent, systematic statement of a theory within which any resource use might be appraised. The test of its value for that purpose lies in its adequacy in providing for accurate and sufficient description of the conditions in which resource use takes place. Since Firey is not attempting to apply the theory in a rigorous way to either planning or the studies necessarily preceding planning his statement should provoke careful testing in depth in diverse situations. Geographers should undertake this promptly. One can ask immediately whether or not the theory seems to encompass factors that are known to be significant in specific situations, and whether or not they are combined in a fashion that lends itself to understanding resource use and to normative judgments about use.

Firey's scheme does not directly deal with the process of decision making by which uses are selected by private or public managers. By not confronting the theories of decision, it omits a crucial aspect of the intricate relationships surrounding resource use. One would like to see this scheme reconciled with the major formulations which Dyckman reviews from the standpoint of planning (John W. Dyckman, "Planning and Decision Theory," *Journal of the American Institute of Planners*, vol. 27, 335–45), and which Kates reviews in connection with resource management (Robert W. Kates, *Hazard and Choice Perception in Flood Plain Management*, University of Chicago, Geography Research Paper no. 78, 1962, 25–28). It also touches only lightly upon basic problems of risk and uncertainty as they affect choice.

There is little attention to the role which resource estimation plays in choice, and particularly to the differences in resource perception within and between social groups. Technology tends to be given a subservient position in the model although we know this can be a major factor in selecting a particular use. Generalizations as to limits of the natural environment, equilibrium, and the efficacy of primitive habitat use are somewhat loose. The reviewer is inclined to think that this is the best formulation of resource use thus far produced, but that it fails to give sufficient weight to factors such as resource estimation and technology. To add to the three-part model is to increase the complexity of relationships tremendously. However, resource management is in fact exceedingly complex and there are numerous feedback situations, such as the effect of the level of applied technology upon the perception of environment. Planners or scholars cannot be satisfied with analysis that does not help explain a major part of present resource use patterns or of the full consequences of a change in use. Explanation seems likely to come only through systems analysis taking some of these other factors into account.

While the internal construction of the theory is tightly woven, and good use is made of mathematical symbols, certain of the distinctions seem strained. The concept of a resource system is rather amorphous and could not be mapped readily on the ground. The criteria of stability for a resource complex, the contrast between development and conservation, and the characterization of resource planning as being either revolutionary or reactionary are not easily preserved in practice. Indeed, there are few of Firey's points which fail to provoke theoretical argument or an interest in checking them in the field.

Firey's analysis is a landmark in the expanding terrain of thought about natural resources. Whether its synthesis of physical, cultural, and economic approaches is supported by later work or is modified into a more complex system, it will serve as a baseline for efforts at refined description. It is required reading for any student of natural resources.

Science and the Future of Arid Lands

The role played by basic research in setting the limits for social development in arid lands is nowhere more apparent than in the struggle to narrow the gap between scientific knowledge and practical action in resource use. If there were miraculously to be an abundance of teachers to work with the Somali grazers and if there were adequate funds to pay for whatever pamphlets, radio sets, slide projectors or posters they might wish, there still would remain the question of what should be taught. For the Somalis, in order to make better use of their camels and grasslands, must develop a new way of life that is consistent with their traditions and religious aims as well as with their climate and soils. The task is far more than adjustment of routine teaching methods to a new language or community; it is a searching appraisal of the culture's capacity to change, and this, in turn, assumes that knowledge of social processes is sufficiently precise to permit accurate appraisal. What is required is appraisal that goes so deep into the value systems and economic and political processes of each tribal grouping that it reveals the conditions in which man now uses the meager resources and the circumstances in which improvements can be made. Fundamental investigation by the anthropologist and geographer thus must precede sound decisions as to effective ways of using the rich store of developed technology to advance the basic cultural aims of these people. No simple transfer of "know how," such as in controlling livestock diseases, will do; no massive allocation of capital investment for new water or housing projects can substitute for ingenuity in reorienting the society with full awareness of the pride and aspirations of their family and tribal groups.

(1960e), 82

Notes

1. Walter Firey, *Man, Mind, and Land: A Theory of Resource Use* (Glencoe: The Free Press, 1960).
2. For a review of major efforts toward integrative and comprehensive concepts of natural resource, see below, vol. II, chap. 3 (Clawson).

12 Vietnam: The Fourth Course

In 1961–62, White chaired a consulting group for the Ford Foundation on the social and economic aspects of developing the Lower Mekong River in Southeast Asia. As the war in South Vietnam and Laos escalated, he attempted to interest the world, but particularly U.S. policy makers, in an imaginative effort to replace hostility within the region and achieve peaceful change by a cooperative harnessing of nature's riches. Thus he published several articles, including a paper in *Scientific American* (1963c) and this selection from the *Bulletin of the Atomic Scientists* on the development of the Lower Mekong. He also made private overtures to government officials that culminated in President Johnson's speech suggesting the possibility of common development of that region.

In the end, the attempt to beat swords into the great plowshares of river development failed in the face of the escalating, "morally reprehensible and desperately dangerous" war that White foresaw. In retrospect, perhaps, the whole idea appears naive, as was Walter Lowdermilk's ambitious effort to ease Arab-Israeli tensions by joint development of the Jordan River. Yet for people of good will, it surely must have been the road not taken that might have averted that terrible war. And, somewhat ironically, the hope nourished in the Lower Mekong Coordinating Committee still persists, as joint consultation on Mekong development continues on a technical level to this day despite the continuing hostility of the nations of the region.[1]

A peaceful and honorable resolution of the conflict in South Vietnam and Laos may be found in a bold plan for land and water development which already unites factions in four nations of Southeast Asia. For seven years, Cambodia, Laos, Thailand, and South Vietnam have been working with little publicity and without disagreement on a huge development program. These four countries, which do not

Reprinted by permission of *The Bulletin of the Atomic Scientists*, a magazine of science and public affairs, 20, no. 10 (December 1964): 6–10. © 1964, Educational Foundation for Nuclear Science, Chicago, IL, 60637.

cooperate in anything else, have reached accord on development of the Lower Mekong Basin.

Work already is under way in drawing engineering designs, moving earth for dams, building power plants, cultivating pilot farms, and training village technicians. Even guerrilla troops have not halted field work.

If the United Nations were to designate this area for international development according to the plan already drawn by the four nations, there is a strong possibility that peace could be achieved in a common pursuit of agricultural and industrial growth. This is a solution to Southeast Asian violence which would make sense to peasants in the rice fields and to American taxpayers.

There are three well-publicized choices for the United States:

The current combination of military, political, and economic assistance to local governments opposed to the Vietcong in Vietnam and to the Pathet Lao in Laos could be continued or stepped up. Many observers argue that the United States cannot hope to settle the conflict or stabilize the two governments if it relies on this course of containment.

The United States could carry the war to North Vietnam in an effort to cut off support of guerrilla activity. This would be morally reprehensible and desperately dangerous. At best it would create another Korea; a new border guarded by armies would contribute new rigidities to the search for world peace. At worst a big war might unfold.

The area could be neutralized under an international agreement that would establish a peacekeeping force. As a political measure, this would oblige the United States to step down from its unhappy military position without improving the dim prospect for continuing political stability in Vietnam, Laos, or neighboring lands.

The Lower Mekong River may be the key to a fourth course of action, a more constructive and humane one than any of the others. For the imaginative scheme to manage the winding streams and alluvial soils of that great basin now provides a framework within which all nations could join their technical, financial, and police assistance under the United Nations flag in working toward a concrete goal. The attractions of taking positive international action based on indigenous plans for the Lower Mekong are obvious. The difficulties of getting agreement are great and have not been fully assessed. But they should be explored with all of the energy and skill now going into war plans. Even if the specific framework suggested by Mekong experience were to prove impracticable, other viable and constructive solutions may lie in this direction.

Planning for the Lower Mekong was begun by a committee of the four countries under auspices of the United Nations Economic Commission for Asia and the Far East in 1957. The basin below the Burmese and Chinese borders includes virtually all the area of Cambodia and Laos, the Korat Plateaus in northeast Thailand, and the Delta and southwest interior of Vietnam. It embraces half the population of the four countries. It is a huge area drained by an untouched river of the dimensions of the Columbia.

The committee set out to gauge streams, explore damsites, chart soils, study farming, and do all the other things needed to design a program of managing water and land for the welfare of its twenty million people, mostly rice farmers. The survey effort was unique in dealing with the entire lower basin as a unit before construction began, in doing this under United Nations guidance, and in calling upon technical help from other nations. Always before, nations have joined in river development, as in the Indus or Rio Grande, under pressure of contending claims. In the Mekong, they started with a clean slate and a generous supply of both water and land. As yet there are no serious conflicts. They lacked essential data and decided to defer construction until more facts were in hand.

More than $14 million has been spent to date on basic studies. These now involve twenty other nations and eleven international agencies. The U.S. Bureau of Reclamation helped with a reconnaissance study. Lieutenant General Raymond Wheeler, former chief of the Corps of Engineers, headed a United Nations mission which recommended a series of detailed investigations. France was the first to offer direct assistance; Canada flew the aerial photographs; the United States studied the hydrology; Australia investigated damsite geology; Japan sent reconnaissance engineering teams; the Philippines did detailed mapping; France surveyed soils and fish life. The studies and organization became more complex. India sent a team to design a large dam; Israel contributed agricultural engineers. The names of the Netherlands, Pakistan, the Scandinavian countries, New Zealand, the United Kingdom, the International Labor Organization, the World Health Organization, and the Ford Foundation began to appear in progress reports.

The Food and Agriculture Organization designed pilot farms, and the World Meteorological Organization advised on flood forecasts and hydrology. These and many other basic surveys are being put together by an international staff and Advisory Board under the Committee's guidance.

With all of these nations in the picture it might seem impossible to get anything done. There was no model to follow. Skeptics in Bangkok and Saigon and Washington called the idea visionary. In fact, there have been delays and administrative snarls, but the work has moved surprisingly well. An Indian survey team uses American hydrology, Philippine maps, and Australian geology to design a dam in Cambodia which also would benefit Vietnamese farmers downstream.

The total U.S. contribution to Lower Mekong planning has cost less than four days of military aid in South Vietnam, now reported to exceed $1.5 million per day.

Burma has such a small and inaccessible part of the upper basin that it has not taken part. The People's Republic of China has at least 74,000 square miles in the upper basin. This lies in a long narrow segment, most of which is steep gorge cutting through high dissected plateaus. While China has not participated in the planning for the lower basin, integrated development of the entire basin ultimately should include the upper reaches. Hydroelectric storage there probably would benefit the regimen of the lower stream, but even if China were to divert its entire share into other basins—a most unlikely prospect—there still would be enough water in the Lower Mekong basin to care for major works now planned. North Vietnam does not have a part in the basin. However, if large-scale hydroelectric installations were to be made along the mainstream above tidewater, its cities would be a possible market for the power.

Much of the planning centers upon improvement of agriculture by supplying water for irrigation of a second crop in the dry season and by preventing drought losses. Stream regulation would cut down the heavy dislocations caused by annual floods. There are massive opportunities to increase output of rice and to diversify crops in the Delta, around the Grand Lake in Cambodia, and in the winding alluvial valleys upstream. Channel, terminal, and shipping improvements would be essential to rebuilding the commercial life of the Delta. Hydroelectric power plants on the tributaries would provide cheap power for urban and commercial growth, but the marketing of large blocks of power would depend upon development of industrial complexes. These have been actively discussed by Japanese interests.

In no case would the simple construction of a dam, irrigation canal, or power plant alone guarantee economic growth. Roads, credit, seeds, and other village improvements are essential if a Thai farmer is to adopt effective irrigation. Astute economic promotion must precede the sale of power. Engineering must march hand in hand with technical workers who will get their feet wet in the rice fields, and who will understand that the preoccupation of the villager is with a way of life adjusted to monsoon rains and the annual drought. The best in social study and judgment must be marshalled along with scientific and technical knowledge.

Funds have come from the four countries, other participating nations, and the United Nations Special Fund. All committee decisions have been unanimous. French and English are the official languages, and the common bond is concern with rice cultivation, water flow, and power output. Diplomatic relations among the countries may be strained or broken, but the Lower Mekong activities continue without interruption. When Cambodia's borders were closed to all Thai commerce and diplomats, the Mekong Committee met in Phnom Penh with its full complement of Thai delegates. Indeed, the Committee is the only agency in which the four countries regularly participate. Enthusiasm for a massive common effort thus far has overridden shorter-term hostilities on political issues.

With the major outlines of a program in hand, work already has begun on construction of several tributary projects. West Germany is loaning funds for a multiple-purpose power and irrigation project in the Nam Pong tributary basin in Thailand. France is helping with two small power and water supply projects in Laos. Part of the United States AID program supports water control and village development schemes that fit into the general plan in Laos and Thailand. In Cambodia, one multiple-purpose undertaking has begun under Australian engineering supervision and a second will be ready to start if funds can be found.

More extensive water and land improvement projects are in early prospect. High priority is given to measures to rehabilitate the deteriorated inland water transport system so essential to the economy of the Delta by rebuilding waterways, barges, terminals, and ship repair yards. A combination power and irrigation dam in the Nam Ngum basin near Vientiane in Laos would bring large-scale rural improvement and cheap electricity to the doors of the capital. To meet the needs of growing population and to raise the level of living above the present per capita income of $60–$100 annually, much heavier investments will be in order. These allocations should be related to national growth requirements in other sectors. The very large main-stem projects take time to design (projects the size of Hoover Dam are on the drafting boards) and no doubt will be de-

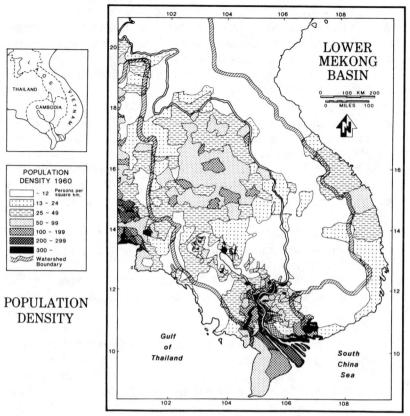

Map 12.1

ferred for some years, but expenditures of the scale of $200 million per year may not be unreasonable over the next two decades. Much of this can be on village improvement schemes that benefit the farmers directly and promptly.

To approach this level of investment, innovations in financing and security are needed. Rather than relying upon separate loans or grants from contributing nations, the Mekong countries should have a central financial agency to receive and supervise expenditure of funds. All four countries probably would gain from unified administration of help from the outside. It is possible that the International Bank for Reconstruction and Development might be enlisted to play this role. Commitment to finance a long-term program would in itself be an earnest of intention by donor nations to support sound social growth along the river.

But heavy financial support will not come unless the security of the area can be assured more confidently. The United Nations might be expected to provide a blue-helmeted watch and ward service for those sectors of the project area where security is threatened. It could do this on the invitation of the country concerned. Cambodia and Thailand would have no immediate need beyond protection of their borders. Laos and South Vietnam would find it essential in the areas where civil unrest has been intense. United Nations measures would have to be launched with the authority of the General Assembly and with guarantee of costs and personnel contingents from interested member nations. In a more basic sense the United Nations presence would be effective only if it were grounded on agreement among the nations concerned that present national military operations should be suppressed. Reduction of violence will take time in the most favorable circumstances, for the guerrillas and army groups will be slow to disband and will have trouble finding and settling into work in their home villages.

This type of agreement would be fundamentally different from the cease-fire that is envisioned under a neutralization treaty. The four countries, the United States, North Vietnam, and other interested nations would commit themselves primarily to advance a great development program for the welfare of the people of the lower basin on which the four governments have agreed. This would mean withdrawal of national military units. It would require support for a United Nations force to maintain security of the program—from Delta canal to Laotian rice paddy. It would substitute a development goal for an indistinct battle line, and it would permit the United States to withdraw gracefully in favor of an international force committed to that goal. Financial obligations of the United States would in the future be linked with contributions of money and people from other nations.

Would this be acceptable to North Vietnam and the Pathet Lao? No one can say for certain what ultimate position they would take. Developments to date are encouraging. None of the important field work has yet been halted by guerrilla activity. It is known that the head of the Pathet Lao is highly sympathetic to multi-purpose water programs in his country. The goal of harnessing a river's resources to serve the common man is widely hailed in both Communist and non-Communist countries, and the People's Republic of China and the USSR have been distinguished practitioners of the idea. Neither has raised formal obligations to its embodiment in the Lower Mekong. And while Cambodia for political reasons has cut off United

States bilateral assistance, it is willing to accept help through the Lower Mekong channel.

A proposal for expanded international cooperation in the Lower Mekong would give a new choice to elements which have pushed guerrilla action in South Vietnam and Laos. By outright opposition they probably could block an enlarged venture toward goals they respect. On the other hand, by withholding opposition they could get the integrity of national governments in the two countries guaranteed by a United Nations presence devoted to advancing those aims. Thereafter, a weak United Nations administration with flabby assistance from the outside no doubt would favor growth of dissident groups, as in Laos in recent years. But a vigorous international effort aimed at bettering the lot of villagers and city workers might stabilize the governments and in time enable the people of Laos and South Vietnam to choose by nonviolent means the government they want.

Is it possible that the vision of a majestic river harnessed for the advance of twenty million people by an unprecedented piece of international cooperation would so command the imagination of the nations that the present grueling conflict could give way to a struggle for more abundant life? Could this mean to a world increasingly aware of its network of mutual responsibilities what the Tennessee Valley Authority meant to proponents of national development thirty years ago? The Mekong Committee's files are beginning to bulge with the necessary technical plans and surveys. Such investigations must continue. What is needed immediately, however, is incisive analysis by the United Nations as well as by the interested nations of practical steps to negotiate on a new basis of a common human cause. Just as the river's planning has called forth a unique degree of collaboration among scientists and nations, new forms of political and economic organization will be required to translate plans into village and city action.

If the Lower Mekong program in its present form does not prove a complete way out of the Southeast Asian dilemma, it at least suggests a route that is challenging. This is a way calling for international collaboration in advancing the economic welfare of peasants who long months ago had enough of terror in the night. It builds on plans shaped by the four nations and commands their support. It draws technical assistance, financial help, and necessary police protection from a score of other nations under the United Nations flag. In such a venture the United States could with honor and deep conviction invest its men, experience, and capital as a member of an international team. The route is broadly plotted, its technical foundations are laid, but its political surface is yet untried. That testing warrants all of the ingenuity, careful logistic planning, and determination on both sides that marks the current undeclared war of attrition in the Mekong swamps and hills. Nothing less than the same searching appraisal now should be made of a constructive solution.

It is just barely possible that out of this most incredible of places— the swirling political caldron of Southeast Asia—may come a new pattern for international action in harnessing nature's riches to achieve peaceful change.

Comparative Analysis of Complex River Development

When international experience with complex river development in sectors of Africa, Eurasia, North America, and South America is examined in the light of historical trends, the mix of human adjustments practiced, and the probable environmental impacts, a few needs stand out. Long-term climatic and landscape changes need to be assessed more carefully. The tactics of manipulating water quality and quantity need to be appraised more fully with a view to employing a broader range, including social measures, of alternatives. Critical post-audits of what in fact has happened in the trail of large-scale development should be expanded. The great rivers of the world provide striking, earthy lessons that can be ignored only at the cost of environmental degradation.

(1977c), 18

Notes

1. For a description of the evolving situation, see below, vol. II, chap. 5 (Day et al.).

13 Rediscovering the Earth

If a field as varied as geography (earth, space, region, human environment) speaks to young people, what is worth telling? Following his term as President of the Association of American Geographers and the agenda displayed in selection 10, White reached out to colleagues, foundations, and school administrators in order to organize the High School Geography Project and the production of new, fresh curriculum materials.[1] His role was one of leadership—bringing together participants from the various branches of the discipline, raising money, and interpreting project results to education officials.

This paper was a latter effort addressed to educators. It answered the question of how to give young people ways of ordering the world around them. The project fulfilled its immediate promise; it produced an interesting, indeed exciting, set of teaching materials used today in perhaps five percent of high school systems that teach geography. It sensitized a generation of geographers to the potential for having something lively to say to young people. But it did not revolutionize earth knowledge or lead to a significant increase in the teaching of geography within American high schools. Even where geography is taught, caricatured versions of it still persist in many schools.

In 1719 the French cartographer Sanson drew a map of the western hemisphere. Using the best reports available from explorers and traders, he showed with considerable accuracy the east coast of North America and of Central America but left blank the space which we now know as the Pacific coast of the United States. Not being certain what was there, he bluntly admitted his ignorance.

Sanson's map was copied by many lesser cartographers in later years, but few had the courage to admit ignorance as he had. Most

Reprinted from *American Education* 1, no. 2 (February 1965): 8–11.

filled in the blank space by drawing imaginary coast lines, straits, islands, and peninsulas.

In a sense, each of us does much the same thing every day as these later cartographers did. Confronted with a reference to a snowy steppe in a Pasternak novel or to a Vietnamese village in a State Department news release, we improvise a kind of image which often fades into a blurred gray view of snow or jungle. We may visualize an Africa that is all tropical rain forest or an Iran that is all rocky desert or an Iraq that stretches out in waves of Hollywood sand dunes.

Much of our view of the world is based on shaky and shabby kinds of improvisation, drawn from the movies and travel booklets. This improvisation is quite unnecessary. Just as Sanson's map told his time what he knew and did not know about the then partly explored hemisphere, our present-day geography from its new discoveries can help people understand what they know and do not know about the character of their changing world.

While the classical period of exploring and mapping the earth's surface is nearing its end with the ascents of remote Himalayan peaks and the establishment of base camps at the South Pole, a different kind of discovery is unfolding. The new discovery comes neither from the probing of outer space with all of its drama of interplanetary travel, nor from the sounding of ocean depths and sediments. It comes from a study of the earth's thin skin with a view to accounting for the amazing diversity and order of natural and man-made features.

The new exploration of the earth centers on understanding how and why the main elements in the landscape—the land forms, water, climate, vegetation, and soils, and the towns, crops, roads, and other works of man—differ in distribution from place to place. It asks what regularities there are in these distributions. And it looks into what kinds of changes may be expected as man expands his numbers and his technological prowess and destructiveness in managing the earth's resources.

The findings of this new exploration have profound implications for education. Fortunately, not only professional geographers but foundations and the federal government are aware of these implications and are moving to improve the content of geography courses and the preparation of geography teachers.

Three examples of geographic study suggest the significance of the insight which modern geography is gaining.

The first centers on the location of buildings and people within a typical American or Western European city area. A very simple kind of relationship which most of us would recognize intuitively is that population density decreases with distance from the downtown center of the city. If we take a sample American city, such as Chicago, and chart the population density on traverses leading out from the city center, we arrive at a kind of relationship that may be stated in an elementary formula, that the density diminishes in a linear fashion with distance (fig. 13.1). If we apply this formula to other American cities, we find that it holds in most instances, but differs with size of city (fig. 13.2).

As long as the current form of urban growth prevails, the density-distance relationship may be expected to hold for United States

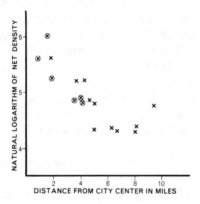

Figure 13.1 Density of population tends to diminish with distance from centers of American cities, as shown on this diagram of the population of Chicago. The circles indicate nonresidential areas.

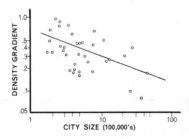

Figure 13.2 The density-distance relationship holds true for most American cities, as it does for Chicago, but it differs with the size of the city. This relationship is used in traffic planning.

cities. It is used in the massive and complicated studies which the Chicago, Pittsburgh, and Detroit metropolitan areas have made in recent years of their needs to provide traffic arteries.

In addition to its practical applications, the density-distance relationship shown in figure 13.1 is important to understanding the character of the city in which we live and leads us to ask how far it applies to other urban areas around the world. We find that for parts of Europe, North America, and South America this relationship holds, but that Asian cities are much more compact. Moreover, the change in density of an Asian central city area over time is radically different from that in the West (fig. 13.3).

Such findings raise questions about why there are these differences in development, differences for which we must seek explanation in the history of a culture and which may or may not prevail as the technology of the West makes its full impact elsewhere.

The second example shows how we can find order in the way a common physical element is distributed on the earth's surface. For example, if we examine the natural input of water through precipitation at any place, we can describe in a rough way the water balance for that place over a period of an average year (fig. 13.4).

Water balance is the difference between the amount of water which falls as snow and rain and the amount which is lost by evaporation from soil and plant surfaces and by transpiration from plants. By taking account of the amount which can be stored in the soil, we can roughly estimate surface runoff.

With this concept of water balance, we can study differences in the water regime from place to place. We find that these differences

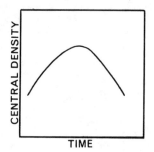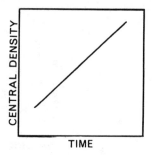

Figure 13.3 Asian cities are much more compact than most cities of Europe, North America, and South America. Moreover, over a period of time the changes in density in Asian cities are sharply different. Diagram on left suggests that in United States cities the central densities first increase and later decrease. Diagram on right shows that in certain non-Western cities they have thus far continued to increase.

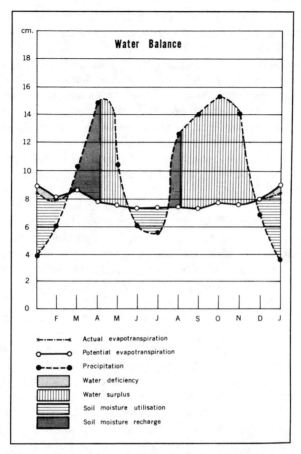

Figure 13.4 During an average year at Mubende, in central Uganda, water for plant growth is deficient only during December and January. Enough water usually is stored in the soil to carry through the dry period of midsummer. Even places with heavy annual rainfall may suffer periodic water deficiency.

have rather clear regularities in relation to the energy balance on the land. With knowledge of how energy input from the sun varies over the earth's surface with latitude and altitude, and knowledge of the general pattern of atmospheric circulation, it is possible to outline what on a theoretical basis would be the distribution of major types of water balance on a continental area if it were everywhere a low plain. Africa, for example, might be thought of as having six types of water balance, which could be plotted schematically.

This theoretical kind of distribution can be adjusted to the real altitudes and, with important exceptions, will be found to conform to the actual distribution. Without memorizing African climatic data but having an understanding of water balance, of major African land elevations, and of atmospheric circulation patterns, a high school student can estimate how the chief types of water balance are distributed in Africa. Such deduction is no mere academic exercise. It applies an insight into basic regularities and their relation to human management of the earth. Understanding of the water balance is essential to wise planning of projects for irrigation and for introduction of new crops, and essential to management of water generally.

My third example of geographic study bears on the human practice of shifting cultivation, a common form of land use in large sectors of the world. It consists of the burning and girdling of trees and the burning of grass, with a view to clearing crudely a place where crops may be planted. After a period of years the extraction of nutrients, the growth of weeds and pests, and other factors lead to abandonment of the field in favor of newly cleared land elsewhere. The period of cultivation ranges from one to ten years; there usually is no livestock.

Although sometimes condemned as a highly destructive use of soil and vegetation, shifting agriculture is, in fact, a delicately adjusted scheme of land use in areas where combinations of climate and vegetation and problems of soil fertility and leaching make a stable sustained agriculture extremely difficult.

Density-distance relationships, water balance, and shifting cultivation are widely observed on the earth's surface, and we can offer generalized kinds of explanations for them. These enable us to see a little order in the tremendous complexity of the earth's surface and to anticipate certain consequences of changing elements in that great web of soil and water and life which almost everywhere shows the hand of man. Other types of relationships for which there are generalized explanations are the variations in climate, soil, and vegetation with differences in altitude, and the spacing of villages, market towns, and cities in a predominantly agricultural area.

It would be comforting if we knew enough about these relationships to confidently predict what we might encounter on any given part of the earth's surface. We do not. We know enough, however, and can pass on enough to enable a high school student of average ability to successfully pass what one geographer, Preston James of Syracuse University, likes to call the "thumb test."

Place yourself in your imagination in front of a globe or a large map of the world. Assume that your thumb finds its way at random to some point on a continent. What would you anticipate would be the character of the climate, land forms, soils, vegetation, major types of land use, and spacing and form of urban settlement?

If you were familiar with the basic notions and relationships which have grown out of modern-day discovery of the earth's surface, you could make a rough prediction of precisely these items. In doing so you would not be able to operate altogether from deduction. You could not, for example, apply the density-distance relationship to all cities of the world. You would have to know that there are situations where for various reasons this rule does not apply, and you would have to learn the exceptions. But with the kind of basic information I have described you could make a moderately effective prediction of what you would find at a selected point.

You then could go on and offer a partial answer to the question of the extent to which this complex of landscape features might be expected to change in time if one of the features were altered. For example, what might happen to the pattern of land-use density in a city with the construction of a new expressway? What might be the shift in the spacing of villages in an agricultural plain if a new, improved road is driven through the middle of the area? Or what might be the consequences for both man and soil of attempting to establish a permanent system of cropping in a region of annual water deficiency in soils which are now handled by shifting cultivation? We know very little about many of the changes which man instigates in soil and water and settlement by his use of fire and livestock and bulldozer, but we know enough to begin to identify them and estimate their magnitude.

A geographer sometimes is greeted with a condescending smile from a person who thinks the world's geography was disposed of neatly after he learned about Japan's tea gardens in grade seven and who is a bit amused at a grown man giving time to the subject. Nor is it still entirely surprising to come across a school child who speaks of geography as the most recent place names he has learned by rote.

What I have suggested certainly is not geography as we have known it in school or in the older and traditional textbooks. But it is very much in harmony with the major traditions in geographic studies. It draws upon a concern with description of spatial variation on the earth's surface which has been shared over the centuries by such scholars as Strabo, the Greek who died in A.D. 19 leaving us a 17-volume geography of Europe, Asia, Egypt, and Libya; Immanuel Kant, best known as a philosopher but also a prolific writer on politics, anthropology—and geography; and the adventuresome German geographer Alexander von Humboldt.

It deals with our persistent curiosity as to how man affects the face of the earth and, in turn, is limited in his activities by limitations in the resources of his particular corner of the earth; it asks what have been the unique historical circumstances shaping particular landscape features of hill or farm or town; and it inevitably is forced into regional generalizations about areas large or small.

Intellectual discipline is required to perceive relationships among several factors which vary in space. It is essential to recognition of associations of factors such as water balance, soil, and vegetation. A sense of scale is involved, and so, too, are skills in thinking spatially and in relating first-hand field observation to broader and distant areas. The best of thinking from geophysics, anthropology, economics, and history is used in explanation. All of this is part of geographic exploration in our day.

For emphasis, we may describe such analysis by what it is not. It is not the routine memorization of capes, bays, and mountain elevations. It does not try to compete with the friend who proudly recites in alphabetical order the area and capital city of each of the fifty states, congratulating himself upon having brought his early learning up to date. It clearly is not a dull review of a miscellaneous assortment of physical facts about particular segments of the earth's surface. Rather, it is an effort to understand the order and regularity of features of the earth. In that effort the student engages in a search which both enables him to solve problems about why things are where they are and leaves him standing in awe at problems which so far have defied human solution.

Attention to unfamiliar places need not inhibit discovery of the familiar. The corner store or local park or the journey to school may provide the base from which a new, disciplined way of looking at the world can grow. Spacing of city shopping centers or of rural stream courses can be seen as related to physical and social processes having wide application.

Better teaching of geography helps students build habits of direct, analytical observation. It uses knowledge of process and it demands precise information where appropriate.

Some of the understanding requires facts which do not yet fit into a refined intellectual framework. There is no simple way other than hard memory of learning where the major types of land forms are on most of the continents. No easy formula is at hand for recognizing the few areas rich in metallic ores or the sedimentary basins where petroleum might be found. Facts of that sort are most effectively learned in the course of explaining exceptions to the other regularities over which we have some command.

The lively, imaginative geography course leaves the student with a whole series of question marks and unsolved problems but enables him to cope with a few more elementary ones. Geographers currently address themselves to a fascinating series of questions. A recent report by a committee of the National Academy of Sciences-National Research Council calls attention to a few of them. The ways in which the processes of erosion, stream transport of silt, climate, vegetation, and human use of resources are related to each other are perceived only roughly in most drainage basins. As nations grow in number at an unprecedented rate, the conditions for effective international boundaries and for regional organization of resource management are matters of urgent inquiry. One of the most fruitful lines of study is in theories that bear on the location of land uses and on the functional relation of urban centers and transport networks.

Whatever its attraction as a research field and whatever its usefulness in practical management of urban growth and natural resources, geography is a major teaching challenge in American society today. From it come approaches to several persistent problems troubling man. Knowledge about the dimensions and character of the earth responds in a concrete way to the concept of one world which gradually has emerged with improvements in modern communications and with the elaboration of international responsibilities. The world of the United Nations is no political theory: it is a mosaic of real people on real land. The pictures which satellites now give us of an entire sector of a continent, with systems of clouds drifting across it and with the main groupings of land forms and land use clearly distinguished, literally confront the viewer with the need for reasonable explanation of the combined phenomena.

Man continues to be deeply disturbed by doubt about the capacity of the earth to support rapidly expanding population at reasonable living levels. He soberly asks how resources can be marshaled to meet new needs, and how he can deal helpfully with the tremendous tensions provoked by deep disparities from place to place in level of living and in resource endowment.

Perhaps most of all, men in the western world are troubled by the individual's search for identity, by the struggle young people are having to see themselves as constructive parts of an intricate and massive set of interlocking systems.

To these problems, exploration of the kind I have suggested brings a combination of understanding and mode of thought from both social sciences and the natural sciences and offers one of the bridges which joins them in the service of man. To supplement the historical approach and the analysis of social process there is much to be said for a mode of thought which views events in their spatial setting and probes their relations to each other in earthly places. However confusing its political and intellectual currents, the world nevertheless has sufficient unity and regularity for a young person to acquire a framework which explains some of the diversity and shows the points at which knowledge of process is wanting. This is a kind of discovery of the earth that can leave the student deeply excited by problems upon whose solution the outcome of the human adventure depends.

To cultivate this mode of thought is no easy task. It is handicapped by lack of basic research and by the slowness of university people to make their research results intelligible to teachers. It is retarded in part by the tradition of routine descriptive study. Two decades ago a college which did not consider such study warranting a place in its curriculum, or a high school which felt that a student had covered geography by the time he finished eighth grade, was theoretically weak but had practical reason for its position. Now only the stodgy or uninformed institution could take this position. We are gaining practicable means for teaching the understanding and skills that are needed.

Much can be done to improve the quality of the elementary school teaching, but probably the key situation is in high schools, where geography has been most neglected in recent years and where, when combined with other disciplines in social studies, there has been a tendency to water down or to handle it in an uninspired fashion. Clearly, every student in his liberal education should be exposed to the geographic mode of thought and have some capacity to exercise it. This does not necessarily imply a long sequence of courses in geography as is found in the Soviet Union and in Great Britain and France. A shorter, more condensed, and more thoughtful set of courses probably could serve the needs I have suggested. Nor does it argue for only one kind of course. The basic skills and understanding of process can be taught in regional study as well as in systematic units or in pursuit of a dominant problem such as man-land relations.

The continuing efforts which the National Council for Geographic Education has made to improve elementary and secondary teaching now are being assisted in part by new high school geography units of instruction designed by the Association of American Geographers. Funds came first from the Ford Foundation's Fund for the Advancement of Education and more recently from the government's National Science Foundation. New materials and course outlines for teachers are being prepared by the association.

Under the National Defense Education Act, as amended last October, groundwork is being laid for institutes for teachers which will get under way this coming summer with funds administered by the Office of Education. It is apparent that improvement will require not only changes in teaching methods but the training of teachers on a much larger scale. A few years ago, when the city of Chicago attempted to give more instruction in high school geography, it found itself frustrated by a fact it discovered overnight—that it was short 100 qualified teachers of the subject. This shortage has never been filled.

Every week's events show some distant place to be an important part of daily life. With a relatively modest outlay of effort, significant changes can be made in the understanding which young people have of both familiar and distant parts of the world. The tools and framework of thought are at hand with which we can help our students to think of the earth's surface as a dynamic complex and to perceive some of the consequences that the actions of human beings will have on it.

Approaches to Weather Modification

Uncertainty attaches to both our understanding of the natural systems and to our confidence in being able to modify them. Were either amenable to more confident prediction, research in techniques as well as in impacts and relationships would move ahead rapidly. Clearer recognition of vital ways in which weather modification could contribute to social improvement would stimulate fresh lines of physical research; greater promise of workable modification techniques would launch enthusiastic inquiries into the possible results. In the face of doubt in both directions it may be that the wise strategy, as Leonid Hurwicz suggests, is in diversification of effort and in directing attention to fundamental understandings that may have the wider implications.

Whatever the appropriate strategy, there are few other fields of natural resources management in which larger uncertainty dogs the path of the researcher. This is at once a discouragement and a tantalizing prod. In the words of Keats, it calls for the man of "negative capability," that is the man who is "capable of being in uncertainties, mysteries, doubts, without any irritated reaching after facts and reason."

(1966e), 21–22

Notes

1. For a fuller discussion of these efforts, see below, vol. II, chap. 7 (Feldman).

14 Optimal Flood Damage Management: Retrospect and Prospect

In what White calls "program notes on a retrospective show," he updates three decades, 1936–66, of effort to understand human behavior and to change public policy on the floodplains of the United States.[1] Along the way, he is most outspoken. The careful reader would note the labeling of the major government agencies as "soggy newspapers," the likening of normative economic analysis to carrying around a useless heavy anvil, and the judging of his own first essay in that genre as the "worst paper ever published on the subject." Characteristically, no one took offense. Behind the scenes, the soggy newspapers began to catch fire in a major public policy change directed at encouraging a broader range of adjustments to floods. This was heralded by the widespread acceptance that same year of the Bureau of the Budget's Task Force on Federal Flood Control report (1966d), for which White served as chairman. The economists, comfortable in their ivory towers of normative assumptions, failed to take notice, and the jury was still retired in the case of his self-judgment.

In the worldwide struggle to manage water resources for human good the gap between scientific knowledge of optimal methods and their practical application by farmers, manufacturers, and government officials is large and generally widening. Man's ability to forecast streamflow, to store water and transport it long distances, to alter its quality and extract it from great depths is growing more rapidly than his skill in putting the improved technology to earthy, daily use. This gap between the desirable and the actual is especially dramatic where peasant societies are exposed to new agronomy and water technology, but the contrast is hardly less striking in sectors

Reprinted from *Water Research,* ed. Allen V. Kneese and Stephen C. Smith (Baltimore: The Johns Hopkins Press for Resources for the Future, 1966), 251–69.

of United States production where we often think of new techniques as being avidly embraced. Farmers who overirrigate, manufacturers who neglect tried water conservation devices, and officials who misuse measures of economic efficiency testify to the difficulty.

At times we placidly accept this gap, as when a projection of future national water needs assumes that industry will not increase the adoption of new technology.[2] At other times we confidently ignore it, as when economic justification for a flood protection scheme assumes all the farmers will change their farming system to take advantage of new conditions.[3] When the failure to employ economically optimal management practices is frankly confronted, it often is charged off to cultural lag, or ignorance, or inept social organization.

In a broad sense we are dealing with the basic and immensely intricate problem of how social change takes place nonviolently, and we must recognize it has strong and diverse components of customary cultural behavior, organized social action, and natural environment. A soothing and academically safe observation is to say we need more research on why the optimum is not achieved. Clearly this is true, but I feel uncomfortable about even that platitude because some of the research on both technology and analytical methods seems to have widened rather than narrowed the gap in the field of water resources. And the gap promises to widen still further if I read correctly the lessons to be drawn from the American encounter with water resources and the shape of federal programs to come. There is reason to suspect that concentration of analysis on testing for economically efficient solutions may have impeded attainment of the efficiency goal. To modify Voltaire's observation, perhaps the search for the best has been, in these instances, the enemy of the good.

The United States' attempt to manage flood damage over more than three decades illustrates the point. From experience with the planning, construction, and study of the national flood protection efforts, beginning with the massive public works projects of 1933, a few conclusions emerge. That record of natural catastrophe, far-flung surveys, and gigantic expenditures suggests that similar though less grandiose events may well unfold on the floodplains of other developing lands. It also suggests that some portion of the myopic view of remedies, of the failure to understand how choices are made, and of the difficulty in interpreting scientific findings—troubles which have dogged federal flood activities to date—may trip up such growing programs as pollution abatement, water supply, and recreation.

I shall try to review the experience with management of flood losses in several steps. In beginning, a confession must be made of our confusion concerning optimal management. Next, the state of our knowledge about the human use of floodplains is summarized. That is sharpened by pointing out those findings that seem sufficiently valid to warrant recommendation for public policy. The obverse, of course, is admitting what we do not know, thereby outlining major research needs. Both knowledge and ignorance are stated more concretely in reference to the Denver flood of June 1965. Finally, what we have learned that may be of value in other sectors of resources management is the subject of a concluding section.

Were this a review of works of art rather than of engineering construction, public policy, and geographic studies it might be called program notes on a retrospective show. I do not attempt to appraise all the thinking which has bloomed in the muddy wake of U.S. floods. It has been summarized elsewhere.[4] The emphasis here is on what some of us associated with University of Chicago studies have learned that commends further public action and on the lessons that may be relevant in distant floodplains or in broader fields of water management.

Concepts of the Optimum

If the optimum management of flood damage is taken as that form of floodplain use that is most satisfactory and desirable in yielding the largest net social benefits over an investment period, it must be admitted at the outset that there are pitifully few grounds for judging where it has been achieved. This is because the methods of assessing optimal use are still crude and have been tried in only a few areas. It is difficult to tell whether a given reach of a valley will best serve the needs of the community in one use rather than any other. Rare is the community that is able to define its land needs clearly. Even if methods were established and readily applied, the political task of using the results in urban planning is fraught with complications. When faced with the question, it is common for agencies to follow a greatly simplified process of judgment, such as that embodied in the Milwaukee and Detroit municipal policies of buying up valley bottoms for recreational use without attempting to weigh the alternatives, or the implicit decision in Los Angeles that any land which can be protected from floods at reasonable cost should be used for residential, commercial, and industrial purposes. In some of these cases and in certain clearly warranted protection projects, the present use no doubt is optimal.

If, on the other hand, optimal management is taken as that combination of adjustments to flood hazard for any given use that yields the largest net social benefits, the task is much simpler. At hand are relatively satisfactory methods of comparing the social costs of protection or flood proofing or insurance with the ever available alternative of simply bearing the loss.[5] Not that there are no differences of professional judgment; there are, as is shown later, but that major definition and approach seem valid.

The federal policy, freely translated and omitting numerous details, is that flood damage is undesirable, and that the government is prepared to prevent it by feasible engineering or watershed treatment so long as the benefits to whomsoever they may accrue exceed the costs and so long as local interests contribute to the costs of nonreservoir projects. Surveys by the Corps of Engineers and the Soil Conservation Service characteristically assess costs and benefits for one or more engineering schemes for loss reduction. As economic critics have pointed out, these comparisons do not select the optimum: they may ignore incremental analysis, and discount practices are in question.[6] Thousands of completed survey reports give benefit-cost findings and their implicit forecasts as to what the future will bring.

Inherent in those surveys are a pair of ambiguities. One hinges on the obvious fact that what is optimal for the community may not be optimal for individual managers of floodplain land, depending upon arrangements for sharing costs and benefits. A second is the policy of cost sharing. The Corps sometimes counts benefits from land value enhancement resulting from more intensive use only if necessary to obtain a favorable benefit-cost ratio, and then seeks to recover part as local contribution; the Soil Conservation Service counts them but has no authority to require reimbursement for them. The Corps is not expected to seek reimbursement for the cost of protection from reservoirs. A change in one of these policies by the Congress or a change in the social arrangements for providing technical assistance to individual managers would alter the estimated optimum.

It must be admitted, then, that hortatory declarations as to the proper national policy for making wise use of floodplains can only come to grips at long distance and in a shadowy way with the question of wisest use. They really are addressed, for the most part, to the more limited question of whether or not, given a particular use, the adjustment selected is optimal.

However, there are several bits of evidence that the uses prevailing in United States floodplains in the 1960s are less than optimal. In the more limited goal of federal legislation, we have not been doing well. It is clear that the mean annual toll of flood losses has been rising for half a century.[7] This in itself would not be evidence of less than optimal use, for it might be that rising total losses would be a characteristic of wise use. Such a conclusion is thrown into doubt by the widespread attempts of flood sufferers to avoid future losses through flood protection works, and by the inability of the federal government, in its expenditure of more than $7 billion since 1936 on protection works, to curb the volume of flood losses. Further skepticism is induced by the readiness of some managers to change their uses of floodplains when presented with new information and opportunities, and by the growing recognition by the Corps of Engineers, itself, that engineering works alone are not fully serving their intended aim.

We know a little about the overall shift in use of floodplains in the United States. Corps of Engineers studies show that encroachment continues to be rapid in most sections of the nation. The invasion is predominantly urban, and in many areas consists in enlargement of existing commercial and industrial uses or the expansion of urban uses along new highways.[8] Loss potential mounts vigorously in reaches where partial protection is offered by dams or levees.

Two aspects of rural floodplain use seem clear from sample studies. One is that the major encroachments of loss potential upon rural floodplains are for new urban use rather than for agriculture.[9] The second is that in selected places where additional protection against flooding has been provided by watershed protection works there has not been rapid intensification of cropping.[10] Insofar as the latter relation is true, it challenges the assumption underlying much of the Watershed Protection program that decreases in flood frequency and magnitude will lead to prompt increases in intensity of farm use. It also appears from Burton's work that where parts of a farm are protected, there is a tendency for farmers to gamble on higher risks in unprotected lands.[11]

In general, the primarily urban protection programs are running fast to keep up with further encroachments; and the agricultural program is failing to satisfy its promise for farmland use changes, and increasingly is protecting or promoting urban invasion of floodplains. Individual managers of the floodplain are working over large areas at cross-purposes with stated federal aims.

How much do we now know as to why and how these managers—both public and private—behave as they do in the face of floods?

Knowledge of Floodplain Use

My first preoccupation with American flood problems was with the primitive methods of economic justification practiced by public agencies thirty years ago. When the first national Flood Control Act was passed in 1936, the simple and in many respects more comprehensive comparison of land values that had been used to justify local drainage and Miami Conservancy investments was being replaced by the more complicated but narrowed benefit-cost computations used in the 308 Studies. From my appraisal of the situation came probably the first and certainly the worst paper ever published on the subject in an academic journal.[12] There also emerged the recognition that engineering protection works are only one of the several adjustments which a floodplain manager may make to flood hazard.

Very slowly it became evident in subsequent years that a benefit-cost calculation has little meaning in seeking optimal solutions unless it predicts with modest reliability the streams of gains and losses that probably will flow from the area in the future. To do this requires sufficient understanding of the behavior of floodplain managers to discern how they may be expected to respond to whatever conditions are postulated for the future. Is a dam likely to reduce net losses downstream? How would a shopkeeper be inclined to use a new map showing his hazard from flooding? Would a change in cost sharing alter a farmer's view of his feasible alternatives? Seeking answers to questions such as these leads to examination of what managers in fact see as their range of choice, how they evaluate the flood hazard, what they regard as available technology for adjusting to floods, what appears to be the economic efficiency of each available technique, and how they are encouraged or constrained by social action. We cannot realistically assume these people to be uniformly well-informed, rational optimizers, nor can we dismiss them cavalierly as stupid, pigheaded, or knaves. More will be said later about models of floodplain behavior.

How Much Choice?

The range of choice open to floodplain users in dealing with flood hazard was outlined systematically in 1942 and has not changed greatly since.[13] It was apparent then, as now, that the heavy federal emphasis on full support of reservoir projects and partial support of other engineering protection encouraged both public agencies and

individual property managers to think of their choice as being between bearing the loss or enjoying federal protection. The early
Tennessee Valley Authority program and the Department of Agriculture Watershed Protection program reinforced this view. Emergency evacuation, while facilitated by Weather Bureau river forecasts,
was not specifically organized. Structural adjustments and land elevation were not canvassed in federal studies, and were not promoted by federal or other public agencies. Relief was available
principally for extreme hardship. Insurance was lacking in all but a
few special cases. Change in land use rarely was considered; a reduction in intensity of use was presumed to meet invariably with
local opposition (a presumption later shown to be false). Prediction
of enhancement in land value by changing to a more intense use was
avoided for a long time, because of fear of open involvement with
speculative land development, and the extreme option of relocation
of entire communities was so abrupt, radical, and lacking in implementation that it was practical in only a few instances.

The effect of the earlier classification, which still appears in discussions of zoning ordinances or floodplain information studies, is
to imply that the range of adjustments applies only to the current
use, and that any change in use will mean reduction in intensity. As
a result, the possibility of a different use or of use regulation often
is ruled out of consideration before it is canvassed: a town rejects
zoning on the ground it would automatically prevent commercial
uses; a property owner opposes land use studies because he thinks
they surely will lead to his relocation or to downgrading of land
value. They may, but they do not necessarily have that effect.

Table 14.1 Typology of range of choice in adjustments to floods

1942	1965
Land elevation	Land use A—
Flood abatement	Bearing the loss
Flood protection	Protection
Emergency measures	Emergency measures
Structural adjustments	Structural, including land elevation
Land use	Flood abatement
Crops	Insurance
Urban relocation	Public relief
Public measures	Land use n—
Public relief	Bearing the loss
Insurance	etc.

Moreover, we have come to recognize that the process of choice is dynamic rather than static. To think of there being a decision only when a major project is under consideration is to ignore the variety of paths along which adjustments move.[14] When attention is focused on these diverse paths, then more importance is attached to critical times of decision and to methods of planning for changing and flexible adjustments.[15]

How Is the Hazard Perceived?

One of the comforting explanations for apparently less than optimal use of floodplains is that there is lack of information as to flood hazard. The obvious corrective—making the information available— was slow in coming in the form of flood hazard maps, and when the first map appeared, it failed to yield the expected results. The U.S. Geological Survey map of Topeka in 1959 did not have a prompt, significant effect upon the thinking of people in Topeka.[16] Experience there and in the Tennessee Valley pointed out the need for understanding how people habitually perceive hazard and for designing means by which hazard information may become meaningful to floodplain dwellers.

Kates's study of LaFollette and five other communities indicates that perception of flood hazard is a function of direct flood experience and of outlook toward nature, and that the mere supply of information, as to where the water reached and when, does not necessarily lead to more precise recognition of the probability and magnitude of floods.[17] He shows how the ambiguity of flood risk and the difficulty of facing uncertainty foster a division between flood-plain dweller and the scientist in recognizing the hazard for what it is to each of them.[18]

The pioneering flood reports of TVA[19] and the wide experimentation of the Northeastern Illinois Metropolitan Area Planning Commission with USGS maps have begun to point out specific steps, such as distribution of hazard reports to financial agencies, which will enhance their reception. Regional studies by the USGS of flood frequency are making possible more reliable estimates of hazard. From the burgeoning number of Corps of Engineers floodplain information reports new experience is accumulating but has not been assessed in detail.

We still have only a rudimentary knowledge of how man perceives the risk and uncertainty of a flood or any other natural disaster,[20] and until this is refined, the judgment of what information in what form will be useful to him will be largely pragmatic.

What Technology Is Available?

It early became apparent that many floodplain managers, while vaguely aware of the possible range of choice and of the flood risk, did not understand the techniques of certain adjustments such as emergency evacuation, structural change, and insurance. From the LaFollette study there is evidence that those who have lived longest on the floodplain are more aware of alternatives, that those most recently flooded perceive emergency action more acutely, and that the times of construction, normal renovation, or disaster repair are more propitious for adoption of alternatives than are other times. Sheaffer's study of flood proofing in Bristol, Tennessee-Virginia showed systematically for the first time the potentialities of combining emergency and structural measures.[21] This led to TVA and state discussions with local authorities and those, in turn, to a significant deviation from usual federal flood control policy. At Bristol and White Oak, the TVA authorized federal contributions to partial engineering works on condition that local government regulate further encroachments, flood proof its own buildings, and take responsibility for advising property managers as to flood proofing of the other structures which would enjoy only partial protection under the feasible engineering works.[22]

Emergency evacuation measures have received little detailed study other than at LaFollette and Bristol, and it is only in recent months that the Weather Bureau has begun to go beyond mail questionnaires to ask how its forecasts are used.

Notwithstanding elaborate engineering studies of flood control spillway and channel design, there is no significant research under federal or any other auspices on methods of using plastics, temporary barriers, electronic control, and similar devices to prevent flood losses, and nothing since 1936 on means of rehabilitating damaged property. Expert services are poured into specifications for a dam, but it still is impossible to get from a federal agency the simplest advice as to technology for flood proofing. This lack becomes especially critical when it is recognized that the number of places receiving unfavorable reports from the Corps of Engineers on local protection projects probably equals those receiving favorable reports; that among the authorized projects more than 200 have not been undertaken because of local complications; and that even where comprehensive reservior programs have been authorized and are under construction, it may be years before the full effect of protection will be felt. Recently, unfavorable reports have tended to include more references to possible alternatives to protection.

Insurance, while a practicable alternative for the few managers who know how to obtain special coverage for flood-proofed property, or who fortuitously are covered by an all-purpose auto policy, is closed to most others. Partly as a result of caution on the part of the insurance companies and partly the consequence of inept beginnings by the administration of the authorized but abortive Federal Flood Insurance Agency, there has been no systematic experimentation with this device. Were the recent explorations by an industry committee to be extended with Housing and Home Finance Agency encouragement and, using the full technical advice of federal agencies, to offer limited coverage in sample areas, it would be practicable to find out precisely how much could be indemnified at costs and in circumstances that would supplement or substitute for protection and other alternative measures. Insurance may well have its most influential uses in being available as a substitute where protection and structural adjustments are not feasible, or in covering the hazard during a period when old uses are phased out.

Is It Economically Efficient?

Although public approaches to flood problems have centered upon benefit-cost analysis of protection projects, little attention has been paid to the economic efficiency of other adjustments and, particularly, to the efficiency of choices by private managers. As a result, we still have only a sketchy knowledge of how far individuals seek to optimize their uses and what they would choose if they were to do so rationally using their preferred criteria. It is plain that differences in time horizon, discount rate, perceived benefits, and most of all, direct costs give many individual managers a view of alternatives that would favor federal reservoir projects and would discourage an alternative such as insurance so long as existing federal policy prevails.

By assuming that managers are both rational and subject to the same efficiency criteria as public agencies, the federal programs predict changes which do not materialize—as with the Soil Conservation Service anticipation of intensification of agricultural land use—and fail to predict other changes that do materialize—as with earlier Corps of Engineers disregard for continued encroachment into valleys partially protected by new reservoirs. Economic research has contributed notably to refinement of public analysis without confronting these problems. The recent study of industrial losses in the Lehigh shows that loss and benefit estimates may be sharpened by improved methods, but that a more important step is to recognize differences in the economic stance of an industrial plant versus its region versus the nation.[23]

How Are Other Places Affected?

A principal deficiency of most economic efficiency analysis has been the handling of external economies and diseconomies. In physical terms, little attempt has been made to assess the full physical consequences of changing the regimen of a stream or altering its channel. These go largely unmeasured.[24] In economic terms, the impacts of a change in land use upon other parts of the same community and region are rarely and imperfectly traced. It is much easier to compare possible adjustments for a store in the Dallas floodplain than to estimate what would be the effects upon Dallas of sharply or progressively shifting that use to recreational or residential. Hopefully, the current Pittsburgh study will throw some light on this. And analysis, such as that attempted at Towson, Maryland,[25] of the social costs of alternative uses will help make choices in the light of impacts upon the whole community.

The early studies of floodplain adjustments tended, as Kates has pointed out, to attribute much of the behavior of floodplain dwelling to ignorance, cupidity, or irrationality.[26] We took a long time—far too long—to begin to test these riverside opinions as to motives and as to the decision process itself. Perhaps we lacked a handy model of how decisions are made if they are not made by rational, economic men. Perhaps the relative simplicity of manipulating a benefit-cost ratio fascinated us. Perhaps we had too much faith that if a government commission charted a path, both property owners and government officials would follow it sheepishly. At any rate, we went many years aiming at what people should do about floods without trying to deepen our understanding of the different paths which people actually take in living with floods.

To do this requires a much more complex model of human behavior than the optimization model. It must be a model which takes sufficient account of differences in perception, decision criteria, and the effect of social guides to permit a reasonably accurate description of what and how people decide to cope with flood hazard.

How Does Society Guide the Choice?

The role of social guides in constraining or encouraging the floodplain manager has been noted at several points. More is known about how public managers respond to pressures[27] than about how individual managers respond to public guidance. We have just begun to identify the ways in which federal policies of flood control, flood forecasting and urban renewal, state regulatory action on stream channels and highway location, municipal building codes, urban plans and renewal programs, and financing agency policy can be integrated to promote maximum choice in adjusting to flood hazard and in use of floodplains. We have learned, for example, that while a flood hazard map alone may have little influence upon the decision of a property owner to build in a floodplain, it becomes powerful when placed in the hands of a professional appraiser who has been instructed in its use, and even more powerful when mortgage insurance officers in the Federal Housing Administration and Veterans Administration are instructed to look for it and use it. We have learned that a cost-sharing requirement for a protection project may change a manager's view not only of the project, but of the alternatives open to him. Reactions to enforcement of encroachment lines, zoning ordinances, and subdivision regulations are better known. The legality of setting encroachment lines now appears well established, but no solid test yet has been made of restrictive zoning of the plain above[28] the channel, or of the liability of a property owner who knowingly exposes others to the damages of flood.

Perhaps the most baffling single problem in achieving any significant advances in public action with respect to flood loss reduction is the set of mind and institutional position of public agency personnel. To ignite the field staff of an agency like the Corps of Engineers or the Bureau of Reclamation with a new approach in contrast to a new technique is like setting fire to a pile of soggy newspapers— time, patience, and systematic ventilation are required. And a new organization can mismanage a promising innovation, as in the case of flood insurance. But there are exciting exceptions, as in the Tennessee Valley Authority where one man gave the leadership for a new federal-state-local cooperative approach to flood loss reduction.[29]

Basic to the action of the administrators and the public pressures they feel are the attitudes toward nature, water, and floods which they and other resource managers share. The sense of man as a conqueror of nature, the view of access to water as a divine right, these and other attitudes combine to condition the position of a city manager who argues that restriction on channel encroachment is a denial of a right to share nature's largess, or of an engineer who after daily travelling a hazardous expressway to his office concludes that the only suitable engineering works are those promising virtually complete safety from loss of life.

Unlike the larger enterprise of river basin development, difficulties in flood loss management cannot be charged primarily to splintering of responsibility among agencies operating under divided congressional authority. True, the number of federal, state, and local agencies having a hand in guiding decisions about floods is large, but they now have little conflict or duplication—except where the Soil Conservation Service bumps into the Corps of Engineers as it moves downstream with larger dams and smaller local requirements—and the greatest opportunity for coordination appears to rest in the application of an integrated federal approach within existing legislative policy.

Side Effects of Normative Analysis

Emphasis upon normative economic analysis by federal agencies and their academic reformers may have retarded the attainment of

optimal goals. Refinement of the benefit-cost practices has directed
attention to what should be engineering investment in floodplains
rather than to explaining their present use. While there is much room
for improvement and while benefit-cost analysis would be enhanced
by insight into managerial behavior, it detours the realistic prediction
of future use. By assuming the managers of floodplain property are
either optimizing, rational men, or narrow, inflexible men, it avoids
difficulties that would arise were account taken of information lacks;
interpretation of information; differing perception of hazard, technology, and efficiency; differing decision rules; and inconsistent social guides. Having prescribed a simple cure, there is little incentive
for the government doctor to return to the bedside to find out what
happened to the patient. Public doubts about economic justification
are soothed by B/C ratios exceeding unity while public side effects
are largely ignored. It should not be abandoned, but the weight of
its authority should not be carried unless it is used to help explain
the vital question of what will be the future use if a given adjustment
or use is adopted.

One day a short, sturdy Maine blacksmith, after long wooing the
willowy village belle, won her consent to marriage as they talked in
his shop. Enthusiastically, he jumped onto the anvil and kissed her.
Then they walked hand in hand out across the meadows. "Shall we
kiss again?" he asked. "Not yet," she said. Farther on he again put
the question. "But why do you keep asking?" she replied. "Because," he said, "if there isn't going to be any more kissing, I'll
stop carrying this anvil." Normative economic analysis is an anvil
that can serve several important uses, including intangible aesthetic
ones, but there isn't much point in carrying its weight unless it is
used properly.

Warranted Public Action

Given optimal use as our distant aim, what do we know sufficiently
well to warrant immediate prescription of public program? I think
we know enough about human use of floodplains to assert that if
the burden of annual losses is to be reduced without concomitant
expenditures for engineering works, federal policy and program
must be revised and supplemented in several ways. These would
help to see to it that managers of floodplain land are supplied with
full information, in intelligible form and at the right times, as to
the range of choice in use and adjustment open to them in using

the floodplain. This means piecing together a genuine program from diverse parts.

Solid data on flood losses would help. There are grounds for believing that sufficient research has been done to enable us to design and carry out a new system for collecting flood loss data which would substitute synthetic for reported losses, estimate the degree to which potential losses are averted, and rely upon a stratified, systematic set of samples.[30] We ought to end our unhappy plight of having to say that mean annual flood losses are "in the neighborhood of $300–900 million."

Technical studies on flood proofing and insurance and on relationships between flood characteristics and flood losses should be pushed and their results disseminated where and when people will use them.

A profound change in approach would result if the Corps of Engineers were to exercise its present authority to require, as a condition of further construction, state agencies to regulate encroachments upon stream channels, and to require local agencies to regulate land uses in reaches of streams where federal protection is to be provided.

Perhaps more important is the opportunity for the Corps of Engineers and the Soil Conservation Service, in collaboration with the appropriate federal agencies dealing with urban problems, to encourage local communities to assure that consideration of flood hazard enters into public and private planning decisions.[31] This would include expansion of reports on flood areas and frequencies, interpretation of the results to interested managers and key people such as architects and mortgage officers, stimulation of flood-forecasting and warning systems, dissemination of information on alternative measures for loss reduction, promotion of improved land use regulations, and support of selective public acquisition of hazard areas. Reconciliation of cost-sharing policy is in order.

In brief, the chief threats from clearly uneconomic encroachment should be curbed. Next to this, the major opportunity now seems to rest in promoting choice by private managers among combinations of protection, emergency, structural, and insurance measures. In the light of our knowledge of decision making and of a few practical experiments, it seems likely that public action with that aim would improve floodplain use and that it would be politically acceptable. Only a few aspects would require new federal legislation.

In contrast to earlier times when proposals for federal policy, whatever their theoretical merit, fell on inhospitable ground, three apsects of the present scene make action seem more hopeful: (1) There is a rather widespread feeling in both federal and state agencies that engineering works are not enough. (2) At the same time practical experience is building respect for new analytical techniques and for such social guides as flood information reports, mortgage insurance policy, and zoning. (3) The public groups concerned with flood damage management have been expanded significantly in recent years. To the flood sufferers have been added citizen and government contingents seeking speedy urban renewal and greater open space, and their support for a broadened approach seems assured.

Now let us list points at which understanding is so lacking that it is necessary to be cautious about recommending public action.

While there are rough methods for weighing alternative adjustments for the same land use, such as protection and flood proofing, there is no satisfactory method of comparing the feasibility of different land uses, having in mind the external economies and diseconomies accruing to other parts of the same urban area. Until this is achieved the public choice of alternative uses in hazardous areas, a choice which at best is subject to highly political pressures, is bound to be vague.

In the case of both urban and rural reaches, much more needs to be known about the precise factors affecting individual choice, particularly the managerial perception of uncertainty and of economic efficiency. Without such refinement of relationships, U.S. Department of Agriculture's predictions of floodplain land use in Watershed Protection projects, and Corps of Engineers' estimates of urban land use changes will remain speculative. We need to find out precisely how many alternatives people can consider and in what conditions.

Denver as an Example

With the June 1965 disaster in Denver in mind,[32] how would these conclusions apply to that inundated South Platte floodplain as its residents dug out of the mud and repaired their bridges behind them? After Red Cross and federal disaster assistance had been assured and people began to look to the possibility of future floods (including some which, they have been told for many years, will bring a flow twice as large as that of 1965), a program of the type suggested could have affected the city in the following fashion.

The Corps of Engineers, while open to discussion about getting funds for an upstream dam, would announce that no federal funds would be available for protection until the state took steps to curb further encroachment on stream channels. Any further protection for Denver would also be contingent upon the city showing it had extended its present regulation of land use from a few of the tributary dry washes to the full South Platte floodplain. Assistance would be offered to the municipal authorities in getting accurate, intelligible reports on the hazard areas and likely frequency of flooding, and in advising them as to how the results could be put in the hands of property managers. Technical assistance would be available to owners, architects, and city officials, not only on how to speed up appropriations by Congress but on how to save on salvage operations, how to floodproof existing buildings against future losses, how to improve the flood-warning service, and how to draw up building codes which would prohibit dangerous uses such as flammable gas installations. Highway officials and the Federal Housing Administration and Veterans Administration would be alerted to hazard areas. Appraisers would state flood hazards, like termite infestation, as a routine part of their reports. The new urban affairs agency would offer to collaborate in planning of renewal for part of the area and in preparing schemes for open land reservation and acquisition for other parts of the area. Insurance would be available, perhaps through private companies at rates reflecting the risk, to property owners who already were within the reach of floods and could not be protected or who must wait years for protection.

What would all of this mean in terms of future occupance of the South Platte? As a minimum, the harmful encroachments on natural channels and needlessly hazardous invasion of adjoining floodplains at last would stop. No further increase in damage potential would take place without both property managers and public agencies being aware of the hazard and the possible choices. Emergency damage reduction plans would become a part of industrial and municipal preparations. Structural changes would receive routine consideration in any rehabilitation or construction of buildings in hazard areas. Insofar as renewal or open space schemes were studied, there would be conscious public judgment of the merits of changing existing uses. As a maximum, Denver might find itself after two decades of progressive adjustments with only as much property in the path of floods as city and owners feel warrants the risk. Thus, the optimal adjustments to flood hazard would be approached. The present policy of permitting dangerous invasion of the channel and of restricting many floodplain dwellers to a choice between bearing flood losses or seeking federal aid would be abandoned.

Lessons from Three Decades

This rich and partly distilled experience of the United States in coping with the risk and uncertainty of floods is relevant to broader efforts at water management in several ways. Wherever growing cities spread into floodplains and whenever governments feel obliged to curb the losses from inundation there is danger that public investment may take the course which has been etched in buildings, earth, and concrete along American river valleys during the past three decades. This need not be repeated in growing countries with rapid urbanization.

Beyond the floodplain, there are three more general lessons that may apply to other sectors of water management. One is to guard against becoming bemused by commitment to a single engineering solution to the exclusion of other alternatives. Just as devotion to dams, channels, and levees obscures the view of possible gains in flood management from changes in emergency action, structures, and insurance, or from other land uses, the preoccupation with stream dilution may divert attention from alternative waste treatment, and the search for new water supplies may prevent the exploration of methods of saving water already available.

Another lesson is to beware of easy, unsupported explanations of why people manage resources as they do. Willingness to believe that floodplain dwellers are rational managers who will seek the optimal economic adjustments to floods can sidetrack our understanding of the complex factors which in fact affect their decisions. In the same way, simple curbstone explanations of manufacturers' preferences may delay measures to improve the efficiency of their water use, and casual assumptions concerning an irrigator's aims may obstruct the public attempts to help him adopt soil- and water-saving practices.

In both the exploration of alternatives and the understanding of the intricate process of resource decisions it is essential, if the results are to be useful, to be continuously alert to ways of reporting and interpreting findings so that they will be employed at the right time by farmer or manufacturer or government official. It is better to know that floodplain dwellers interpret the same sequence of flood events differently than to believe erroneously that a single government statement of hazard will have similar significance to all. It is even more important to learn by both study and trial what form of information will be meaningful to them and in what circumstances decisions will be made. The manager must not only see the choices, but see precisely how he or others can carry out each alternative.

When we address ourselves to lessons such as these, we are confronting—whether in floodplain or upland or city or farm—the fundamental problem of how man, in the face of diverse cultural tradition, social rigidity, and resource disparity, manages peacefully to gain a more fruitful living from the earth.

The Limit of Economic Justification for Flood Protection

At the present writing there is no uniform method for estimating the limit of economic justification for flood protection among the various private and government agencies interested. An honest and competent technician, using the same basic data for a given protection scheme, can in a large number of cases conclude either that protection is justified, or that it is unjustified, depending upon the premises which are assumed at one or more of the several points at issue. The precise arithmetic ratios of benefits to costs in which estimates may be summarized do not reflect the rough approximations and arbitrary assumptions involved in arriving at those ratios. It seems probable that by taking one extreme conservative interpretation of the problems noted in the foregoing pages almost no additional flood protection could be shown to be justified, and that by taking an extreme liberal point of view it could be shown in number ratios that some protection is justified on almost every floodplain. The technician, lacking any guiding policy, establishes premises at his own discretion, although it is rarely that a technical report states all the premises in lucid fashion.

(1936a), 147–48

Notes

1. For a current update, see below, vol. II, chap. 2 (Platt).

2. Senate Select Committee on National Water Resources, *Report*, 86th Cong., 1st sess., Report no. 29 (Washington, D.C.: Government Printing Office, 1961), 24.

3. Ian Burton, *Types of Agricultural Occupance of Flood Plains in the United States,* Department of Geography Research Paper no. 75 (Chicago: University of Chicago, 1962).

4. Gilbert F. White, *Choice of Adjustment to Floods,* Department of Geography Research Paper no. 93 (Chicago: University of Chicago, 1964); Tennessee Valley Authority, Technical Library, *Flood Damage Prevention: An Indexed Bibliography* (Knoxville: Tennessee Valley Authority, 1964); Robert William Kates, *Hazard and Choice Perception in Flood Plain Management,* Department of Geography Research Paper no. 78 (Chicago: University of Chicago, 1962); William G. Hoyt and Walter B. Langbein, *Floods* (Princeton: Princeton University Press, 1955).

5. Burton, *Types of Agricultural Occupance,* 83–92, see above, note 3.

6. Maynard M. Hufschmidt, John V. Krutilla, and Julius Margolis, *Report of Panel of Consultants to the Bureau of the Budget on Standards and Criteria for Formulating and Evaluating Federal Water Resources Developments* (Washington, D.C.: U.S. Bureau of the Budget, 1961); *Policies, Standards, and Procedures in the Formulation, Evaluation, and Review of Plans for Use and Development of Water and Related Land Resources,* Senate Document no. 97, 87th Cong. (approved by the President on May 15, 1962).

7. Howard L. Cook and Gilbert F. White, "Making Wise Use of Flood Plains," *United States Papers for United Nations Conference on Science and Technology,* vol. 2 (Washington, D.C.: Government Printing Office, 1963).

8. Gilbert F. White, et al., *Changes in Urban Occupance of Flood Plains in the United States,* Department of Geography Research Paper no. 57 (Chicago: University of Chicago, 1958).

9. Gilbert F. White, *Rural Flood Plains in the United States: A Summary Report,* Department of Geography mimeograph (Chicago: University of Chicago, 1963).

10. Burton, *Types of Agricultural Occupance,* see above note 3.

11. Burton, *Types of Agricultural Occupance,* 39–41.

12. Gilbert F. White, "Limit of Economic Justification for Flood Protection," *Journal of Land and Public Utility Economics* 12 (May 1936):133–48.

13. Gilbert F. White, *Human Adjustment to Floods,* Department of Geography Research Paper no. 29 (Chicago: University of Chicago, 1945), table 2.

14. White, *Choice of Adjustment to Floods,* 18–21, see above, note 4.

15. White, *Choice of Adjustment to Floods,* 17–18; Douglas James, *A Time Dependent Planning Process for Combining Structural Measures, Land Use and Flood Proofing to Minimize the Economic Loss*

of Floods, Report EEP-12 (Stanford: Stanford University Institute in Engineering Economic Systems, 1964).

16. Wolf Roder, "Attitudes and Knowledge in the Topeka Flood Plain," *Papers on Flood Problems,* Department of Geography Research Paper no. 70 (Chicago: University of Chicago, 1960).

17. Kates, *Hazard and Choice Perception,* see above, note 4.

18. Robert W. Kates, "Variation in Flood Hazard Perception: Implications for Rational Flood-Plain Use," *Spatial Organization of Land Uses: The Willamette Valley* (Corvallis: Oregon State University, 1964).

19. Senate Select Committee on National Water Resources, "Flood Problems and Management in the Tennessee River Basin," *Committee Print no. 16,* 86th Cong., 1st sess. (Washington, D.C.: Government Printing Office, 1960).

20. Ian Burton and Robert W. Kates, "The Perception of Natural Hazards in Resource Management," *Natural Resources Journal* 3 (1964).

21. John R. Sheaffer, *Flood Proofing: An Element in a Flood Damage Reduction Program,* Department of Geography Research Paper no. 65 (Chicago: University of Chicago, 1960).

22. *Plan for Flood Damage Prevention at Bristol, Tennessee-Virginia* (Bristol: Bristol Flood Study Committee, 1962).

23. Robert W. Kates, *Industrial Flood Losses: Damage Estimation in the Lehigh Valley,* Department of Geography Research Paper no. 98 (Chicago: University of Chicago, 1965).

24. M. Gordon Wolman, "Downstream Changes in Alluvial Channels Produced by Dams," U.S.G.S. Professional Paper, in preparation.

25. Ian McHarg and D. A. Wallace, *Plan for the Valleys* (Towson: Worthington and Green Valley Planning Commission, 1964); Walter G. Sutton, *Planning for Optimum Eonomic Use of Flood Plains,* A.S.C.E. Environmental Engineering Conference, Atlanta, 1963.

26. Kates, "Variation in Flood Hazard Perception," see above, note 18.

27. Arthur Maass, *Muddy Waters: The Army Engineers and the Nation's Rivers* (Cambridge: Harvard University Press, 1951); Francis C. Murphy, *Regulating Flood Plain Development,* Department of Geography Research Paper no. 56 (Chicago: University of Chicago, 1958).

28. Edward W. Beuchert, "Recent Natural Resource Cases: Constitutional Law-Zoning-Flood Plain Regulation," *Natural Resources Journal* 4 (1965).

29. The work, which has been carried out under the leadership of James E. Goddard, is reported briefly in two papers: James E. Goddard, "The Cooperative Program in the Tennessee Valley," *Papers*

on *Flood Problems,* Department of Geography Research Paper no. 70 (Chicago: University of Chicago, 1960); James E. Goddard, "Flood Damage Prevention and Flood Plain Management Improve Man's Environment," *Proceedings of the American Society of Civil Engineers* 89 (1963).

30. White, *Choice of Adjustment to Floods,* 111–14, see above, note 4.

31. White, *Choice of Adjustment to Floods,* 124.

32. White et al., *Changes in Urban Occupance of Flood Plains,* 127–33, see above, note 8.

15 Formation and Role of Public Attitudes

This is an extraordinarily rich paper, one of White's seminal papers, which documents the launching of a new research field. It is not, of course, the first in the field of environmental perception. It comes after the genetic papers of Kirk[1] and Lowenthal,[2] linking geography to psychology; the historical work of J. K. Wright,[3] Glacken[4] and Tuan[5]; the work of White's students Burton and Kates,[6] Lucas,[7] and Saarinen[8]; and the initial academic celebration of the new field at the Association of American Geographers meeting in 1965. It draws directly on White's first studies in perception, conducted by a group of social psychologists and White himself rather than his students.

The studies fell short in the sense that an integrated report of them was never published. The separate pieces by the social psychologists and this paper by White are the only written records of the studies; but the stimulus of those studies is apparent, reflected in some of the best and clearest writing concerning the relationships between attitudes, values, and behavior and between individual and public action.

The paper does four things very well. It expands on the primitive notions of perception found in "Choice of Use" (see selection no. 9), incorporating the best available thinking of social psychology and geography. It introduces readers to the many approaches available for the study of attitudes. It synthesizes the few generalizations that were then available, and it makes a very strong argument, replete with detail (even to the counting of water managers), of why this field of study is important for the understanding of the emerging field of environmental management.[9]

Reprinted from *Environmental Quality in a Growing Economy: Essays from the Sixth RFF Forum* (Baltimore: The Johns Hopkins Press for Resources for the Future, 1966), 105–27. The author is grateful to Wesley C. Calef, Robert W. Kates, W. R. Derrick Sewell, Fred L. Strodtbeck, and Meda M. White for critical comments on an earlier draft of this paper.

At the heart of managing a natural resource is the manager's perception of the resource and of the choices open to him in dealing with it. At the heart of decisions on environmental quality are a manager's views of what he and others value in the environment and can preserve or cultivate. This is not a conclusion. It is a definition: natural resources are taken to be culturally defined, decisions are regarded as choices among perceived alternatives for bringing about change, and any choice presumes a view of the resource together with preferences in outcome and methods.

Little can be said by way of conclusion as to how the public's evaluation of environment and methods does in fact affect the decisions which are reached in polluting air and water, or defacing the terrain, or destroying habitat or in efforts to prevent or repair such damage. The terms themselves are slithery. Pollution or defacement of a physical landscape can be measured only against a human preference. Human perception and preference are related to environment and personality in ways which are not well explored. Much of the public discussion is masked by a rough plaster of horseback judgments that hide the structure of action and opinion formation. The difficulties, pitfalls, and opportunities are illustrated in a recent decision by a community of 50,000 people.

The voters of Boulder, Colorado, went to the polls in July 1965 to cast their ballots for or against two proposals certain to alter the environment which they will occupy in the years ahead. One was authorization for a bond issue to pay for construction of an additional sewage disposal plant to reduce the waste that the rapidly growing city pours into Boulder Creek below its municipal limits. The second was to approve an agreement under which city water would be delivered to a mesa south of the city and thus permit residential invasion of a high, scenic sector of the piedmont adjoining the Rocky Mountain front. Voting was desultory, the totals were close, but the practical outcome was plain. The treatment plant gained approval, the water extension met defeat.

To understand how these decisions were reached and how public attitudes influenced them would require answers to a series of questions beyond the voting returns. The answers needed for Boulder exemplify those needed wherever public agencies alter environment. What is the decision-making network in Boulder, not only as it shows in the electoral decision but in the public and citizen agencies that framed the issues and sought to influence public action? What attitudes toward water quality and landscape enjoyment were held by those who took any part in the decision? How did their attitudes shape the public choice? What factors affect these attitudes? To what extent and in what circumstances are the attitudes subject to change? These questions will be reviewed in a broader context at the close of this paper.

The Boulder decisions were relatively simple in the concreteness of the issues, the number of people taking part, and the narrow area affected.

Consider the difficulty of trying to understand the genesis and implications of a declaration from the Department of the Interior[10] that:

> Slowly, there is dawning in man an understanding of the intertwined cause and effect pattern which makes him subject, in some small way, to every slightest tampering with his total environment. If he is to enjoy the fruits of a truly Great Society, he must be willing to work for quality everywhere, not just in his own back yard; he must consider not just the fumes from his own car, but the total exhaust cloud from the Nation's vehicles; he must wonder not just where the next drink of water is coming from, but what is being done to keep the world's taps from going dry.

How is it known what is dawning on man, and what is the quality he enjoys? Some observers argue that a consensus is emerging as to what is good quality in the environment. Looking to problems of policy formation, Caldwell argues:

> The need for a generalizing concept of environmental development that will provide a common denominator among differing values and interests is becoming clearer. And the concept of "good" environment, however one defines it, is certainly no less concrete, tangible, and specific than the concepts of freedom, prosperity, security, and welfare that have on various occasions served to focus public policy.[11]

How can these felt needs be recognized and weighed? How do they or should they affect public decision? In what ways may they be expected to change?

The number of people and organized groups directly involved in decisions that affect the quality of environment can be stated with some accuracy for any area. Water management to alter the water cycle in the United States provides one example. It is known that approximately 306,000 farmers irrigate land independently or through

8,750 districts, that farmers organized in 8,460 districts drain off
excess water from the land, that 18,150 incorporated urban governments provide water and dispose of it for their residents, that each
of 7,720 manufacturing plants withdraw on the average more than
20 million gallons of water daily for industrial purposes, that agencies
and organizations generate hydroelectric power at 1,600 plants, that
at least 3,700,000 rural homeowners provide their own water and
waste disposal facilities, that every one of the farmers affects the
movement of water on and in the soil, and that numerous state and
federal agencies exercise some kind of influence over these choices.

For each of the direct decisions to make or not make one of these
alterations in hydrologic systems there may be presumed to be a
network of relationships. Where one manager is directly involved,
as with a farmer who decides to install his own septic tank, his
choice may be influenced only by a county sanitary requirement, by
technical information supplied by state agencies and local merchants, and by the expressed preferences of his neighbors. Where
an organization is involved, the whole process is complicated by the
character of internal choice and by the number of exterior conditions
that influence the process. Networks for decisions to modify air,
vegetation, and urban landscape are even more complex.

Adequate models are lacking to describe the intricacies of decision
making and, thereby, to indicate critical points in the process. There
seems no doubt that an individual manager of a sector of the environment takes into account in some fashion the range of possible
uses, the character of the environment itself, the technology available to him for using the environment, and the expected gains and
losses to himself and others from the possible action. His perception
and judgment at each point is bound to occur in a framework of
habitual behavior and of social guidance exercised through constraints or incentives. When the decision is lodged in an organization
there is added the strong motivation of its members to seek equilibrium and to preserve the organization while accommodating its structure to changes required by shifts in preferences, environment, or
personnel.[12]

Just how much of a role attitudes play in the final outcome is a
matter for speculation at present. Little evidence is in hand. A few
case studies of decision making suggest points at which they might
be expected to be especially significant. Thus, Gore's examination
of selected government operations suggests that "formal organization accounts for only a part of surface behavior," and that informal
organization, with its sensitivity to motivation, communication,

sanction, habituated behavior, and threat symbol, help explain the remainder.[13] Certainly, individual goals and beliefs figure importantly in the organizational behavior.

Less systematic observations of the course of environmental management show a few obviously critical situations. Early efforts to develop public water supplies in the United States encountered serious inertia because many people did not believe in the germ theory of disease: until attitudes toward disease were altered proponents of the new, more sanitary supplies had hard sledding. A large city electric utility corporation recently installed precipitators in the stacks of a thermal fuel plant because it thought conspicuous smoke plumes would impair its relations with its customers. The classic case in the water field is the decision of the Board of Water Supply of New York City in the 1950s to pass over further study of the Hudson River as a water source because of its belief that the water-users would object to anything other than pure, upland sources.

Attitudes enter into decisions in three ways. First, there are personal attitudes of the people sharing in the decision. Second, are their opinions as to what others prefer. Third, are their opinions as to what others should prefer. The three need not, and rarely do coincide, although there probably is a tendency for personal and normative attitudes to merge. In Boulder, an influential citizen favored a clean stream, thought that most of his fellows did not, and urged them to adopt waste treatment. Another Boulderite regarded the mesa subdivision as an abomination, thought his fellows did not, and supported a negative vote against it. In a national park not far away, a government official who would prefer to walk and camp in a sector where no roads are present or may be seen, ignored his own preferences in advocating a new highway which would gash the mountain slopes in making the landscape accessible to tourists who, he believed, like it that way. The Outdoor Recreation Resources Review Commission was the first public agency to attempt an orderly canvass of consumer demand for qualitative use of environment, and it was careful in reporting the findings as to trends in outdoor recreation to note "A projection of these trends cannot foretell the future, but there are important clues here indicating the new order of needs."[14]

The literature of resources management and conservation is rife with assertions of what the people want. These range from sweeping declarations that "So far, our history has recorded two great threats, or attitudes, with relation to our natural resources, and now we are beginning a third,"[15] to closely-reasoned arguments that the government actions in a democratic society reflect the ultimate resolution of conflicting preferences.

Far more influential in the daily course of environmental modification is the assessment of public attitudes that goes unsung and largely unrecorded. This is the assessment that is lodged deep in the engineering design of a new Potomac River dam or in the administrative decision to specify standards for land use in a Cascades wilderness. An engineer judges that people will be satisfied with a given taste of water or with a certain monotony of wayside design or with a stream bank that is deprived of algae growth. Once plowed into a design or office memorandum, the assumption may never reappear in its original seminal form, but it may bear profuse fruit in the character of daily action.

Strictly speaking, there is no single expert opinion about attitudes toward quality of environment; there are the opinions each person holds, the opinions he thinks others hold, and the opinions he thinks they should hold. Many public administrators get mixed up about this. Perhaps the greatest confusion arises from their not knowing what others do believe and from lacking means of finding out. What follows is a preliminary attempt to outline what is known and not known about attitudes toward quality of environment and the ways in which they vary from person to person, place to place, and time to time.

Before touching on some of what is known and not known about attitudes affecting environmental decisions, caution should be offered at four points. The first and basic reservation is that the analysis is made by a geographer who, while seeking help from social psychologist colleagues, has not absorbed their scientific lore, and who reports the use of their conclusions but may not fully comprehend the grounds for them. During the past year a few social psychologists and geographers at the University of Chicago have joined under an RFF grant in an exploration of attitudes towards water.[16] Much of what follows stems from that investigation, especially the remaining cautions.

The term "attitude" is used interchangeably with "belief" or "opinion" to describe a preference held by a person with respect to an object or concept. It does not in itself constitute a value or mark of value; it is the result of a valuation process of some kind and always involves a preference. Insofar as it applies to an aspect of the environment it requires perception of that environment. By perception is meant the individual organization of sensory stimulation.[17] Apparently, there is no perception which is not organized on the basis of social experience. All of the evidence indicates that the same mountain landscape may be perceived quite differently by two people, to one as lowering and ominous, to another as refreshing and uplifting; that one man honestly terms clean a stream which another labels dirty; and that the same size coin looks larger to one boy than to another.[18] A common feature of perception is distortion of unfamiliar phenomena to adjust to familiar orientation, as when a geometrically minded American sees a skewed window frame in a different perspective in order to make it appear rectilinear. Or it may obscure painful reactions, as when a loving father does not observe a marring twist in a daughter's face. The term cognitive dissonance, as used by Festinger, describes the transformation.[19] There can be no thoroughly objective perception of the environment, only degrees of distortion which are minimized in rigorous scientific description. If this is true, then there can be no absolute standards of aesthetic experience, only standards which vary with experience and personality. What is perceived as reality may differ from person to person, and it seems likely that in such elementary ways as viewing abstract designs people vary in their spatial styles, in their preference, for example, of vertical as against horizontal lines.[20]

The next caution is against equating quality of environment with quality of life. When people speak of a high quality of an environmental vista, they often mean that the stimulus which it offers has led to a perception and accompanying response by the viewer which they regard as good for the viewer. The proper test is not the landscape itself but the response of whoever is stimulated by it. The response of an ardent ecologist who is inspired by exposure to an almost wholly undisturbed ecosystem may be like that of a mathematician in reverence before a perfect proof: the object inspires joy in the recognition of something which satisfies a particular human yearning for perfection. If this is true, then it is misleading rather than helpful to distinguish between quality of natural environment and quality of social environment. Quite aside from the fact that virtually no bit of the earth's surface is wholly undisturbed by man, it is important to remember that what commonly is called natural environment has meaning solely in a social setting in which the preferences are those of man interacting with man and nature.

The suggestion that the natural-social distinction be dropped may offend a few environmental engineers, but there is not likely to be solid objection from engineers or architects who attempt to design new buildings and communities. They subscribe in principle to the idea that they are shaping the total environment. The chief difficulty arises in trying to carry out the theory. Thus far, there are no instances where this has been done through rigorous application of what is known about human environmental stimuli.

A final caution has to do with the tendency to explain part of man's use of environment as rational and part as irrational. This is an attractive and convenient dichotomy, particularly when attention is directed toward economic optimization. It is said, for example, that if farmers were rational they would adjust their operations, within whatever constraints are set by social institutions, so as to maximize their net returns. Quite aside from the baffling task of recognizing social incentives and constraints, there are two difficulties in trying to pursue the distinction of rationality. One is that human goals rarely if ever are clearly defined; generally they are ambiguous. There is not a single program or single policy in recent United States resource management that displays a unitary, unambiguous aim. Several aims are fused, and the most ardent administrators revel in the flexibility afforded by the resulting ambiguity: flood control is to save lives and protect economically efficient development; highway beautification is to enhance the landscape and make it more accessible by concrete expressways; waste treatment is to reduce health hazards and to render streams more useful for a variety of purposes. As it is with organizations, so it is with individuals. It is sanguine to expect neat, unambiguous aims and decison criteria.

A second difficulty is that the factors of personality and environment are so complex that to speak of a rational process is to ascribe a clarity of action and observation that rarely is attained. It is enough to struggle for rational, accurate description without seeking or claiming to find rationality in the action itself.

To sum up the cautions: Do not regard this as comprehensive from the standpoint of the social sciences. Remember that every attitude toward environment involves perception that is organized by individuals. Avoid equating quality of life with quality of environment, for the latter is judged only by the former. Abandon any early claims for rationality, and look at the way in which living people behave.

Five different avenues are followed by those who would discover attitudes. They twist and sometimes cross, but rarely merge.

The first and more traditional method is to analyze the interpretations which articulate man has made of his environment. The central route here is scholarly, sensitive appraisal of what man has felt about nature through his writings and his graphic art. Lowenthal gives the most comprehensive introduction to this approach,[21] and

Glacken presses it searchingly in his examinations of attitudes toward nature held by scientists and other observers.[22] Tuan has called attention to landscapes which in literature have taken on special symbolic significance.[23] Travel diaries, the notes of explorers, or rock paintings may reveal the terrain as humans observe it.

Content analysis has the same purpose and uses the same material—the written word—but applies a more rigorous method. The Boulder decision can be subjected to analysis by examining the entire printed discussion that preceded the election according to prescribed categories of form, direction, authority, and value.[24] On the water question, the supporters stressed the need for planned development and for improvements which would enhance the city's growth. They argued that the mesa service area would enlarge the tax base, and appealed to the city's reputation and the responsibility of its citizens. Opponents stressed a prospective shortage of water and higher taxes. They argued that the city officials had acted unwisely and that the citizens should concern themselves only with the welfare of the city itself. Generally, the arguments against were not the obverse of those for, and the opposition was oriented around individual cost while support was oriented around community gain. The opposition tended to place more emphasis on effects which could be stated quantitatively than on quality and to criticize details of design. The printed statements about the sewer issue were similar in character.

An intermediate variant of content analysis is represented by Elson's investigation of more than 1,000 textbooks used in the first eight years of American schooling during the nineteenth century.[25] It leads to the following type of conclusion as to attitudes toward nature:

> Thus the nineteenth-century child was taught that nature is animated with man's purposes. God designed nature for man's physical needs and spiritual training. Scientific understanding of nature will reveal the greater glory of God, and the practical application of such knowledge should be encouraged as part of the use God meant man to make of nature. Besides serving the material needs of man, nature is a source of man's health, strength, and virtue. He departs at his peril from a life close to nature. At a time when America was becoming increasingly industrial and urban, agrarian values which had been a natural growth in earlier America became articles of fervent faith in American nationalism. The American character had been formed in virtue because it developed in a rural environment, and it must remain the

same despite vast environmental change. The existence of a bounteous and fruitful frontier in America, with its promise not only of future prosperity but of continued virtue, offers proof that God has singled out the United States above other nations for His fostering care. The superiority of nature to man-made things confers superiority on the American over older civilizations. That Uncle Sam sooner or later will have to become a city dweller is not envisaged by these textbook writers, although their almost fanatical advocacy of rural values would seem to suggest an unconscious fear that this might be so.

All of these appraisals raise the problem of how representative were the artists or pedestrian textbook writers upon whose work they are based. Was Corot's view of the forest in any sense indicative of the attitude of French foresters who managed the state land? Did McGuffey speak for the tillers of soil in Indiana? Were local newspapers which said of the Indiana prairie land in 1830 that the soil was suitable for cultivation a more accurate measure of the contemporary farmer's perception of his environment than the latter-day historians who spoke of avoidance of the prairies? McManis shows that they were.[26]

If the artists' interpretation is to be verified, a second course of inquiry is to go directly to the people. Here enters the opinion pollster. An expression of attitude may be solicited by questions, and this may be checked for internal consistency and structure. Beginning with the Department of Agriculture Program Surveys in 1940,[27] the opinion surveys of the National Opinion Research Center[28] and the Center for Survey Research[29] have canvassed segments of the American population from time to time as to its preferences concerning environment. The most extensive effort, sponsored by the Outdoor Recreational Resources Review Commission, inquired into the current habits of use of outdoor lands, and attempted to forecast the likely shifts in demand which would result from changes in population, income, and transportation.[30] Several recent studies of the survey type dealt with opinions toward air pollution in the Clarkston, Washington,[31] and metropolitan St. Louis areas,[32] and toward aviation noise at an Air Force base.[33] From the array of responses to those surveys it is possible to describe certain articulated attitudes toward environment, ranging from air quality to juvenile delinquency, and to correlate them with social status, location, and views of the community.

Assuming that the samples are representative, complications in the use of results of opinion surveys arise from the degree to which the interview situation reflects conditions which would be at work if the respondent were faced with a decision in real life.[34] A man who, sitting in his living room, says he would favor a waste treatment plant may behave somewhat differently in a voting booth where the question is posed in terms of authorizing a bond issue, and still differently when the issue is a matter for discussion by a neighborhood group where he is open to new information and to interaction with peers and authorities. There is a lesson in the account of the professor who said he didn't know what he thought of a complex issue because he had not started to talk about it. An election with a yes or no choice lends itself especially to polling prediction and verification. Where the issue is more complex and the range of answers susceptible to wider interpretation, the procedures require supplementing, as was revealed by the forecasts of consumer preference for a paragon of a car called the Edsel. Ideally, the polling should follow after a much more searching investigation in depth which would isolate the factors of environment and personality that may be expected to figure in the final choice.

Although the polling techniques have been popular for a quarter of a century there seems to have been no attempt to find out trends in stated opinion toward the environment during that period. Nor has there been published any successful correlation of opinion survey and content analysis for environmental attitudes at the same place and time.

Akin to the opinion poll, though more concrete in its findings and more provocative in its interpretation, is the examination of actual consumer choices. This is a third avenue of study. The school of thought that argues that the public generally "gets what it wants" asks where people go for recreation in order to find their taste in recreational facilities, and asks what they are willing to pay for water as a measure of the value they place upon it.[35] Such analysis, where its use is practicable, raises basic and disturbing questions for it may well challenge accepted beliefs as to preferences. The trouble here is that so few aspects of environment are subject to free pricing, and that so few past decisions have been made without the encumbrance of extensive social guides which impose constraints and offer incentives. It is much easier to work out shadow prices and comparative valuation of uses for purposes of benefit-cost calculation than it is to trace out the effects of different value judgments along trails of practical choices that are hedged with public prods and carrots. The opportunities to refine estimates of this sort are large and increasingly recognized. Herfindahl and Kneese point out, for example, that "preferences for pollution-free air can perhaps be inferred from relative land values, expenditures for air purifiers, and commuting costs people are willing to incur to avoid polluted air."[36]

Rather than to ask the citizen what he wants or to deduce his preferences from what he ends up taking, it is possible to look into how he goes about making his choice in daily life. Because the models of decision making are far from satisfactory and because the task of sorting out all of the factors bearing on a decision is intricate at best, this fourth avenue has been pursued only a short distance. Thus far, the studies of organization decision making have given little attention to broad environmental considerations.[37] Geographic studies have tended to focus on perception of particular elements in the environment. Lucas traced out the concepts of wilderness, with its elements of beauty and solitude, as held by users of the Boundary Waters Canoe Area.[38] Situations of distinct hazard from natural phenomena may present problems of perception in a clear light.[39] Kates studied the perception of flood hazard as it related to adoption of loss-reduction measures and, with others, compared fresh-water and coastal situations.[40] Saarinen investigated the perception of drought hazard by Great Plains farmers.[41] Meda White examined the perception of tornado disaster by persons responsible for taking relief measures.[42] The Ohio State Disaster Research Center has pursued the problem of how group interaction affects response to a disaster situation.[43]

A fifth and possibly more revealing method of assessing attitudes is found in subjecting people to experimental situations in which they are asked to voice opinions after being exposed to a variety of information and persuasion and to interacton with peers seeking answers to the same problem. In the Chicago study of attitudes toward water, groups of young adults of relatively homogeneous education and age were given the opportunity to learn, discuss, and take positions toward a series of problems involving pollution and other conflictive uses of water. This more nearly approximates an actual decision situation, and makes it possible to observe the effects of changes in experience and in the opinions of their fellows. So far as is known, this is the first venture along that path in assessing attitudes toward environmental quality.

From the scattered evidence accumulated along all five avenues, a few conclusions seem warranted. Generalizations are difficult because there are no adequate models of personality and attitude formation to which to relate the empirical findings.

Perhaps the most obvious observation, as might be deduced from the earlier assertion as to distortion in perception, is that different people may view the same segment of the environment differently. In two neighboring Georgia towns taking water from the same stream, one group of citizens regard the taste as satisfactory, the other as unsatisfactory.[44] A landscape which seems friendly and inviting to one traveler is austere and hostile to another.[45] Perceptions of "dirty" water, "ugly" landscapes, "barren wastes," "murky" hazes, do not appear to conform to any universal aesthetic. Without commenting on how these perceptions differ or how their variance is related to other factors, a few other conclusions can be stated.

Judgment as to the severity of a perceived aspect of the environment varies greatly from person to person. Two people in the same metropolitan area may see air pollution as high or low, while agreeing that it constitutes an impairment of the habitat. Two dwellers on the same floodplain may regard flooding as frequent or infrequent, severe or benign.

So also may the city dwellers' concern with environmental quality differ. Even when the degree of severity is seen similarly, their expression of anxiety over its occurrence may vary widely. Thus, their ability to perceive niceties and complexities of the same phenomena of clouded stream or disfigured mesa-top is diverse.

Closely linked with concern is the sense of capacity to change or adjust to the environment. This ranges from the fatalistic acceptance of any feature—pleasurable or obnoxious—to confidence in individual or collective competence to correct the perceived faults. The view of capacity to deal with environment may be expressed in relation to particular aspects of land, water, plants, and air, or it may show in a general attitude toward nature.

The broad value orientations found by Kluckhohn and Strodtbeck in their study of five communities—Mormon, Texan, Spanish-American, Zuni, and Navaho—in the U.S. Southwest seem to apply more widely.[46] Their man-nature classes conform to the commonly held theory that people in their orientation toward nature may be grouped as seeing man in a position of: (1) mastery over nature, (2) subjugation to nature, and (3) harmony with nature. These orientations are seen as related on the one hand to other cultural behavior, motives, and perception of reality, and on the other hand to the social structure and process of groups.

But why are there these great and persistent variations? The circumstances in which attitudes may change or be open to change may offer some clues, although it is unlikely that the variations can ever be wholly accounted for. Ingrained in the mythology of resources management are a number of explanations that apparently do not hold water in contemporary American society. For example, no close relation has been shown between physical setting and attitudes. In the semiarid landscape of the Southwest, the value orientations toward man in relation to nature do not reveal homogeneity with regard to the same landscape: subcultures have different orientations toward the same physical phenomena but there are not absolute differences between distinct cultures; the same components appear in different rank orders.[47] In the recent Chicago studies of attitudes towards water, American young adults do not appear to vary accordingly to the aridity or humidity of the environment in which they spent their earlier years. Neither do their attitudes seem to differ with religious training and membership. If these negative findings are correct, much of the belief popular among government officers that people raised in dry areas have distinctive attitudes toward water in contrast to those raised in the humid East is challenged. The public expression of concern about water may be different in Nevada than northern Maine, but explanation must be sought beyond the physical aspects of childhood environment.

Four sets of factors do appear to play some kind of part in attitude formation: the decision situation, the individual's experience with the environment, his perception of his role, and his competence in dealing with its complexity. A different classification of the factors no doubt could be more systematic but these groupings are convenient.

Perhaps the most careful studies of the circumstances of public choice of environmental quality have centered upon the issue of whether or not public water supplies should be fluoridated and on the social situations in which public action is taken. The decision to fluoridate as a measure against tooth decay often appears as a single question for popular or council vote. Studies made of a few of the communities in which it has been proposed and adopted or rejected since 1944 throw light on the generality of knowledge as to attitude formation.[48] Paul, Gamson, and Kegeles, in reviewing what is now a large literature of social studies, point out that people who feel deprived or alienated by society tend to express resentment by voting against fluoridation, and that the local leaders of opposition are moved by feelings of the "remoteness and impersonality of the sources of power and influence affecting the daily life of the individual.'' When the histories of fluoridation campaigns are examined, it is observed that the same action which would be advantageous at one stage may set back the effort at another stage, and that often the leaders of campaigns, some of them professionals, may work against their own purposes because of their inability to recognize the roles expected of them. Although there has been considerable analysis of demographic, educational, and economic characteristics of voters and leaders, the more critical points for further investigation seem to center on personality traits and on the local social and political situation in which the decision is made.

Few issues offering choice of environmental quality have been investigated in as much detail as the fluoridation disputes. The aims are less clear and the range of possible means is much wider when an ordinance to ban billboards is up for a vote, or when the farmers' use of pesticides is at stake. Less is known about effects which the circumstances of social organization and personal interaction may have. In an interesting review of studies of why American farmers have been slow to accept recommended soil conservation practices, Held and Clawson examine the scattered evidence on farmers' attitudes toward erosion control measures and find that while certain factors, such as reluctance to change old methods and age of operator, may be significant they should be examined in the context of the tenure, farm management, and cost-price relations in which the farmer acts.[49] The parts played by perception of soil conditions and by personality have received only passing attention.

Probably the greater part of studies to date have dealt with the intricate set of relations involved in human response to environmental stimuli. An excellent sampling of representative work is given by Kates and Wohlwill in a recent issue of the *Journal of Social Issues*.[50] They note the paucity of psychological study of the effects of physical environment on man's behavior, and call attention especially to its significance for the professions that design new rural or urban environments.

At the most elementary level, it was shown that in the two Georgia towns noted above the town which found its water taste unsatisfactory received water from time to time from an alternate source having less pronounced taste, whereas the town regarding the taste as satisfactory drew only on what the other branded an obnoxious source. Experience counts, but does not have a simple linear relation to either perception of the environment or willingness to deal with it. Thus, Kates finds, floodplain dwellers with direct exposure to floods have different perceptions of the hazard and a greater propensity to cope with it. Lucas shows that the canoeist views the same wilderness differently than does the motorboater, and that the responsible government officials have perceptions conforming to neither. Saarinen demonstrates that wheat farmers on the Central Great Plains become more sensitive to drought hazard up to late middle age, and that then their awareness declines sharply. People who are more annoyed by noisy aircraft tend to be those who fear air crashes, who are less convinced of the importance of the air base to the area's welfare, and who are also annoyed by automobile noise. Only a few cross-cultural comparisons have been made. Each such finding throws light on the wisdom of public measures to manage an aspect of environment by sharpening the understanding of how individual citizens and officials view the same physical landscape.

If a major aspect of the individual's perception of the world around him is related to his sense of his own role in that world, then it becomes important to seek out his identification of himself. Oftentimes students of environmental problems like to categorize themselves and others as behaving according to a professional stereotype. The economist optimizes net returns, the engineer gets the right things built, and the conservationist stops the *wrong* things from being built. There are niceties and colorful elaborations that need not be repeated. Just how much the individual's identification by training, professional status, and interaction with his peers leads him to behave in particular ways toward his environment has not been demonstrated. By analogy with other professions, such as medicine,[51] it might be expected that the sense of vocational role would be strong and that it would be reinforced by a high degree of self-selection among students who, sharing certain stereotypes, choose a profession that they hope will be congenial.

A fundamental line of inquiry is followed by Strodtbeck and his associates in examining sex identification. Because it has relevance both to the individual's view of the world around him and to his sense of role in interaction with other people, sex identity may be a powerful means of recognizing personality traits that are significant in formation of attitudes toward environment. In the study of attitudes toward water, it was found that young American men when confronted with situations in which water problems were presented as either severe or not severe, and in which the possibility of taking positive action was seen as either promising or not promising, responded very differently according to the sex role with which they consciously and unconsciously identified themselves. Thus, the man who was both consciously and unconsciously strong in male characteristics was more likely to take action if he was told the problem was capable of solution, whereas the one with strong identification of himself as having female characteristics had the greater propensity to act when the problem was not likely of solution. Other findings are given in the report already noted. The point here is that role identification may turn out to be a highly significant factor in attitude formation.

The fourth set of factors may be grouped under the heading of what Henry and Schlien call "affective complexity."[52] This refers to the personality attributes that permit an individual to be aware of complexity in the world around him, and to respond to them without being entirely defensive or threatened. It implies openness to impressions from the outside and ability to confidently incorporate them in guiding his own behavior.

Although no studies have been made of this set of attributes as they relate to natural resource uses, the approach seems worth noting as aiming at situations that often attach to management of the environment. Quality decisions always refer to environmental change, to individual preferences for change, to a complex environment, and to programs with ambiguous aims. Much of the effort to change the urban scene presumes the response of persons who, finding the new city in conflict with the value orientation of their culture and the preferences cultivated by past experience, have the personality attributes to be able to explore modifications of both the city and their own behavior without merely launching war upon the city.

The scholar who predicts what future preferences the public will express for the quality of water or air, and the administrator who wonders how far a later constituency may tolerate a current decision as to standards, may ask how likely are the underlying attitudes, once tagged, to change. Also, judging from the number of students of resource management who appear to feel that what the people should prefer coincides with what they themselves prefer, there are some who brood over how they can manipulate public attitudes in what they regard as the right direction. Indeed, a considerable part of the public information expenditures of the federal and state departments dealing with natural resources is based on conviction that a flow of facts about resources and their use will influence public action either by changing attitudes or by providing information on which people can act more intelligently. The Chicago study on attitudes toward water finds pronounced differences among managerial groups in their belief in the degree to which their actions can modify public attitudes.

Insofar as the attitudes are related to role identification and affective complexity they may be regarded as largely inflexible. The decision situation—the time of the vote, the leaders who force the decision, the way they phrase the question—is much more subject to alteration. Among the numerous aspects of environmental stimulation, probably the one most susceptible to change is the information about environmental conditions and ways of managing them. Although certain findings suggest that the information alone, as in the case of a flood hazard map or a government pamphlet about wind erosion, may have little effect upon attitudes toward those phenomena, other studies indicate that if the situation in which choice is exercised is modified or if the individual's sense of efficacy in dealing with the confusion of the world is changed suitably, the information takes on different significance. Given favorable circumstances (and this qualification may be crucial) there is no reason to think that some amount of shift in attitude would not follow the receipt of new information about the environment.[53] Just how far the shift will go, and just how much personality traits, such as role identification, will have to do with it, is far from clear.

The most difficult question remaining is whether a shift in attitude would have any perceptible effect upon the decisions reached about environmental quality. Experimental evidence seems wholly lacking, and most of the observations of a curbstone character are made without a rigorous scheme of analysis. Generalizations must come from definition or from casual reflection on a few past decisions.

One striking fact is that a large number of environmental quality decisions are made by people who feel a strong professional identification. Their view of themselves as conservationists, economists, sanitary engineers, foresters, etc., may be expected to shape their perception of the environment and their competence to handle it. In these roles they not only inherit customary ways of defining significant parts of the environment but they are disposed to distort or ignore phenomena that they regard as beyond their professional responsibility or competence. (If you can't measure a diseconomy, sweep it under the rug.) Their perceptions and preferences become the implicit and usually unchallenged determinants of plans presented for public choice.

A second fact is that these professional judgments often involve assessment of public preferences that go largely unchecked. An engineer's view of public valuation of a polluted stream or a soot-ridden sky rarely is tested by investigation and commonly enters into public decision in situations in which individual citizens can express a disapproval of the plan but not of its assumptions as to their preferences. When the New York Board of Water Supply decided against using Hudson River sources it had no generalized scientific evidence on the way in which citizens of those cultural groups regard water sources, it lacked any findings on New York preferences, and it passed on its judgment in a form which eliminated any public expression of those preferences: since it was concluded without verification that the people wanted upland sources, they were asked to vote for upland sources as the best solution. Their favorable vote could neither confirm nor deny the conclusion.

In the absence of a more adequate model of decision making, the testing of the influence of attitudes on both officials and the related citizen groups remains largely conjectural. It may be useful, nevertheless, to outline two hypotheses that grow out of experience with water resource management debates and that do not appear to be inconsistent with observed relationships. These indicate the kind of question calling for systematic examination, and are selected from more than twenty appearing in the report of the Chicago study.

1. Because of the complexity of systems of water management, the ambiguity of their ownership, and ignorance of the natural processes explaining their behavior, there is a strong tendency to rely upon exterior authority for judgment as to management of water quality. In simple decision networks, where individual control of environment is large, the unknown is explained by myth: in complex networks the judgment is referred to professional experts.

2. Perception of the effects of water management upon others is a function of the degree to which the individual regards man as a master of nature and to which he resents manipulation of himself by others.

In each case the disposition to take public action is seen as related to the personality traits of the individuals involved. How much these are confined or magnified by the social setting in which the decision is reached is difficult to say.

Until recent years a high proportion of public decisions on resource management in the United States were taken following great natural disasters or in anticipation of serious human deprivation. Flood control legislation often followed in the muddy wake of major floods, and soil conservation measures sprang up in the lengthened shadow of dust clouds. Timber management was promoted in part by the anticipation of a future timber shortage. Much of the rhetoric hinged on the fear of a reported crisis or of a new one looming in the years ahead. Such appeals are still strong, as with the brooding concern for the human effects of pesticides and fungicides and with the dark prediction of national water shortage. However, the rising level of per capita production and the accelerating pace of technologic change enlarges the conditions of choice in the direction of greater freedom. As shown by a recent report by the Committee on Water of the National Academy of Sciences,[54] the unfolding opportunities for water management are in exploring the whole range of possible alternatives in transactions with the environment, and in weighing their relative social impacts.

This is a turn away from the customary promotion of single solutions in an atmosphere of present or impending crisis. Consideration of alternatives for changing the environment implies less reliance on choices by a technical elite and more confidence in a base of citizens who have the maturity to deal with complex and probabilistic conditions. To the extent such a shift occurs, sensitivity to the direction of public attitudes, as well as to their limitations, may be expected to take on greater importance.

Faced with trying to understand public decisions on a stinking stream and a scarred landscape such as those presented to Boulder's voters, social science can offer a few sturdy methods and a larger set of questions that remain unanswered. By content analysis it is possible to define the issues and the attitudes toward them as articulated in the public argument. By opinion survey of a representative sample of the population, the expressed attitudes of sectors of the electorate can be assessed, and a prediction of voters' behavior can be checked against the vote itself. By analysis of the voting situation a rough judgment can be reached on the extent to which conflicts or compacts among public officials and political groups in the community may so operate as to obscure any voter preference as to the kind of environment the people prefer. There are slight but provocative grounds for expecting that the stated attitudes would vary in some degree with length and type of the respondents' experience with stream and mesa, that perception of severity would vary with social status, and that propensity to act would be related to role identification and affective complexity. These and other data provide an initial base from which to speculate on the situations in which various types of information might promote changes in attitudes if given to the public at the appropriate time.

Because there are no satisfactory models of the decision process or of the interaction of factors affecting attitude formation, the speculation for Boulder or any other community could, at best, only explore pragmatic relations or tentative theory. Basic research is needed on decision processes and attitude formation, particularly in settings where resource management produces nonvendible benefits. The network of decisions should be described in sufficient detail to permit recognition of power relations among individuals and groups. The typology of attitudes toward environment invites much more precise analysis. These attitudes deserve searching examination in experimental conditions where the personality traits of the subjects are known and where the inputs of information and the decision situation are partly controlled. From those investigations may be expected increased understanding of human response which would permit a more incisive wording of questions for opinion surveys and, in turn, more intelligent interpretation of their results. An auxiliary step would be thoughtful appraisal of environmental quality features of opinion surveys of the past twenty-five years in order to recover data that have been lost from sight.

As these studies proceed they will throw light on how decisions in truth are made, on how the professional's own preferences figure in the proposed solutions, on what he thinks the citizen prefers, on what the citizen, given a genuine choice, does prefer, and on how all of these may shift with the circumstances and experience surrounding the choice. In a time when many types of environmental change are little suited to precise definition or quantitative expression, when there are few market checks of value judgments, and when professional judgment obscures assumptions as to preferences, the future public management of the environment's human satisfactions has growing need for discovery of the delicate process by which individual preferences find their way into public choices of vista, taste, odor, and sound.

Environmental Perception and Its Uses: A Commentary

It is pertinent to ask how much and in what way the processes and consequences of land use or similar problems are better understood as a result of the studies since 1965, and what lessons have been learned as to the comparative merits and demerits of the available means of studying environmental perception. . . .

It is noteworthy that apparently there has been no systematic comparison by geographers of the results from applying and replicating a variety of methods to the same set of phenomena. Would the findings be greatly different, and if so how, were reliance placed on individual interviews, nondirective interviews, content analysis, or any of the score or more types of tools that might be applied? And how much confidence can there be in studies which have not been replicated? This suggests the second step that might now be taken toward a fuller appraisal of what has been learned about method since the environmental perception concept began to gain popularity. Until this is achieved the answer to the question of relative effectiveness will remain speculative.

(1984a), 94, 96

Notes

1. William Kirk, "Historical Geography and the Concept of the Behavioral Environment," *Indian Geographical Journal, Silver Jubilee Edition,* George Kuriyan, ed. (Madras: Indian Geographical Society, 1952), 152–60.

2. David Lowenthal, "Geography, Experience and Imagination: Towards a Geographical Epistemology," *Annals of the Association of American Geographers* 51, no. 3 (1961): 241–60.

3. J. K. Wright, *Human Nature in Geography: Fourteen Papers, 1925–1965* (Cambridge: Harvard University Press, 1966).

4. Clarence J. Glacken, "Changing Ideas of the Habitable World," *Man's Role in Changing the Face of the Earth,* William L. Thomas, Jr., ed. (Chicago: The University of Chicago Press, 1956).

5. Yi Fu Tuan, "Attitudes Toward Environment: Themes and Approaches," paper presented at the 61st Annual Meeting of the Association of American Geographers, Columbus, Ohio, April 20, 1965.

6. Ian Burton and Robert W. Kates, "The Perception of Natural Hazards in Resource Management," *Natural Resources Journal* 3, no. 3 (1964): 412–41.

7. Robert Lucas, "Wilderness Perception and Use: The Example of the Boundary Waters Canoe Area," *Natural Resources Journal* 3, no. 3 (1963): 394–411.

8. Thomas Frederick Saarinen, *Perception of the Drought Hazard on the Great Plains,* Department of Geography Research Paper no. 106 (Chicago: University of Chicago, 1966).

9. The subsequent evolution and direction of the field may be found in volume II, chap. 9 (Whyte).

10. *Quest for Quality:* U.S. Department of the Interior Conservation Yearbook (Washington, D.C.: Government Printing Office, 1965), 13.

11. Lynton K. Caldwell, "Environment: A New Focus for Public Policy?", *Public Administration Review* (1965): 138.

12. Herbert Simon and J. March, *Organizations* (New York: John Wiley and Sons, 1958).

13. William J. Gore, *Administrative Decision-Making: A Heuristic Model* (New York: John Wiley and Sons, 1964).

14. *Outdoor Recreation for America,* U.S. Outdoor Recreation Resources Review Commission (Washington, D.C.: Government Printing Office, 1962), 27.

15. *Quest for Quality, op. cit.,* 6.

16. The report currently is in preparation under the title of *Attitudes Toward Water: An Interdisciplinary Exploration.* Principal participants have been Fred L. Strodtbeck, Meda White, William Bezdek, and Don Goldhammer from the Laboratory of Social Psychology, and W. R. Derrick Sewell, David Czamanske and Richard Schmoyer, from the Department of Geography. Several of the cautions as to definition were suggested by Fred Strodtbeck.

17. David Lowenthal, "Geography, Experience, and Imagination: Towards a Geographical Epistemology," *Annals of the Association of American Geographers* 51, no. 3 (1961): 241–60.

18. Jerome S. Bruner and Cecile C. Goodman, "Value and Need as Organizing Factors in Perception," *Journal of Abnormal and Social Psychology* 42 (1947): 33–44.

19. Leon Festinger, *A Theory of Cognitive Dissonance* (New York: Harper and Row, 1957).

20. Robert Beck, "Spatial Meaning, and the Properties of the Environment," in *Environmental Perception and Behavior,* edited by David Lowenthal (Chicago: University of Chicago Geography Research Paper no. 109, 1967).

21. Lowenthal, *op. cit.*

22. Clarence J. Glacken, "Changing Ideas of the Habitable World," in *Man's Role in Changing the Face of the Earth* (Chicago: University of Chicago Press, 1956), 70–92; and "Man's Attitude Toward Land: Reflections on the Man-Nature Theme as a Subject for Study," in *Future Environments of North America,* the papers of the Conservation Foundation Conference of 1965 (New York: The Natural History Press, 1966). See also Alexander Spoehr, "Cultural Differ-

ences in the Interpretation of Natural Resources," in *Man's Role in Changing the Face of the Earth*, 93–102.

23. Yi-Fu Tuan, "Attitudes Toward Environment: Themes and Approaches," in *Environmental Perception and Behavior*, edited by David Lowenthal (Chicago: University of Chicago Geography Research Paper no. 109, 1967).

24. Bernard Berelson, *Content Analysis in Communication Research* (Glencoe: Free Press, 1952). David Czamanske applied this type of analysis to the Boulder data, using material printed in the *Boulder Camera* during one month and compiled by Mary B. White.

25. Ruth Miller Elson, *Guardians of Tradition: American Textbooks of the Nineteenth Century* (Lincoln: University of Nebraska Press, 1964), 39–40.

26. Douglas R. McManis, *The Initial Valuation and Utilization of the Illinois Prairies, 1815–1840* (Chicago: University of Chicago Geography Research Paper no. 94, 1964), 49–58, 89–95.

27. The Program Surveys were initiated by farseeing officials and social psychologists who sought to understand more precisely why farmers accepted certain government measures and rejected others, why some were concerned about eroding soil or sustained woodland management and others were not. Political exigencies of the period soon drove the effort into assessment of responses to international policies and wartime controls, and other data collected then has received little notice. It would merit reexamination for its revelations of attitudes prevailing in the 1940s and for its suggestions of factors which then seemed relevant. Comments on the work are to be found in Rensis Likert, "Opinion Studies and Government Policy," *Proceedings of the American Philosophical Society* (1948): 341–50.

28. Studies published by the National Opinion Research Center involving assessment of environment: Paul N. Borsky, *Community Reactions to Sonic Booms*, Oklahoma City Area, part 1, University of Chicago, National Opinion Research Center Report no. 101, 1965; Community Conservation Board, City of Chicago, *The Hyde Park–Kenwood Urban Renewal Survey*, University of Chicago, National Opinion Research Center Report no. 58, 1956; and Community Conservation Board, City of Chicago, *The Near West Side Conservation Survey*, University of Chicago, National Opinion Research Center Report no. 63-B, 1957.

29. Studies of the Survey Research Center, Institute for Social Research, University of Michigan, Ann Arbor, published by the Institute unless otherwise noted: Angus Campbell and Charles A. Metzner, *Public Use of the Library and Other Sources of Information*, 1950; Eva L. Mueller and Gerald Gurin, with Margaret Wood, *Participation in Outdoor Recreation: Factors Affecting Demand Among American Adults* (Washington, D.C.: Government Printing

Office, 1962); Eva L. Mueller, Arnold A. Wilken, and Margaret Wood, *Location Decisions and Industrial Mobility in Michigan,* 1961, 1962; John B. Lansing, Eva L. Mueller, William M. Ladd, and Nancy Barth, *The Geographic Mobility of Labor: A First Report,* 1963; John B. Lansing and Eva L. Mueller, with Nancy (Morse) Samuelson, *Residential Location and Urban Mobility,* 1964; John B. Lansing and Nancy (Morse) Samuelson, *Residential Location and Urban Mobility: A Multivariate Analysis,* 1964; John B. Lansing, *Residential Location and Urban Mobility: The Second Wave of Interviews,* 1966.

30. *The Future of Outdoor Recreation in Metropolitan Regions of the United States,* vol. I, *The National View—Present Conditions and Future Prospects of Outdoor Recreation for Residents of the Metropolitan Centers of Atlanta, St. Louis, and Chicago,* A Report of the Outdoor Recreation Resources Review Commission, Study Report 21 (Washington, D.C.: Government Printing Office, 1962). *The Quality of Outdoor Recreation: As Evidenced by User Satisfaction,* Report to the Outdoor Recreation Resources Review Commission, Study Report 5 (Washington, D.C.: Government Printing Office, 1962).

31. Nahrum A. Medalia, *Community Perception of Air Quality: An Opinion Survey on Clarkston, Washington,* Environmental Health Series (Cincinnati: U.S. Department of Health, Education, and Welfare, 1965).

32. *Public Awareness and Concern with Air Pollution in the St. Louis Metropolitan Area,* Public Administration and Metropolitan Affairs Program, Southern Illinois University (Washington, D.C.: U.S. Department of Health, Education, and Welfare, 1965).

33. Paul N. Borsky, *Community Reactions to Air Force Noise,* WADD Technical Report 60-689, parts 1 and 2 (Dayton: Distributed by U.S. Department of Commerce, Office of Technical Services, 1961).

34. E. J. Baur, "Opinion Change in a Public Controversy," *Public Opinion Quarterly* 26 (1962): 212–26.

35. Nathaniel Wollman, et al., *The Value of Water in Alternative Uses* (Albuquerque: University of New Mexico Press, 1962), 6–19.

36. Orris C. Herfindahl and Allen V. Kneese, *Quality of the Environment: An Economic Approach to Some Problems in Using Land, Water and Air* (Washington, D.C.: Resources for the Future, 1965), 29.

37. Gore, *loc. cit.* An interesting description of opinion in relation to land and water use organization is given in: Charles K. Warriner, "Public Opinion and Collective Action: Formation of a Watershed District," *Administrative Science Quarterly* 6 (1961): 333–59. He concludes, "the 'need' arises as the organization comes into being

and thus is as much a creator of the need as the need is the creator of the institution" (p. 358).

38. Robert C. Lucas, "Wilderness Perception and Use: The Example of the Boundary Water Canoe Area," *Natural Resources Journal* 3 (1964): 394–411.

39. Ian Burton and Robert W. Kates, "The Perception of Natural Hazards in Resource Management," *Natural Resources Journal* 3 (1964): 412–41.

40. Robert W. Kates, *Hazard and Choice Perception in Flood Plain Management,* Geography Research Paper no. 78 (Chicago: University of Chicago, 1962).

41. Thomas Frederick Saarinen, *Perception of the Drought Hazard on the Great Plains,* Department of Geography Research Paper no. 106 (Chicago: University of Chicago, 1966).

42. Meda M. White, "Role Conflict in Disasters: A Reconsideration," paper presented at American Sociological Association, 1962. See also Harry Estill Moore, and F. L. Bates, J. P. Alston, M. M. Fuller, M. V. Layman, D. L. Mischer, and M. M. White, *And the Winds Blew* (Austin: Hogg Foundation for Mental Health, 1964).

43. Studies published by the National Academy of Sciences-National Research Council, Disaster Research Group (formerly Committee on Disaster Studies), Washington, D.C.: Lewis M. Killian, *A Study of Response to the Houston, Texas, Fireworks Explosion,* Disaster Study no. 2, Publication no. 391, 1956; Anthony F. Wallace, *Tornado in Worcester: An Exploratory Study of Individual and Community Behavior in an Extreme Situation,* Disaster Study no. 3, Publication no. 392, 1956; Fred C. Ikle and Harry V. Kincaid, *Social Aspects of Wartime Evacuation of American Cities, with Particular Emphasis on Long-Term Housing and Re-Employment,* Disaster Study no. 4, Publication no. 393, 1956; George W. Baker and John H. Rohrer (eds.), *Symposium on Human Problems in the Utilization of Fallout Shelters,* Disaster Study no. 12, Publication no. 800, 1960; *Field Studies of Disaster Behavior: An Inventory,* Disaster Study no. 14, Publication no. 886, 1961; Raymond W. Mack and George W. Baker, *The Occasion Instant: The Structure of Social Responses to Unanticipated Air Raid Warnings,* Disaster Study no. 15, Publication no. 945, 1961; George W. Baker (ed.), *Behavioral Science and Civil Defense,* Disaster Study no. 16, Publication no. 997, 1962; F. Bates, *The Social and Psychological Consequences of a Natural Disaster: A Longitudinal Study of Hurricane Audrey,* Disaster Study no. 18, Publication no. 1081, 1963; Harry E. Moore, *Before the Wind: A Study of the Response to Hurricane Carla,* Disaster Study no. 19, Publication no. 1095, 1963.

44. Robert S. Ingolds, "Taste Test Taxes Theories," Engineering Experiment Station, Georgia Institute of Technology, Atlanta,

Georgia. Reprint no. 176 from 1964 *Water Works and Wastes Engineering*.

45. Joseph Sonnenfeld, "Variable Values in Space and Landscape: An Inquiry into the Nature of Environmental Necessity," *Journal of Social Issues* 22 (in press).

46. Florence Rockwood Kluckhohn and Fred L. Strodtbeck, *Variations in Value Orientations* (Evanston: Row, Peterson and Company, 1961), 1–48, 363–65.

47. Kluckhohn and Strodtbeck, *op. cit.*, 341–42.

48. Benjamin D. Paul, William A. Gamson, and S. Stephen Kegeles (eds.) "Trigger for Community Conflict: The Case of Fluoridation" (eight articles), *Journal of Social Issues* 17, no. 4 (1961): 1–81, quotation on page 7. See also: Fluoridation (special issue of 22 articles), *Journal of the American Dental Association* 65 (1962): 578–717; Robert M. O'Shea and S. Stephen Kegeles, "An Analysis of Anti-Fluoridation Letters," *Journal of Health and Human Behavior* 4 (1963): 135–40; Arnold Simmel and David B. Ast, "Some Correlates of Opinion on Fluoridation," *American Journal of Public Health* 52 (1962): 1269–73.

49. R. Burnell Held and Marion Clawson, *Soil Conservation in Perspective* (Baltimore: The Johns Hopkins Press for Resources for the Future, 1965), 254–62.

50. Robert W. Kates and J. F. Wohlwill (eds.), "Man's Response to the Physical Environment," *Journal of Social Issues* 22 (in press).

51. Everett C. Hughes, *Student's Culture and Perspectives* (Lawrence: University of Kansas School of Law, 1961).

52. William E. Henry and John M. Schlien, "Affective Complexity and Psychotherapy: Some Comparisons of Time-Limited and Unlimited Treatment," *Journal of Projective Techniques* 22 (1958): 153–62.

53. For a review of changes in attitudes toward culture groups, see E. E. Davis, *Attitude Change: A Review and Bibliography of Selected Research,* Social Science Clearing House Documents no. 19 (Paris: UNESCO, 1965).

54. "Alternatives in Water Management," National Academy of Sciences Research Report no. 1408, August 1966 (Washington, D.C.: National Academy of Sciences-National Research Council).

16 Strategies of American Water Management

In this excerpt from his volume of Cook lectures given to the Law School of the University of Michigan, White begins with his five major strategies of U.S. water management (listed in fig. 16.1), the evolution from single means–single purpose water management to the current multiple means–multiple purpose management. This attempt to place a structure of a simple means-ends contingency table over the history of water management has been well received and is widely used.[1] Whether it serves as a neat summary or implies an evolutionary process is never quite clear—another example of implied rather than explicit theory.

This, the concluding chapter of his lectures, draws heavily on his participation in two activities: the work of the Senate Select Committee on Water Resources and the Northeastern Illinois Planning Commission. In both of these activities, he correctly identified the critical role of projections and tried to direct these studies to project water "use" by considering realistic alternatives of development as well as unconventional solutions to both supply and demand problems.

VI. Regional Integration: Linear Projections and Finite Resources

The five major strategies of water management show signs of coalescing into a sixth and much more complex strategy, embracing features of each and thus far defying precise description. Vaguely defined, this transformation is in the direction of regional integration. Within the framework of either metropolitan areas or groups of drainage basins the public agencies are groping toward an illusive

Condensed from *Strategies of American Water Management* (Ann Arbor: University of Michigan Press, 1969).

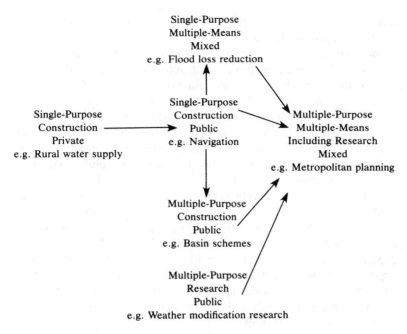

Single-Purpose
Multiple-Means
Mixed
e.g. Flood loss reduction

Single-Purpose
Construction
Public
e.g. Navigation

Single-Purpose
Construction
Private
e.g. Rural water supply

Multiple-Purpose
Multiple-Means
Including Research
Mixed
e.g. Metropolitan planning

Multiple-Purpose
Construction
Public
e.g. Basin schemes

Multiple-Purpose
Research
Public
e.g. Weather modification research

Figure 16.1 Schematic diagram of changes in water management strategies.

style of planning which would continually explore multiple means of reaching multiple goals. One way of stating the trend and the prospect is to say that the traditional analysis of regional opportunities in terms of linear projections of demand leading to long-range plans of public agencies to regulate a finite water supply is moving toward appraisal of a wider range of alternatives, including scientific research, and that the outcome is not a single 50- or 100-year plan but a set of guidelines within which short-term projects by both public and private agencies may be undertaken. While meeting certain of the objections to multiple-purpose public construction, a serious effort at regional integration invites confusion as to administration by drawing in private managers and a large number of local governments and relies more upon public education than upon neat administrative organization to provide continuity and aim.

Linear projection of demand against a finite resource symbolizes a main theme of American water management as it has unfolded in single and multiple-purpose construction programs. To depart from that approach is to invite complexity of analysis, more flexible plans, more systematic use of scientific research, and greater diversity and looseness in administration.

Single-purpose construction by private managers tended to move to either public single-purpose construction or public multiple-purpose construction (fig. 16.1). Navigation, irrigation, and flood control followed the path of single to multiple-purpose construction; hydroelectric power, while predominantly single-purpose private, moved directly to public multiple-purpose construction. Municipal water supply and waste disposal are edging in the same direction while showing an interesting diversion toward canvassing multiple means. A few of the changes were in the opposite direction, as with a reversal of irrigation development toward single-purpose private construction. The shift to single-purpose multiple-means in the case of flood loss reduction inevitably brought a mixture of public and private activity. The effect of applying public research to water management was largely to refine particular techniques to serve multiple purposes. However, both the strategy of multiple means and of multiple-purpose research when added to multiple-purpose construction so expand the character of drainage basin planning as to suggest that it will be basically different.

A Substitute for Multiple-purpose Construction?

While there is no clear outline of what this new strategy may be as it looms among gigantic basin plans, a few of its charateristics emerge from the theoretical relation of the several strategies and from planning that is under way. Forms it is denying may be easier to discern than forms it is achieving.

Because of the growing concern for exploration of alternative means and for continuing assessment of the nature of public preferences, the fashioning of a discrete, long-term plan for a basin is fading in prominence. Like the once-popular master plan for urban growth, the comprehensive plan for river basin development loses allegiance among its practitioners. It should do so if for no other reason than the rapidly changing technical suitability of engineering works by comparison with the long time horizons required for their economic justification. To speak of a system of storage reservoirs on the Potomac River costing at least $160 million as justified by benefits of low water regulation flowing over 50 years when the pace of improvement in waste treatment and the spatial pattern of urban living promises radical changes in 20 years is to run the risk of built-in obsolescence.

There also seems little doubt that while the complexity of problems canvassed calls for a stronger coordinating role by one agency—the Corps of Engineers, the Bureau of Reclamation, or a new drainage basin commission—it also prohibits vesting full public responsibility for planning in any such entity. The diversity of interests drawn together in water planning for any one area precludes their being served adequately by the operations of one agency, particularly as the range of devices considered by public agencies is widened. As means of coping with water problems multiply, the idea of finding a solution by a single construction program seems less and less suitable.

It is especially cumbersome to translate a program relying on construction plans and water allocations into an interstate compact which will satisfy national goals and regional equities, or to settle interstate allocations by going to court. The Great Lakes Diversion case is an example of a tragic waste of public funds and time centered on a single engineering solution. Instead of building up elaborate evidence for or against a relatively small diversion of Lake Michigan water to supplement Illinois River flow asserting a local proprietary control of the nation's water, the cities and states of the basin could have joined in exploring the full range of possible ways of handling waste and regulating water quality in the Lakes system. (An abortive attempt was made to bring the four Lake Michigan states together in 1929.) Ultimately, they will be obliged to do this cooperatively.[2] Meanwhile, the wisest judicial decision is a poor substitute for joint appraisal of the technical, economic, and administrative alternatives open to them.

Another way of describing the type of strategy which might replace multiple-purpose construction is to outline trends in aspects of the decision process. Aims have become progressively more numerous and ambiguous. When means have multiplied they have sharpened the recognition of aims, and have tended to develop around limited purposes. During the expansion of multiple-purpose public construction the time horizons lengthened, but with more consideration of alternative means and of scientific research as one of those means, the horizons have begun to shorten. From a situation in which heavy risks were taken in private resource management, there has been a gradual reduction in hazards of shortage and excess, accompanied in recent years by doubts as to the degree that reliance should be placed upon engineering measures in contrast with other methods of spreading risk. The spatial extent of planning has spread from local navigation and irrigation projects to stream reaches, entire basins and then groups of basins, but with attention to economic analysis and to multiple means the scope has enlarged on the one hand to consideration of genuinely national efficiency and has contracted on the other hand to assessment of the complex of possible measures within metropolitan areas. Concern for recognition and measurement of environmental impacts has increased at an accelerating pace. The long-term trend toward heavier public investment has been modified modestly by increasing attention to specialized and dispersed management, such as nuclear power and household waste disposal.

A strategy which meets the principal objections to multiple-purpose public construction in a theoretically viable fashion would have the following characteristics. Its aims would be multiple and consciously recognized as evolving with public preferences which in themselves would be partly shaped by the process followed. Its means would be multiple, and would take account of a full range of alternatives, including scientific research as a tool for devising new technologies affecting both demand and supply. For that reason, it would utilize a mixture of public and private administrative instruments, encouraging as much decentralization of choice among individuals and local agencies as consistent with broad guidelines supported by public consensus. Standards for water quality, hazard reduction, and social valuation would not be rigid, but decisions would be based on criteria of keeping the range of choice as wide as practicable and of working toward short-time horizons within frameworks describing long-term human needs and physical limits. There would be intensive investigation of resources and the theoretical possibilities and social consequences of altering them. Because the populations served are likely to be in dispersed metropolitan areas and because the locus of political decision will rest there, the planning would need to take account of metropolitan organization. It would be alert to the distinctive hydrologic unity of drainage basins and aquifers, but would not attempt to conform the social process of choice to the physical entities of watershed lines. Whether or not such a strategy may be achieved is conjectural. However, several trends support it.

With the spread of sprinkler irrigation, the diversification of flood loss reduction measures, the increasing independence of residential and industrial water installations, the creation of drainage basin commissions such as the Delaware, the launching of special surveys such as the Northeast Water Supply Survey, and the prominence of cities in regional water controversies, as in the Great Lakes, the direct role of individual and local managers in water planning is more explicit. The part played by the Los Angeles metropolitan area in exploring opportunities for linking desalting with nuclear reactors is symptomatic of growing organization of water management in the expanding structure of metropolitan areas. While a few years ago the negotiation of interstate arrangements for water was thought of primarily as a state responsibility, the reapportionment of legislative votes among urban and rural constituencies, the strength of urban planning agencies, and the declining relative importance of agriculture have raised the importance of the city clusters in dealing with other states as well as with federal agencies.

Indicative of a radical departure from the tradition of authorizing specific water projects to meet vaguely stated needs was the Appalachian program of 1965. Water development was treated as only one of several activities, including highway and education development, which might serve a broad aim of income redistribution in assisting a chronically depressed area.[3] Not only did the customary goal of economic efficiency as measured by national productivity give way to redistribution aims similar to the authorization of special irrigation projects in drought-stricken areas during the late 1930s, but the evaluation of projects was seen as comparing water with other types of investment. . . .

Attitudes Toward Nature

At base, this emerging strategy reflects a shift in man's attitude toward nature and his concomitant role in society. The view of man the transformer or man the conqueror that permeates so much of the single and multiple-purpose construction and that shows itself in the chart of the future as a contest between rising human demands for water and bounded natural supplies is replaced by another. In the view of man as the cooperator, man the harmonizer, construction is only one means of coming to terms with an environment he never fully explores and that is constantly changing under his hand. With the adoption of this view, the means and instruments of handling water become increasingly complex, the concern with tracing environmental impacts more acute, the adjustments to human preferences increasingly sensitive, and the demand for citizen participation heavier. The emphasis shifts from construction to scientific probing, and from long-term commitment to short-term flexibility.

Water and Choice in the Colorado Basin

The myriad alternatives possible in the Colorado Basin make clear that there is no single panacea for regional economic development or even for water development. Building dams or "making the desert bloom" by bringing in additional water are no more than alternatives. The best cure for a threatening water shortage is not necessarily more water; savings in water use, or transfer of water use to less-consumptive, higher-yield applications, or discovery of new techniques of water management may offer better solutions. Indeed, if objectives are clarified, water development *per se* may not be the desired solution. It is time for the nation to draw upon its great reserve of scientific capability and consider how it can best meet the different objectives that people seek, instead of expecting new water projects to be the solution to all water problems.

(1968c), 100

Notes

1. See the discussion below, vol. II, chap. 5 (Day et al.).
2. Since the lecture the Supreme Court has released the report of Albert B. Maris, its Special Master on *Wisconsin, Minnesota, Ohio and Pennsylvania v. State of Illinois and the Metropolitan Sanitary District of Greater Chicago, December 8, 1966* (Philadelphia: Legal Intelligencer). This is an admirable review of the points at issue in the case.
3. U.S., 90th Cong., 1st sess., *Report of the Committee on Public Works on Revising and Extending the Appalachian Regional Development Act of 1965*, Senate Report 159, 1967, 14–15. The Corps of Engineers is given the difficult task of harmonizing regional development and national river basin planning.

I7 Preface to the Second Edition of Integrated River Basin Development

In this preface to the second edition of the Report of a Panel of Experts on *Integrated River Basin Development,* White took the opportunity to update information on international river basin development in the ten-year interim since his last report.[1] He brings to it the enrichment of his experience as advisor on man-made lakes to the United Nations Development Program and as consultant on the Lower Mekong river basin development. On the international level, this brief report continued his long-term interest in the three types of integrating regions—river basins, arid lands, and metropolitan regions—and introduced to international audiences some of the thinking evident in previous papers on the American scene. Thus it sums up what he considers important to share internationally: new perspectives on development and environment, new research methods and technologies, and new international administrative arrangements.

During the decade that has passed since the publication in 1958 of the report of a Panel of Experts on *Integrated River Basin Development*[2] the pace of public investment in water management has accelerated and United Nations involvement in promoting both national and international action in river development has expanded widely. Even in the light of that mounting experience with studies, construction and operation of works on the world's rivers, much of the 1958 report still seems sound and timely. However, the march of science, the increasing demands for water and the growth of government commitment make it seem likely that if a new Panel

From *Integrated River Basin Development,* 2d ed. (New York: United Nations, 1970). The writer wishes to acknowledge the help he has received from a number of experts within and outside the United Nations Secretariat who have shared with him their current appraisal of the 1958 report.

were to be convened it would place different emphasis on some points and add others. The events of the decade suggest that several fresh aspects of water policy would be stressed, that certain recent advances in scientific research and training would be noted and that account would be taken of major changes in the activities of the national and international agencies involved in water management.

That most of the suggestions in the Panel's report seem appropriate ten years later testifies to the breadth of the earlier appraisal of river basin problems. The changing emphasis and activity reflect basic changes in the technical tools of water planning and in the views of the role of such work in advancing human welfare.

As with most national ventures in river basin development, there is relatively little critical appraisal of work accomplished in terms of the economic returns, engineering reliability or the effectiveness of administrative organization. Each regional economic commission of the United Nations seeks to keep informed of the efforts at water planning in its region, to assess the availability of basic data and to promote the exchange of experience among the member countries.[3] Beyond a few assessments of completed projects, descriptions of projects such as the Kitakami in the area of the Economic Commission for Asia and the Far East,[4] and a series of evaluations by the International Bank for Reconstruction and Development of selected power and irrigation enterprises financed by the Bank,[5] the emphasis is heavily on the planning of new ventures rather than on lessons from the past. For this reason much of the judgment offered by the Panel still can be compared only with the judgment of other experts: systematic examination of the actual results of earlier work on river development is largely lacking.

New Perspectives

One major shift in perspective of river basin planning has to do with the view of what constitutes integrated development. In many parts of the world attention continues to center on single-purpose control of water for hydroelectric power, irrigation, navigation or domestic supply, but where multiple purposes are sought there is a tendency to broaden consideration of the means taken to reach those ends. Human factors loom larger than heretofore.

As the Panel noted, engineering measures are not likely to bring the desired improvements in level of living unless they are accompanied by secondary measures affecting other aspects of resource use. The essential storage and canal facilities of an irrigation project must be supplemented by alterations in credit, marketing, transport, fertilizer, seed supply and similar services if they are to bring genuine gains in farm production. An electric power generator and transmission line must be tied into adequate facilities for distributing power to the customers. The importance of the secondary measures is emphasized by the unhappy experience of completed projects which failed to provide expected benefits. In a larger sense, however, water control activities are seen as only one aspect of natural resource development, and it is recognized that there may be other, more effective ways of promoting economic growth and social welfare than by storing or conveying water.[6]

When the enlarged perspective is adopted, the process of water planning changes. It becomes more important to relate basin plans to the national economic plans, as in the Lower Mekong where efforts are made to have regular consultation among the international planning group and the central planning agencies of the member countries.[7] A scheme for waterway improvement is examined in comparison with other possible programs to provide low-cost and effective transportation, as in the Soviet Union.[8] A proposal for reducing water pollution by stream flow dilution is compared with possible measures for additional waste treatment, for diversion of the waste effluent or for oxygenating the stream, as in the Potomac Basin. A national program for reducing flood losses, as in the United States of America[9] and in Japan, is designed to include, along with reservoir and embankment protection, measures to promote flood-proofing, emergency removal, land use planning, flood insurance and flood hazard information. Under this view of water planning as encompassing analysis of alternative means of reaching similar economic, social or political goals, greater weight is given than before to defining precisely what goals are held and to canvassing a wide range of water management and other development techniques. So far, the concrete examples of this approach are few, but the interest is wide and the necessity for giving it more attention is clear. Measures to prevent floodplain encroachment and heavy pollution loads from expanding also may be important in developing countries.

A second major shift in perspective is towards keener recognition of the full network of ecological impacts incurred by construction of water projects. These impacts include the effects of storage impoundments and water diversion on resettled human populations, terrestrial and aquatic ecosystems, stream sediment, groundwater supplies, disease vectors and water-borne diseases, as outlined in the earlier edition of the report.[10] With the initiation of many huge reservoirs, particularly in tropical areas, the potentialities of those projects both to enhance and injure the life around them, quite aside from their primary purposes of power generation and flow regulation, have claimed the interest of a wide number of scientific disciplines.

Drawing upon the experience with reservoir construction in Europe, Asia and the Americas, a series of investigations is under way on the Kariba, Volta, Kainji and Sadd El Aali (High Aswan) projects to find the probable consequences of those new man-made lakes and to explore methods of managing them so as to increase social returns and minimize their social costs. Ways of enlarging fish production, cultivating the seasonally flooded reservoir margins, preventing the spread of schistosomiasis, and curbing the social dislocations from population resettlement are among the aims of the investigations carried out by the Food and Agriculture Organization of the United Nations and the World Health Organization under financing through the United Nations Development Programme. The problems of maintaining public health in these conditions are outlined in a report presented in annex 4 by the World Health Organization and the Food and Agriculture Organization of the United Nations. Problems of life in the lakes and on adjoining lands are examined in the selections from the report by the Food and Agriculture Organization in annex 5.

It seems increasingly clear that many of these problems could be minimized by more careful study while the projects are being planned, and this suggests the need for a kind of early warning system which would alert the competent agencies to the possible hazards before construction gets under way. Such warning may not in itself suffice to prevent later difficulties, for the press of time and shortage of funds may lead to neglect of desirable supplementary investigations even when they have been prescribed in advance. Thus, the monumental preliminary studies for the Volta River project identified questions of health, fisheries, and resettlement which were given inadequate attention during the construction phase because political negotiations to finance the project took several years and the government then became preoccupied with the urgency of building power generation and aluminum-smelting facilities.[11]

The past decade has also seen a pronounced change in public concern for reducing the growing pollution of streams from the wastes of city, farm and factory.[12] As pollution loads increase through rising population, new agricultural technologies and complexity of industrial processes, and as the standards of public health and of recreational and aesthetic uses of water are raised in industrial countries, the demands on water management schemes to take account of opportunities to eliminate, dilute or treat effluents become more exacting. These demands show themselves in enlarged attention to pollution abatement in basin development schemes, and in strengthened national programs to cope with pollution problems. In Europe, waste management is strongly linked with plans for industrial and urban expansion. In North America, it is closely related as well to the restoration or maintenance of a natural habitat for recreation and aesthetic enjoyment. In developing countries, such considerations may appear less urgent by comparison with economic growth, but they are commanding increasing attention.

Where water management is not restricted to single-purpose projects, the tendency continues to be to seek to deal with river basins as a whole.[13] Often, this is honored more in theory than in practice. However, there are concurrent tendencies to investigate river development within the framework of the interests of major metropolitan and industrial areas or within areas having common groundwater resources. Improved methods of prospecting for groundwater, pumping it and artificially recharging it have enlarged the opportunity to plan for ground and surface supplies jointly. Thus, unified treatment of the Khabour Basin in Syria requires management of both.

The Dubrovnik resolution of the International Law Association of 1956 was revised in 1966 at Helsinki to restate the general principles that were then regarded by the Association as deserving recognition by nations entering into action on international streams.[14] The Helsinki Rules are reproduced as annex 7 to this report. Among other changes from the earlier version, they recommend the use of the term "international drainage basin" in place of "international river" and, in the spirit of the Panel's report, define the basin as including the whole system of waters—surface and underground—flowing into a common terminus.

Scientific Research, Assistance and Training

To an increasing degree the international scientific community is mobilizing to probe and to train personnel to assess the basic and global problems of water behavior. The most comprehensive of the efforts to enlarge fundamental knowledge about water on the earth is the International Hydrological Decade launched in 1965 under the auspices of the United Nations Educational, Scientific, and Cultural Organization (UNESCO) with the participation of other United Nations agencies and the International Council of Scientific Unions. Covering the whole gamut of questions from precipitation, evapotranspiration, groundwater distribution and chemical quality to watershed relationships and world water balance, the International Hydrological Decade is making available scattered data and research findings that will facilitate the study of individual river basins and the identification of regional or world trends in quantity and quality of the basic resource. Its publications vastly extend the range of scientific problems listed by the Panel in annex 4 of its report, and that annex has been omitted from the present edition. Individual nations gain from the Decade in the stimulation it gives to the improvement of data collection and research and in the information they are able to draw from other areas.

The United Nations specialized agencies extend their activities to foster international collaboration in several of the fields essential to sound water development. The World Meteorological Organization is involved in hydrometeorology and networks for surface hydrology.[15] The Food and Agriculture Organization of the United Nations promotes cooperation in studies of aquatic biology, watershed influences and, in cooperation with UNESCO and the World Meteorological Organization, agro-climatology.[16] The United Nations Educational, Scientific, and Cultural Organization sponsors collaboration in basic investigation of geology, geochemistry, landforms, pedology and ecology. It has experimented with the design and conduct of integrated surveys for small areas and with the stimulation of research stations relating to broad problems of arid and humid environments. The World Health Organization gives detailed support to problems of epidemiology and control of water-related disease. The United Nations takes special responsibility for studies of groundwater and of the costs and value of water in different uses.[17] The World Health Organization has launched a spearhead program to improve community water supplies.[18] Two of the surveys under the Secretary-General's program on the development of natural resources deal with the potential for development of international rivers and with the needs and resources in potentially water-short developing countries.[19]

The United Nations and all agencies with responsibilities in water resources development enter into training programs to enhance the quality of technical personnel for water management. It is in the middle levels of the data collection and study agencies of developing countries that the shortages are most acute, and bilateral aid has been directed at remedying some of them.

The net effect of extensions of specialized services is to make available to interested nations, or groups of nations having a river basin in common, the results of data collection and research from other areas, technical assistance in designing their own activities and aid in training personnel to make basic measurements, analyze the data, and join in research on unsolved problems. Thus, both data and professional skill are more widely shared than was the case a decade ago.

Techniques

Among the several advances in techniques for handling river development perhaps none is more significant than the refinement of methods for appraising the social consequences of development. Economic analysis provides somewhat more rigorous ways than formerly were available to examine prospective flows of gains and losses, and to compare them for suitable time periods and discount rates.[20] A new example of national evaluation criteria is given in annex 3. While much of the sophisticated analysis is aimed at tests of contributions which water projects would make to national economic efficiency, it permits helpful calculations to be made in several other directions.[21] By estimating more accurately the benefits and costs of a given scheme in efficiency terms, public bodies are enabled to count the cost of adopting other schemes which may be less attractive in economic returns but more attractive from the standpoint of satisfying particular groups within the society, as when a project is located where it has greatest political support rather than greatest return. By quantifying the future gains and losses from relatively intangible uses, such as recreation or wildlife conservation, improved economic analysis may give a rough measure of their significance.

Such economic analysis may not be heeded by the government authorities, who may place political aims above economic efficiency. Even where it is taken seriously as a basis for judging the wisdom of heavy public investment, as in the case of the Zuider Zee scheme in the Netherlands, it may lead to diverse policies for financing and reimbursement; it may suggest placing the repayment burden on readily identified beneficiaries; or it may show that the gains are so widely shared as to warrant charging the costs to general tax revenues.

Whereas ten years ago the process of comparing the tentative estimates of probable economic effects of various alternative schemes for management of a river basin was highly cumbersome and time consuming, computer capacity and simulation models make it possible to examine thousands of schemes that differ in location, type and rules for operating the proposed structures. Such comparison of a huge number of proposals for one river basin can only be as precise as the basic data and the assumptions as to value and social aims; but rough as it is, it can encourage sound assessment of the perceived possibilities and aid in selecting individual projects for detailed examination.

In addition to improvements in techniques for mass earth moving and for long-distance transmission of electric energy, several advances in technology over the past decade have altered the prevailing ideas as to the practices that can be applied in basin development. One advance is the perfection of waste-water treatment processes to the point where the effluent from an urban sewage disposal plant can be returned in a quality suitable for human consumption. By these means the technical possibilities of preventing downstream pollution as well as of reusing water in areas where natural supplies are meager are increased immensely.

A second advance is in the methods of pumping and distributing water for overhead irrigation, a technique that enables many individual farmers to practice supplementary irrigation without depending upon heavy capital investment in common storage and canal facilities and that reduces the cost of land preparation.

Desalting of brackish water is the subject of elaborate research and development in a number of countries: in recent years more efficient and cheaper devices have been developed, but even when the desalting process is combined with use of waste heat from large nuclear reactor power-generating stations, the unit costs of producing water are barely within the price range of supplies for municipalities and thus are far above current prices for delivered irrigation water.[22] The earlier estimates are brought up to date in annex 6.

Sufficient improvements have been made in weather modification so that in certain favorable sites with orographic precipitation, annual rainfall may be augmented as much as 15 percent by cloud seeding.[23] This is not generally practicable, however, and the social complications are formidable.

Numerous other technical advances extend man's capacity to deal with water. Automated gauges, water quality monitoring and flow forecasting centers enhance hydrologic operations. New means are available to reduce water losses in agricultural, industrial and domestic uses.

Quite aside from the expansion of scientific assistance by the specialized agencies noted above, the organizations capable of dealing with river basin planning, construction and operation have increased notably, particularly at the international level.

At the national level, both old and young countries tend to add to the numbers of agencies responsible in some fashion for water management, without moving to the device of comprehensive regional authorities or to national organizations encompassing all water-related activities. Valley authorities, where they are tried, do not spread to more than one for a country. Under its new legislation, the United Kingdom of Great Britain and Northern Ireland provides for a national board treating with individual drainage basins as units. In the Federal Republic of Germany the emphasis shifts to maintenance of water quality, and the areas of study are those having common problems of industrial development. Similar changes are occurring in other western European countries. In the United States of America, a new water pollution control agency has been established and a Water Resources Council set up to coordinate the efforts of the special-purpose agencies. The Council encourages the joint undertaking of broad regional water studies and the organization of regional commissions having state and federal representation.

Among the younger countries, administrative organization continues to be largely for stated functions of power, irrigation, navigation, and water supply. In a few cases, such as the Dez Valley of Iran and the Comisión Coordinadora de los Proyectos Multiples para la Gran Lima in Peru, a single agency has been given major responsibility, but the prevailing pattern is one of basin study under a national ministry which handles some but not all of the water-related functions. United Nations assistance is given to countries, such as Afghanistan, seeking unified water policy and administration. Some of the reasons for the persistent emphasis upon single-purpose administration in many areas were suggested by the Panel and still seem valid.

At the international level, an impressive number of new organizations now operate to support river basin development. Beginning with the establishment of the Committee for Coordination of Studies of the Lower Mekong in 1957, a series of basin-wide efforts has taken shape under United Nations auspices. The Lower Mekong Committee acts under authority of a treaty among the riverine nations of Cambodia, Laos, Thailand and the Republic of Vietnam, with staff provided initially by the Economic Commission for Asia and the Far East, and with support for its studies and recent construction coming from outside donor nations as well as from the United Nations Development Programme. Similarly, cooperative investigations are under way in the Senegal Basin and the Chad Basin. In each case, basic support is provided through projects financed by the Special Fund component of the United Nations Development Programme and executed through the United Nations, the Food and Agriculture Organization, and other specialized agencies. Special studies have been undertaken by the United Nations in the Logone and Mono Basins of West Africa.

A joint hydrometeorological investigation by the World Meteorological Organization in the drainage area of the Nile above the Sudan is collecting data that will be essential to any later, more detailed planning for that part of the basin. Other hydrological work is in progress under the United Nations Educational, Scientific, and Cultural Organization in the Upper Paraguay Basin, and a larger study is taking shape in the Plata Basin. The Economic Commission for Europe has established a body to deal with regional water resources and pollution control problems. All of these have in common the assembly of basic facts and understandings that will assist in later development decisions.

During the same period the international agreements affecting river development have increased in several critical areas. The Rhine Treaty of 1963 has added regulation of water pollution among five nations to the control of navigation on the waters of that basin. Treaties, chiefly relating to pollution abatement, are in effect for the Drava, Lake Constance, Lake Geneva and Moselle drainage areas. Canada and the United States of America operate in accordance with a new joint program of development for the Columbia Basin. The grounds for cooperation in navigation and transport were established in the Niger Basin in 1964 with the assistance of the United Nations.

The Indus waters treaty brought a legal and engineering solution to a vexing problem created by the partition of the subcontinent. Yet, while it divides the waters of the Indus Basin to the satisfaction of India and Pakistan, it avoids integrated construction and operation.

In these international ventures, in addition to the regional economic commissions referred to above, two United Nations financing agencies play dominant roles. At the preliminary investigation and reconnaissance stages, the United Nations Development Programme (UNDP) is highly influential in funding and shaping the character of the survey work. Where completed projects run into serious problems of coping with their secondary impacts in the ecology and human organization of adjacent areas, the UNDP provides essential assistance.

In implementing the construction and operation of major projects, the International Bank for Reconstruction and Development frequently assists through its studies of particular projects or of national economic development programs, and through provision of loans for constructions. Whereas many projects are financed through government borrowing in the bond market or through bilateral loans and grants, the Bank affects the process of economic and financial evaluation in a powerful way by virtue of the standards it sets.

In general, national administrative capabilities for dealing with river problems are increasing, and the international machinery to facilitate both national and international basin studies is being strengthened. A coordinating committee of the United Nations organizations meets annually, and while the Department of Economic and Social Affairs provides the Secretariat for those meetings, there is as yet no special office with the full duties recommended by the Panel.

As indicated, the annexes to the report have been changed in this edition to reflect the new developments. The body of the report remains the same except for minor editorial changes.

Evaluating the Consequences of Water Management Projects

One effect of benefit-cost analysis is to give any respectable engineer or economist a means for justifying almost any kind of project the national government wants to justify. (You tell me the project you want, and I'll give you a favorable benefit-cost analysis.) It's thrown a cloud of respectability over sets of analysis which have ignored whole sectors of impacts and have misread other aspects, all in the most earnest, conscientious form. It's made it possible for us, as a technological fraternity, to feel clear and easy in mind in dealing with the political officials in saying this is a good project and has a benefit-cost ratio exceeding unity. Exclusive reliance on benefit-cost analysis has been one of the greatest threats to wise decisions in water development. I wouldn't eliminate it; but I would try to see it in proper perspective in terms of other sorts of guidelines.

Unpublished paper, Columbia University, 21 March 1971, 9

Notes

1. See the discussion below, in vol. II, chap. 5 (Day et al.).
2. *Integrated River Basin Development,* United Nations publication, Sales no.: 58.II.B.3 (New York: United Nations, 1958).
3. See, for example, reports on *Major Deficiencies in Hydrologic Data in Africa* (Geneva: WMO and Economic Commission for Africa, 1966); and on *Multiple-purpose River Basin Development* (Bangkok: ECAFE, 1955, 1956, 1957 and 1960).
4. United Nations, *A Case Study of the Comprehensive Development of the Kitakami River Basin,* ECAFE Flood Control Series, Sales no.: 62.II.F.7.
5. John A. King, Jr., *Economic Development Projects and their Appraisal: Cases and Principles from the Experience of the World Bank* (Baltimore: Johns Hopkins Press, 1967).
6. *Water and Choice in the Colorado Basin: An Example of Alternatives in Water Management* (Washington, D.C.: National Academy of Sciences, 1968).
7. See the annual reports of the Committee for Coordination of Investigations of the Lower Mekong River, Bangkok.
8. A. A. Mitaishvili, *Economic Indices and Advantages of Inland Water Transport of the USSR, Its Place in the Single Transport System and Conformity with Plans and Tasks of the National Economy,* United Nations Symposium on Inland Water Transport, Leningrad, 1968.
9. *A Unified National Program for Managing Flood Losses,* House document no. 465, 1966, 89th Cong., 2d sess., Washington, D.C.
10. R. H. Lowe-McConnell, editor, *Man-made Lakes* (New York: Academic Press, 1966).
11. *The Volta River Project: Report of the Preparatory Commission* (London: HMSO, 1956).
12. President's Science Advisory Committee, *Restoring the Quality of Our Environment* (Washington, D.C.: Government Printing Office, 1965). See also *Waste Management and Control* (Washington, D.C.: National Academy of Sciences, 1966).
13. Ludwik A. Teclaff, *The River Basin in History and Law* (The Hague: Nijhoff, 1967).

14. International Law Association, Report of the 52d conference held in Helsinki, 1967, 484–85. See also *Legal Problems Relating to the Utilization and Use of International Rivers* (A/5409, 15 April 1963).

15. United Nations, *Hydrologic Networks and Methods,* Bangkok, WHO and ECAFE. Sales no.: 60.II.F.2.

16. *An Agroclimatology Survey of a Semiarid Area in Africa South of the Sahara: FAO/UNESCO/WMO Interagency Project* (Geneva: WMO, 1967).

17. United Nations, *Large-scale Ground-water Development,* Sales no.: 60.II.B.3.

18. Bernd H. Dieterich and John M. Henderson, *Urban Water Supply Conditions and Needs in Seventy-five Developing Countries* (Geneva: WHO, 1963).

19. *Development of Natural Resources: Implementation of a Five-year Survey Programme, Report of the Secretary-General* (E/4302, 1967).

20. Arthur Maass et al., *Design of Water-resource Systems* (Cambridge: Harvard University Press, 1962).

21. United Nations, *Manual of Standards and Criteria for Planning Water Resources Projects,* ECAFE Water Resources Series, Sales no.: 64.II.F.12.

22. United Nations, *Water Desalination in Developing Countries,* Sales no.: 64.II.B.5.

23. Special Commission on Weather Modification, *Weather and Climate Modification* (Washington, D.C.: National Science Foundation, 1966). See also a report of the same title by a committee of the United States National Academy of Sciences, 1966, and World Health Organization.

18 Unresolved Issues

This paper, like the previous one, provides an occasion for retrospection, written fifteen years after White organized and edited the first international conference on arid lands research and management (1956a). In this later volume, for which this paper serves as conclusion, he considered the changes over the fourteen intervening years. Some of these appear in the retrospective on river basins—interests in a broadened view of development, concern for the environment, new tools and techniques, new international administrative arrangements and initiatives. But at a gross level, arid lands are poorer in renewable resources than humid area river basins, although they overlap—thus White introduces issues in arid land development that affect population, urbanization, food, and poverty. As a conclusion for the book, he dips into the many chapters, pulling forth the key and unresolved issues.[1] But perhaps the most interesting note is a brief aside. Arid lands, like polar regions and perhaps the high mountains, have a mystique that is lacking in river basins, one that draws together its students in ways enabling them to transcend the narrowness of discipline and the provincialism of national origin.

Each person from the twenty-three nations represented in the assessment of experience with arid environment that is reported in this book is bound to have his own, unique set of impressions and illuminations. The test of the long-range effect of the reports and recommendations recorded here will be in the degree to which these new views are translated into wise action in desert laboratory, university classroom, grassland range, or project-planning office. Immediately, three questions can be asked. What changing aspects of the world now seem to be especially pertinent to the future development of arid lands? Where did man in 1969 stand in his efforts to

Reprinted from *Arid Lands in Transition,* Publication no. 90 (Washington, D.C.: American Association for the Advancement of Science, 1970), 481–91. ©1970, American Association for the Advancement of Science.

live in, and maintain, these lands? What basic issues still seem to be both broadly significant and unresolved?

Any scientist who is concerned with the resources of a developing land can conduct the following experiments. Go to the new university and ask the students individually "What are you doing here?" One replies, "I am preparing to hold a job." Another says, "I am learning to be a scientist." Still another says, "I am helping to build a nation." Obviously, the individual's view of his role in the society of which he is a part shapes his mode of action. And it is becoming more and more apparent that the way he perceives the natural and cultural world around him is not simply a matter of the facts that are available to him in a government report.

Go to dwellers on the floodplains of a fluctuating stream and ask, "What is the risk of flood?" One replies, "There will never be another; things are better now." Another may say, "The floods are getting worse." Another says, "We have floods every seventeen years, like locusts." Still another declares, "My house has no hazard at all; I am selling it to someone else next week." Most of the people who are familiar with the semiarid lands know the climatic assessment that is attributed to a Great Plains wheat farmer, "We have had three good rainfall years: 1917, 1947, and next year."

On the basis of the same scientific evidence, different people perceive the risks and opportunities differently. Previous experience, age, their sense of efficacy in being able to cope with the hazard, and other factors affect what is perceived.

Similarly, the views of scientists and of government administrators regarding arid-lands resources and opportunities may vary according to their worldviews, their reading of the earthy record, and their feeling of competency in coping with the unforeseen.

Aspects of the Changing World

Our perception of the continents of which arid lands form a sparsely and spottily settled one-third of the area is changing rapidly, subtly, and profoundly. In comparison with the situation fourteen years ago, when a previous international group gathered at Albuquerque and Socorro, New Mexico, and reported on the status of arid lands (1956b), we see the world differently, and it is different. In at least eight ways, the global conditions that affect a land-use plan for Tamanrasset or a research program at Jodhpur or an irrigation scheme in Arequipa are shifting significantly. Although a global condition is not limited to arid lands, each condition is bound to influence the course of public action in arid lands. Each is likely to do so the most fully by shaping the ways in which scientific problems are stated and the administrator's sense of efficacy in dealing with them.

1) With all of the clarity and emotion that come with the sudden unveiling of a skeleton long suspected but never seen, the growth of world population now is confronted as a legitimate concern of anyone who would find new sustenance in the earth. Only within recent months has it become entirely respectable in international circles, and particularly in banking circles, to talk of the need for conscious population policies. We not only know but openly say that the 3.5 billion inhabitants of our earth will—short of catastrophe—double their numbers before the end of the century.

2) These multiplying residents of a still finite globe are moving to the city at an accelerating rate. The Bedouin's son and the teacher who could show him desert tradition are seeking the crowded amenities of Cairo. Urbanization is a pervasive current that runs strong, draining from remote sections along the improved highway of many lands. The proportion of total population in agriculture is declining.

3) In finding ways to feed his swelling numbers, man has devised new methods of plant-breeding and fertilizer application that promise heroic jumps in production and nourish new dreams of massive improvements in food production. Drought-resistant, high-yielding wheats change the whole food supply of Mexico or West Pakistan. Rice improvement leaps forward with breathtaking speed under international auspices. And visionary plans for gigantic nuclear-powered desalting plants inflame the imagination of those who despair of either curbing population growth or applying the conventional techniques in any reasonable length of time.

4) The gap between scientific knowledge and its application to the husbandry of soil and water, nevertheless, is large. Whether it is thirty years or fifty years, it still is great. Whether it is narrowing or widening is less clear. The "green revolution" can be viewed either as an aberrant leap forward or as the herald of a new era in which the arts of education and extension—for example, the gaining of acceptance for improved maize seed or water-conservation devices—are deployed more broadly and more effectively than ever before.

5) There can be no doubt, however, that the absolute gap in level of living between the rich countries and the poor countries is widening continuously. Even when countries like Kenya or Iran have made unusually rapid gains in gross national product per capita,

their positions in relation to Australia, Canada, the Soviet Union, or the United States of America are less favorable than they were fifteen years ago.

As a world community, we do not yet know how to cope with this growing disparity in income. Indeed, we are not certain just how the rich countries could effectively help the poor countries to better their lot, even if they were prepared to share much more generously than they are at present in bilateral and multilateral aid programs.

There are bright spots in the panorama of international aid efforts—the Khuzistans and Geziras—but the color of economic growth is disappointingly drab, the effectiveness of many of the common measures is in doubt, and there is no fully satisfactory theory of how and why growth takes place. We are learning, for example, that investment in education may pay much higher returns than investment in structures. A sober, patient diagnosis of modes of economic assistance is in progress. And patience may emerge as one of the less comforting prescriptions.

Perhaps the recognition that simple aid programs have not been yielding the expected fruit in new institutions, higher incomes, and improved resources has strengthened two other changes in views of the world scene.

6) There is a rising tendency to look for alternative and multiple solutions to resource problems, to avoid betting all capital and technical reserves on a groundnut scheme or a massive concrete dam or a single institutional reform of the tax system. The importance of integrated action along several lines is more evident: an irrigation project with the finest dam is useless in the long run without appropriate drainage, seeds, fertilizers, application schedules, transport, marketing, health measures, and credit. Because of the increasing number of technical improvements, the same approach would favor public measures with shorter time horizons, which preserve greater flexibility for the future.

7) Linked in many ways with the heightened interest in alternatives is a quickened sensitivity to the ecological impacts of man's intervention in the environment. There is now more interest in knowing the full consequences for man and landscape of a new highway or manufacturing plant or chemical compound. There is greater willingness to say, "Hold up until we learn more," and there is stronger disposition to sacrifice advantages of low-cost power or transport to preserve a natural landscape or an esthetic opportunity.

8) Finally, improved channels for international scientific collaboration have been created: although their arrangements are ponder-

ous and awkward and the flow of information is often sluggish, they do offer ways to spread experience and scientific insight. New instruments such as FAO and UNESCO and the International Hydrological Decade (IHD) are gaining experience, and, in spite of internecine competition, are building habits of cooperation.

Each of these changes has particular meaning for those who look to the present and future of arid lands. Population pressure is increasing on all sides. Cities are growing at breathtaking rates, both inside and outside the arid zones. There seems to be little doubt that the major new developments on arid lands will be urban rather than agricultural. New technologies and the promise of more shape the sense of the possible in coping with thin soils and sparse water. Yet, the ways to speed up economic development are tantalizingly clumsy, and the ominous problem of deepening disparity looms unsolved, with some of the arid countries persistently outdistancing the others. The stance of the scientist or administrator in confronting resource management is increasingly one of canvassing a wide variety of alternatives and of being sensitive to the likely consequences of each.

Tentative Balance Sheet

At the risk of being charged with flagrant overgeneralization and ignorant misuse of evidence, I venture a rough assessment of how well man is doing to maintain the arid lands. As others have done, I think of the arid zone in three parts: extremely arid, arid, and semiarid. I pass quickly over the extremely arid.

In the arid areas, the thin vegetation almost everywhere is either deteriorating or barely holding its own. Irrigation is spotty; improvements in some well-designed and efficiently operated undertakings are offset by the failure and low return of others. The total area under canals expands rapidly. Tourism blooms vigorously in suitably accessible locales. Cities sprawl, and with them come polluted air and defaced terrain. Nomadism declines; city slums grow. Few of the ephemeral or exogenous streams are fully used, but many are approaching that state and their salt loads are increasing.

In the semiarid areas, the grasslands present a disjointed picture. In North America and southwestern Asia, mechanized cereal culture is pushing more deeply into these lands of highly variable rainfall. The remaining areas either are being restored and improved on a major scale, as is the case in Canada and the U.S.S.R., or are deteriorating rapidly under the pressure of overgrazing or of intensified shifting cultivation, as is the case in the sahelian zone and throughout North Africa. In North America and Australia, the deterioration of the uncultivated grasslands is just now being halted. Irrigation is economically marginal in large project operations but is spreading rapidly through smaller pump schemes. Urban growth in semiarid areas is more dispersed than it is in the arid lands; stream pollution is more widespread; and major river-regulation schemes are common.

A backward glance to the symposium fourteen years ago (1956b) shows that this assessment differs in several ways from the temper of that time. Estimates of the practicability of halting the destruction of natural resources are somewhat less confident. Investigators are more sober and restrained in their feelings about the inevitability of economic growth. Greater stress is laid on urban expansion, but undue weight is still placed on agricultural problems. Less importance is attached to the vision that weather modification or the desalting of brackish water or the harnessing of solar energy will make the desert bloom. The choices seem to be more complex, the outcomes less certain. Most of all, there is heightened emphasis on man, his motives and preferences and life styles, and less emphasis is placed on technological solutions. Without diminishing the importance of probing the basic land-soil relationships, the human factor is commanding greater attention. At the same time, however, the interest in new techniques tends to obscure concern for the understanding of the fundamental processes that are involved.

Techniques for clearing mesquite (*Prosopis*) and burroweed (*Suaeda moquini*) from beef grazing land divert attention from ignorance of what accounts for the spread of mesquite or of what adjustments other animals make to the thorny vegetation. Plans for new means of importing a river's flow conceal the fact that the planners and others do not know what effect additional water has on economic growth.

One of the most remarkable features of this period is that, despite the intense interest in methods of developing arid-lands resources, there has been little systematic appraisal of what actually happened and all the ramifications after a scheme was set in motion. There are snatches of information, much of it anecdotal. Even for small, sample areas, an impartial, analytical record remains to be compiled in a form that permits comparison of results.

Arid-Zone Mystique

The participants in the symposium reported here should face up to the "arid-zone mystique" that binds them together. Why is it that the workers in this area gather together to share findings and problems, ignoring the usual boundaries of discipline and nation? Whoever heard of a temperate-lands conference? Humid-tropics fraternal activities survived for a time in the image of the arid zone and then expired. The polar scientists are the only other group with similar inclinations and consistency.

I think the answer lies in a common perception of arid lands as a place of high risk, with thin margins of survival and a special intensity and fragility of processes. Here, man seems to have few choices, to play for high stakes, to lay bare the elementary network of which life consists. The problems of destruction and regeneration are sharp; the capacity to denude completely a sparse terrain is everywhere near. It is in the sense of taking up a specially intense and hazardous relationship with other elements in the man-land ecosystem that a joint approach to arid lands commands allegiance and enthusiasm.

Seven Issues

In the assessment—represented by the papers in this book—of the state of knowledge of this special province in the changing world scene, seven issues emerged and remain unresolved. I begin with a relatively local question and end on a global scale. And, although each question may be answered by different degrees of emphasis, I try to state it as an exclusive choice by way of sharpening the issue.

I assume that the long-run aims are to enhance social growth without undue destruction of natural resources. At this scale and time, we should not be concerned initially with the salvage of local economies at the heavy expense of national welfare or with programs that serve primarily to meet political commitments or to glorify bureaucratic traditions.

Irrigation

What is the role of irrigation? In the answers to this question, there is a basic cleavage in opinion. In some quarters, it is argued with the conviction of a religious credo that rapid expansion of irrigated lands is essential to meet world food demands by the year 2000. New projects have a major share in several national development programs. In other quarters, it is asserted that many of the remaining opportunities for irrigation would be unduly costly by comparison with other means of increasing crop production; that the limited supplies of water have higher value in the expansion of industrial, municipal, and recreational use; and that investment in alternative programs, such as education and transport, would better advance economic growth.

There is no disagreement that all new irrigation should be designed and operated to avoid the flagrant failures and disappointments of the past. The scale and timing of new investment are the major issues. In this controversy, it should not be forgotten that, for some countries, more irrigation, however costly its capital demands and however meager its capacity to swell national exports, may serve to feed a growing population and to buy time while readjustments in industry and growth rates are brought about. Nor should it be overlooked that in some high-income areas, such as central Arizona, great new projects may serve chiefly to put off the day of public reckoning for the conscious destruction of water resources and are luxuries, which could not be bought by less wealthy nations, to protect regional interests.

New Development

Should new development expand or intensify? Including, but not restricted to, the question of irrigation is the issue of whether it is better to improve already occupied lands or to expand into new ones. It is put as vertical versus horizontal expansion.

Major savings come when agricultural production is improved in places where transportation, schools, and other public services are already supplied. Through work in settled areas, such as north-central India, environmental conditions may be understood more accurately, and the difficulties of setting up new agencies may be avoided. On the other hand, it is argued that some undeveloped areas may be more promising, and that by starting fresh it is possible to leap over all the encumbrances of sticky administration, competitive bureaus, outmoded education, and pedestrian thinking.

Amiran (this symposium) and others suggest that, with increasing urbanization and with the concentration of agriculture on high-value products, an oasis-like pattern may be strengthened. There would be more cities and fewer people between the cities.

Size and Number of Projects

Should the single big project or diverse smaller measures be stressed? Whether to have one big project or several small projects raises the issue of the quick technological "fix." It is attractive to think that a few dramatic changes in technology—a new variety of grain or the longest canal system in the world or a nuclear-powered desalting and agro-industrial complex—will alter the whole course of a nation's agriculture. This is possible.

Some large projects may require less social reorganization than would smaller steps. Yet, many scientists feel that the focusing of attention and funds on certain of these measures has deprived other equally promising activities of vigorous support. They suggest that, although infinite pains are taken with the design of a spillway for a great dam, comparatively little attention is given to improvements in irrigation-water delivery or to the efficiency of small watershed design. Numerous experiments are run on beef cattle, but virtually no one lavishes any study on the much-cursed goat.

Amenities

What is the role of amenities? The part that the amenities will play in guiding the future use of arid lands has been suggested but has not been explored sufficiently. Insofar as increased population is attracted by qualities of climate and landscape for recreation and industry, fresh criteria for the assessment of resources must be devised. And new means of preserving the advantages unimpaired must be exercised, for both clear, dry air and rugged scenery are subject to defacement. If the prospect for recreational and urban growth is as great as many people predict, then a new set of priorities for research and administration is in order.

Basic Relationships versus Problem Solving

Should basic relationships or problem solving take precedence? Whether to emphasize basic relationships or problem solving is a question that dogs education and research in the developing countries. At the elementary level, it shows in the choice between teaching and materials aimed at the distinctive resource needs of the area and the teaching and materials to prepare students for broader views and more advanced studies. At the higher levels, it appears in the old debate between general education and vocational education. In the field of research, it is the most acute in the design of research stations.

For the effective study of arid lands, more than the choice of priorities among research problems is involved. If, indeed, some minimum size and some combination of scientific personnel are essential to the effective probing of the unknowns of arid lands, it is a disservice to encourage national ventures in arid-lands research that fall short. The result may be that neither basic nor applied science is well served. Several arrangements have been tried during at least a dozen years, and it is appropriate to ask candidly what combinations of scientific orientation from the developing, as well as the more highly developed, countries are paying dividends in significant new knowledge.

World Linkages

How can world linkages be achieved? Time and time again, it is pointed out that the wise strategy for development of an arid land depends on its linkages with other parts of the world. Contributions from new crops to national income will depend on overseas markets. Tourist expenditures reflect transportation and fiscal arrangements in distance lands. Where educational levels are low, long-term programs of foreign aid may be important. The establishment of new industry often rests on movement of foreign capital and management and on market prospects. Most of the new research institutions will be nourished by continuing collaboration with older laboratories overseas. These are less obvious but probably are more fundamental than technical assistance and funds for public construction projects.

Thus far, the national development plans have only begun to cope with the complexities of integrating these diverse elements in the laying out of resource studies or improvement programs. The difficulties are particularly acute in low-income, arid-lands nations, which have a narrow range of choice and may be required to stake a large part of available capital and personnel on one irrigation or livestock improvement scheme. There is little good in moving ahead with some of these enterprises unless stable linkages with world institutions and markets can be assured. As is true in the case of some bilateral agricultural research institutions, the fashioning of such links is largely in the future.

Integrated Study and Action

Can integrated study and action be achieved? Throughout its life, the program of international cooperation in arid-zone research has stressed the vital importance of integrated studies and action. An institute has been established in the Netherlands at Delft. Detailed surveys have been made in Australia, Pakistan, and a few other places, but the methods for such surveys have not yet caught on widely. Water planning has begun to deal with entire basins. However, genuine integration in the study of new resource-development programs is the exception rather than the rule. Arrangements in which scientists of different disciplines work in the same laboratory or field or gather periodically for discussion do not usually achieve this.

How to establish integrated study and action solidly in either the developing or the developed countries is not known. If this integration is as important as many people assert, it should be the subject of rigorous analysis and promotion.

These issues are being resolved against the background of a swiftly changing world in which the scientific perceptions of arid lands are shifting as rapidly as the technology of using them.

Problems of Communication Between Scientists and Decision Makers in Resource Management

In most types of public policy decisions the use of scientific information and judgment is obstructed by at least five conditions. The scientist fails to establish credibility. His findings are necessarily incomplete and probabilistic. Some of the findings are important to understanding process but have no apparent relevance to the immediate decision. The scientific definition of the problem differs from that of the policy maker. And new research always enlarges the uncertainty of future events.

A few conditions apply especially to decisions affecting natural resources. Resource decisions involve externalities that do not lend themselves to economic efficiency analysis. The time horizons extend beyond those of private enterprise or political regimes. Political and administrative officers may perceive both the environment and environmental impacts differently than do the immediate resource users. And the decision maker tends to place heavy weight upon early and dramatically visible environmental alterations.

These considerations suggest three questions which may be asked of any attempt to marshal science in resource management. Is there early and continuing effort to reconcile the definitions by decision makers and by scientists of the problem to be addressed? Do those expected to use the findings feel any sense of responsibility for the design and conduct of the research? Does the scientist look for concrete albeit simple ways of demonstrating the relevance and necessary limitations of the findings?

(1979e), 79

Notes

1. For a current report on many of these issues, see below, vol. II, chap. 4 (Johnson).

19 The Meaning of the Environmental Crisis

It is always difficult, standing amidst social change, to discern its source, depth, and persistence. So much of what fills the media and overwhelms our senses will be seen in retrospect as momentary fad, pendulum oscillation of values, or recurrent wave of reform. Normally White prefers a long view: a decade is desirable, two decades better; long looks backward and short, cautious peeks forward. In this unpublished paper, presented the year of the first Earth Day, he uncharacteristically proceeds to dissect the environmental movement in the very middle of the movement's major passage.

We now know that the movement was not simply transitory. The changes in values that were becoming evident by 1970 have persisted, whether measured by the consistency of public opinion polls or the constancy of congressional support. If the wave of intense popular concern has subsided as White speculated it would, it has left on the public shore not only new policies, organizations, and methodologies, but institutionalized public concern as well.

White's thoughts about the meaning of the crisis are still cogent today.[1] It remains true that the sources of public concern transcend the "facts" of the environmental crisis. The trends in scientific perception continue—a focus on meso-scale systems, a search for nth-order impacts, a recognition of the global unity of some environmental issues. Perception of the environment as hazard has flourished—a quasidiscipline of risk assessment has been created in the interim.[2] Many universities now have interdisciplinary environmental centers displaying weaknesses that White feared and few of the strengths that he hoped for. Finally, the expected North-South confrontation evident on the eve of the Stockholm conference has diminished. Many developing countries now recognize the importance of environmental issues, but few developing countries have effectively dealt with them.

Paper presented at the University of California, Los Angeles, 7 December 1970. The author is grateful to Kenneth E. Boulding, Ian Burton, Henry Caulfield, Jr., Stuart Cook, Kenneth R. Hammond, Thomas Heberlein, Richard Jessor, Robert W. Kates, and Fred Strodtbeck for discussion leading to many of the ideas presented in this paper.

Is the current intensity of concern over environmental quality among large, articulate sectors of American society another excrescence of public enthusiasm that will wane like the two earlier conservation crusades? Its strength matches the anxiety of the Governor's Conference of 1907 to prevent uneconomic destruction of forest lands and profligate exploitation of fuel and metal reserves. Its national outreach exceeds the vigorous efforts of the mid-1930s to curb soil erosion on farms, prevent the misuse of public lands, and assure multiple use of the rivers of the country.

Each of those earlier waves of popular concern subsided, leaving on the public shore new policies such as multiple water use, new organizations such as the Soil Conservation Service, and new methodologies such as benefit-cost analysis. Are we now at the peak of a third great wave of interest in environmental matters which also will recede, having established new standards of cleanliness, an Environmental Quality Council, and a method of technology assessment? Or are we in the midst of defining a changed relationship between postindustrial man and his physical world?

The current concern about environment has a crisis dimension which most of us would recognize. It shows in the popular expressions of fear that pollution will render the planet uninhabitable to man and that exhaustion of resources of soil and water in the face of burgeoning population will bring massive famine. It is voiced in the Earth Day of 22 April, 1970, in the campaign against phosphates in detergents, in the rash of grass roots protest provoked wherever a bulldozer scratches the site of a new dam or a nuclear power plant, in efforts to extend wilderness areas, curb signboards, outlaw the private automobile, and ban DDT. It finds institutional form in a set of new federal agencies, *Time's* section on the environment, and in the Michigan law and federal cases recognizing a citizen's right to sue on behalf of the earth and posterity. It is expressed, too, by the girl who says, "I will never have a child: there are too many already," and by the public utility executive who observes, "We can't find any place to build a power plant. I don't know what the hell has come over our people."

The crisis took shape more rapidly and dramatically than any previous one. (Even canny federal administrators who characteristically have their ears close to political ground were caught out in such exposed positions as the Grand Canyon and the Everglades.) The corollary movement is nonpolitical in the sense that it lacks personified villains and carping criticism by one political party against another. While it shows sensitive concern on a wide front—ranging from quality of air, water, and visual qualities in the landscape to sober assessment of population growth in relation to available energy resources—it is without a specific program of action around which there can be an easy and clear grouping of public support such as was involved in the reservation of forest lands or the mounting of aid to farmers in soil conservation. Perhaps most significant, it is the first crusade to have genuinely global implications, although its awareness and anguish are not shared by the developing world any more than it is supported by black and Chicano groups who say other issues overshadow it.

Scholars and public agencies have been so busy responding to this new and urgent demand that they have given only casual attention to the dynamics of its rapid change. Not comprehending the dimensions of shifts in public perceptions or their implications for social structure and process, most of what can be said about the movement to deal with the environment in new ways should be taken as hypotheses we should strive to test and refine. We have precious little understanding of how it took shape and where it is going.

In the heat of attention to zero population growth and pollution abatement, some of the likely meanings of the sense of crisis are neglected. Among these, at least five are worth considering. One is that the popular concern may spring from social factors only indirectly related to the natural environment. A second is that scientific perception of environmental problems is changing drastically. Third, the environment when viewed as risk has a special set of lessons for those who would improve man's adjustment to it. Fourth, the implications for social transformation in the university are far more profound than usually recognized. Finally, the confrontation it provokes on the world scene is seen as less acute than it promises to be.

Sources of Public Concern

It would be easy to assert that the effects of accelerated rates of population growth, resource depletion, and technological development have led to acute public awareness and the demand for rational methods of coping with them; but it is misleading to believe this coincidence of events is the sole cause of heightened public awareness. There is no denying the exponential growth of population in certain sectors of the globe, although the flattening out of the population curve in the United States is coming much more rapidly than many people anticipated as recently as a year ago. Nor is there any denying that technological changes have brought man's intervention in the environment to a new order of magnitude within a period of several decades, as witnessed for example, by the great outpouring of organic compounds, by enlarged earth-moving capacity, and by an increase in per capita demand for energy in the United States that has outstripped population growth four-fold. We recognize, albeit reluctantly, that even if total population were to level off entirely tomorrow, the present rates of growth in technology, if continued, would bring as much technological encroachment upon the landscape in the next decade as observed in several decades previously.

These trends are sufficient reason for alarm but do not give fully convincing explanation for the rapid growth of public awareness and action. In the United States the actual deterioration of environment has not been as dramatic as the soil destruction of the 1930s: water pollution has actually abated in some streams such as the Ohio, no air pollution disasters of the magnitude of Donora have occurred, the areas of protected wilderness have expanded, and the threats of early materials shortage have diminished. While the basic picture has on the whole worsened, sensitivity to environmental quality has increased immensely, and willingness to act has become robust, even frenetic. Other sources must have contributed to the public concern.

Without solid ground for assigning any causative weight to these other factors, they can only be enumerated as associated in some fashion with the emergence of the crisis.

It sometimes is suggested, mostly in cynical jest, that the current environmental crusade is a studied product of public officials who sought to divert attention from the ills and unrest accompanying the Vietnam tragedy. It is thus seen as a magnificent diversion in which the embattled establishment found ways of enlisting disenchanted youth and intellectuals in a campaign which shifted their eyes away from the South Asian horizon and focussed them on a nearer, more manageable and less stressful battleground. It is true, as Henry Caulfield points out, that use of environmental quality terminology by government officers followed a conscious decision to do so reached in the White House in October 1964. However, any such tactic, even had it been carefully planned, could not have succeeded unless it fell upon a receptive sector of the population. There is some reason to think that this was the case, and we may ask why.

A few relatively simple circumstances come to mind. The rate of technological change became sufficiently rapid to permit young people to observe in a few years deterioration which formerly would have claimed a lifetime. The dissemination of chlorinated hydro-carbons was now so wide that people could feel personally threatened by them. The spread of communications made it feasible to see the effects shared with others and to hear scientific assessment of the invisibles. Satellite imagery gave graphic reality to the globe as a discrete and self-contained system for the first time. These circumstances came at a time when, in city slum and in the reflection of the Lincoln Monument, sizable groups of the electorate were convinced that individual actions outside the conventional channels of voting box, petition, and debate were required to achieve desired social change.

We know a little about how these circumstances may affect perceptions of the environment and actions toward it. For example, there is evidence that perception of environmental quality varies with socioeconomic status, that more visible characteristics such as particulate matter are more threatening than the invisible nitrous oxide, and that certain people, such as engineers, tend to see a problem as severe to the extent they feel themselves capable of dealing with it. Nevertheless, most environmental behavior is understood imperfectly, and much of it seems to reflect patterns of value.

Some part of the heightened concern for the environment may be ascribed to the experience of highly frustrated Americans who had been presented with a series of problems of unprecedented severity and magnitude for which no easy solutions were forthcoming. The Vietnam war had dragged on for six years with little immediate prospect for resolution. The distress of racial discrimination had apparently intensified, notwithstanding fifteen years of nonviolent activity in restaurant, schoolroom, and workplace. The gap between rich and the poor grew larger while the federal mechanism showed itself conspicuously clumsy in responding to demonstrated incidents of hunger, disease, and insecurity among the lowest stratum of the wage earners. Faced with these failures in stopping the war, building viable cities, and preventing poverty, affluent and conscientious people turned to a problem of similar import—the environment—which they felt better able to cope with.

Coupled with this frustration and disenchantment with the expanding, dim regions of the bureaucracy is a long-term trend toward ascribing causality to the outside world rather than to inner leading. How much this has to do with inclinations to set the environment straight we do not know.

There are strong religious strains in current environmental awareness. Some proportion of the people who take part in discussions of the crisis feel that in doing so they become part of a wider fellowship. They prefer holistic views of the universe and of natural phenomena. They join in reviling villains who are characteristically impersonal organizations—either government agencies such as the Corps of Engineers or business giants such as Procter and Gamble—and they feel that commitment to the broader cause implies change in individual life-style. For a generation which is conspicuous in discarding conventional religious terminology, neglecting ecclesiastic connections, and in suspicion of the formal trappings of church and secular hierarchy, environmental matters offer a meeting ground and a set of symbols and formulations satisfying many of the needs previously met in church organizations, practice, and theology. Under this view, the new high priest is the ecologist, the brotherhood is worldwide, the symbol is the earth and its wholeness, the dogma is modest but cosmic in its inclusiveness, and the demand for individual fidelity to a universal principle of harmonious living among fellow creatures in the great ecosystem is inspiring but not rigorously exacting in its relation to daily life.

Thomas Heberlein suggests that the shift to an earth morality, as forecasted by Aldo Leopold, is a product of two factors. One is a new recognition by the individual of his personal responsibility for the negative consequences of his actions. The other is his perceived opportunity to prevent such harm by drawing on the range of alternatives now made available to him by technology. Thus, the technology which threatens the landscape may also generate moral values and norms to replace primarily economic values.

Scientific Perceptions of the Environment

In terms of scientific perception of man's role in the environment, there have been three major changes over recent years. The first and perhaps the most important, as pointed out by the University of Colorado Commission on Environmental Studies, is the shift in attention to intermediate systems. Man has achieved relatively refined understanding of the processes of nature at the microscopic scale and at the cosmic scale. He knows a great deal about the structure of the molecule and the motions and growth characteristics of the heavens but very little about the workings of water, air, soil, and organisms in a mountain hillside or an ocean shore. These have been largely neglected, in part because the methods for dealing with their complex interrelationships are so crude, and in part because their analysis inevitably involves an interworking of disciplinary skills, a process running counter to the trend of reductionism. Yet, they hold the answers to most of the questions raised by the impacts of human intervention, and it is recognized that nothing short of investigation of all interrelated facets of an environmental system, including man, will yield the needed results. Examples are comprehensive studies of the grasslands and desert biomes.

A second trend has been the increasing interest in tracing out the impacts of a technological change in the total environment. This involves not only the recognition of changes perpetrated by a dam or a highway on local weather or wildlife or stream flow, but also appraisal of the significance of these alterations in the lives of people. The efforts at "technology assessment" which might be more properly called "social assessment" find wide acceptance but relatively little practice. The practice is difficult because of lack of understanding of the intermediate systems and also because administrative measures for carrying it out are so difficult to sustain. It is significant that one of the most ambitious efforts at this type of assessment was undertaken following the decision to construct the Grand Coulee Dam on the Columbia River. A comprehensive investigation tried to anticipate all of its impacts on the environment—both social and natural—and to take steps to guide and cope with them. Many of the studies launched in the 1940s led to changes in public action in the area, but the vision of comprehensive assessment in advance of a technological change never caught hold. It was not until the early 1960s that financial support and administrative enthusiasm for such enterprises began to take shape.

Interdisciplinary investigations of a large range of impacts from building the Aswan, Kainji, and Volta reservoirs were launched in the mid-1960s. These, however, are essentially salvage operations. They represent the kind of work which ought to be carried out well in advance of construction. The important aspect of the change is that governments now realize that scientific work of this sort ought to be done and are willing to put up funds for it.

A third major change is found in the practical recognition of global unity with respect to many of the phenomena. Pioneered by the international geophysical year and by the collaborative program of arid zone research, a series of international efforts now bring together the interests of scientists in many countries and many disciplines to deal with complex processes having global scope. Typically, these are along the lines lending themselves to a cluster of scientific disciplines, as in the case of the Global Atmospheric Research Program and the International Biological Program, but increasingly they deal with problems cutting across the traditional boundaries. The most recent, and in a theoretical sense the most ambitious, is the effort initiated this autumn by the International Council of Scientific Unions to assess problems of the environment on a global scale. This will draw upon all of the traditional disciplines and will call for collaboration of social scientists as an essential part of the enterprise.

In little more than a decade, the scale of environmental studies has shifted, the concern for impact assessment has been validated, and global cooperation has bloomed. To a large degree the sense of crisis has generated these changes, but the time span is still too short to permit the new findings and methods to in turn show themselves fully in the way environmental problems are defined or treated. That influence may be expected to show itself in the years immediately ahead.

The Environment as Risk

Every intervention of man in the environment around him incurs some risk as to both favorable and unfavorable consequences. Every intervention is taken in the face of partial ignorance as to what its effects will be and involves uncertainty as to the ultimate outcome. As we analyze man's experience in dealing with natural hazards such as hurricanes, floods, earthquakes, and drought we find several lessons that may have significance for all concerned with crisis in the environment. It already has been noted that people tend to have a more accurate perception of the severity of a crisis according to the efficacy they feel they have in dealing with it.

It also is evident that man in industrial society has a strong inclination to fix on a single technological solution for any problem which appears, and having done so, he may exacerbate rather than improve the very situation he sets out to remedy. The effect of building a great series of flood control dams and levees to reduce national flood losses has, thirty years and seven billion dollars later, left us with a larger toll of flood losses than when we began. The concentration on a single means for solution diverts attention from other means and encourages people to incur still greater risk in anticipation of public efforts to bail them out in times of distress. The tragic losses from coastal flooding in East Pakistan in the past month have been produced in no small measure by the earnest efforts of engineers to provide what became inadequate protection against the extremely rare event.

Enchantment with the quick technological fix is illustrated admirably by current action with respect to risks of air and water pollution. Much of the debate about water pollution has in recent months centered on the level of federal appropriations for construction of waste treatment works. Methodological changes in dealing with waste at the source, including reduction in the consumption of materials, and in social devices for spreading pollution costs, such as by effluent charges and by taxes on material consumption, will be far more important in the long run. So long as we cultivate the pleasant myth that spending more money on bigger works will solve the problem, it is likely to become worse rather than better. The same applies to air pollution from automobile exhausts. Technological changes in the internal combustion engine or in its substitutes may reduce the threats to public health in part, but in the long run, we face radical adjustments in life-style and the pattern of urban life if the waste hazards are not to grow worse.

The University Response

A similar kind of interest in simplistic solutions runs strong in university responses to environmental problems. Faced with irrelevant research and unsuitable teaching, the tendency is to set up new departments or schools labeled exclusively for environmental purposes. Such efforts no doubt may serve a useful purpose in a transition period, but from experience with schools of public health and agriculture, it seems probable that in the long run the solutions will come from a fusion of broad environmental interests in the established departments and teaching mechanisms rather than by setting up new enterprises for either research or teaching. To say this is to call for something different than adding a few new slots in existing departments. It calls for basic reform in the university's structure for teaching and research purposes. An approach to intermediate systems in a social context demands a structure within which both students and researchers can operate across conventional disciplinary lines without sacrificing quality of scientific thought or losing recognition. Most universities penalize the man who works across disciplinary lines unless he is a smashing success. And most disciplines—essential to academic morale and aspiration until they are replaced by new ones—discourage the student from working on problems beyond the traditional range of a sector of the discipline.

More than new departments or new degrees we need new approaches and a willingness to radicalize the university by altering departmental affiliations and reward structures.

Confrontation with the Developing World

In the United States we slowly are saying farewell to the traditional conservation resource people who are primarily concerned with physical development rather than with wider ranging human outcomes. Yet, these are the people, trained in our universities, who dominate resources work in the developing world. A recent meeting of the Lower Mekong Committee—a group which continues to foster international collaboration in developing the resources of that basin in the face of a futile war—illustrates the point. The four countries are basically concerned with improving the welfare of the 30 million peasants who live in the basin. Their emphasis at first was on building dams which would generate power, control floods, and provide irrigation water for periodically dry lands. The largest single study expenditure was contributed by the United States through the Bureau of Reclamation for design of a single dam which would cost a billion dollars to build. Recently, they shifted their emphasis to comprehensive organization of agricultural community development so modest water control can be translated into genuine changes in distribution of land, income, and opportunity in the lives of the villagers. They are disturbed and baffled by the visiting ecologist, also from the United States, who blandly says they should put a halt to any further construction because of possible injuries to the environment. It is absurd to think of saddling countries with per capita annual incomes of less than $100 with a mammoth structure. It is equally absurd to tell them to avoid technology that promises any alteration in an environment that has felt the hand of man over a millennium. There is willingness to work for population planning, but there is no disposition to place that aim above the welfare of the young.

This is now a divisive issue on the world scene. The high-income countries are enthusiastic about convening in 1972 in Stockholm a world conference on the human environment. The developing countries are far from enthusiastic and have lagged in expressions of interest. Why, it is asked, are the rich countries calling for caution in the use of fertilizers and pesticides and the development of power plants and mineral extraction for a growing population after they, having polluted their environment in the process, arrive at a point of economic and political dominance? The environmental issue places in sharp focus the question of what stance the rich will take toward the poor as the income gap widens and export sources are depleted. It should not be a stance of condescension, or of moral condemnation of the spawning poor to an eternity of suffering. A new level of understanding and of scientific and economic cooperation will be required.

In looking to the future in both rich and poor countries, there is a deep disposition to discount the direct consequences of either ecological or economic disaster. I am reminded of the survey made by one of my Chicago colleagues at State and Madison Streets in which people were asked whether or not they thought there would be an atomic bomb attack on the city in their lifetime. Over half felt there would be. However, when asked as to what they thought they might be doing two days after the attack, more than ninety percent were clear what their activity then would be: they would be helping bury the dead. In dealing with global environmental problems, the high-income countries are beginning to see that there may be no place for the survivors to stand aside.

The environmental crisis is real but it may be more significant as symbol than as substance. Attention to the environment reflects in some degree the need to deal effectively with other troubles in the national and world societies. Deterioration of water and air is threatening, but the festering rot of the cities may cause the decline of our civilization before famine from shortage of rice.

In both cases, however, it seems likely that man can cope effectively only through an unprecedented integration of natural and social science that will be generated and in turn reflected in educational process and organization. To do so he must move courageously away from reliance on simple technological measures to mixed strategies of social, technological, and scientific enterprise. These efforts will accompany a better resolution of the frustrations which people have in the face of man's theoretical capacity to feed himself and live in peace and a better resolution of anxiety over the mounting hazard of destruction from the hand of military man or of laboratory man.

To the extent concern for the environment is a manifestation of dissatisfaction with the state of man and society, constructive efforts to deal with the environment may be a banner for a crusade for fundamental changes in those societies. But even though it is not, to take the environmental crisis seriously is to adopt a revolutionary stance that searches for a new formulation of human values and calls for social creativity and an immensely exciting experiment in education.

As the Rich Grow Richer

The manned spaceship trip behind the moon was like a successful Roman circus on a planetary racecourse. Catching the imagination and admiration of people in the United States and overseas and executed with consummate technical skill, it comfortably diverted public attention and scientific inquiry from tougher and far more fundamental problems than that of which nation first will land on—and return from—the moon. Although no distant national frontiers are crumbling while pictorial lunar bread is distributed over the television waves, the issues that are obscured by such exploits are of graver consequence to the family of man.

Probably the main issue that is blurred by these high technical feats is whether or not the human race, whatever the nature of the moon, can maintain the earth as a habitable place. At stake are the numbers who can be supported and the level of living at which they will subsist. It has been common in some sectors of the scientific fraternity in the United States to criticize the undue support given to space exploration at the cost of more basic research just as it is popular to lament funds going into the hopeless and tragic Vietnam war rather than into a war on domestic poverty. Getting its priorities right is a difficult and never ending task for any society. A society at war is bound to have things backwards because destructive force always is the enemy of the good and reasonable. Yet, the issue of world environment has a special kind of urgency. It overrides the perennial medical aspiration to let men live a little longer, a little less painfully, and the urge to live at peace rather than war. It deals with the minimum terms of survival—physically and spiritually—in what Barbara Ward calls the "lop-sided world."

The issue is one of rich peoples and poor peoples, of the growing gap between the two, and of the rich fouling their own nests.

Commencement Address, Earlham College, Richmond, Indiana,
20 January, 1969, 1–2

Notes

1. For a parallel analysis of the climatic crisis, see below, vol. II, chap. 8 (Hare and Sewell).

2. For these developments, see below, vol. II, chap. 9 (Whyte) and chap. 10 (O'Riordan).

20 Drawers of Water: Domestic Water Use in East Africa

Drawers of Water is another seminal study, perhaps the best field study White (with his colleagues David Bradley and Anne White) has done. It is simple in concept, clever in design, broad in approach, convincingly large, and attractively presented. It opened a new field—serious study of the oldest of human resource uses, drinking water; developed new concepts—a major classification of water and related disease; and encouraged major changes in thinking about ways in which rural water supplies could be improved. It also updates *The Choice of Use in Resource Management* and the work on public attitudes (selection no. 15) by presenting a concrete analysis of how East African women choose their water source, providing a convincing example of the use of White's decision matrix and the universality of its application. And it marks his and Anne White's growing interest in Africa, stimulated earlier through participation in a committee on African-American studies (1967a), an interest that has persisted in major work on man-made lakes, African river basin development, and domestic water supplies and sanitation in developing countries.

An Elementary Choice

Man has many transactions with nature at various times. Among them, the need to obtain water is universal, and most people have some choice over source and volume used. A study of this rudimentary decision therefore should not only help predict people's reactions to attempts at improving their water supply but also throw some light on the essential character of human choice. One aspect

Reprinted from *Drawers of Water: Domestic Water Use in East Africa,* by Gilbert F. White, David J. Bradley, and Anne U. White (Chicago: University of Chicago Press, 1972), chap. 8. © 1972, The University of Chicago.

of this, as explored in chapter 5, is the volume of water the household elects to draw each day. A second aspect is the source to which the family goes for water. In this chapter we examine the preferences which are shown for different sources, as affected by community norms, and relate these to the choices public bodies make in guiding the domestic use of water.

As populations almost everywhere increase, and people expect better water supplies while suffering the consequences of bad ones, governments attempt to improve supply as cheaply as is feasible. They use whatever methods promise acceptable and effective results, but the response of the users is sometimes unexpected, and persistent and widespread problems may arise.

On the practical side, an immediate problem is how users will respond to an improved supply in their choice of source and in daily use. In rural areas, the record of boreholes lying unused and of improved but neglected springs is ample testimony that builders may miscalculate what will be accepted or maintained by the people of the area. Why does a family prefer a polluted spring to a rather pure borehole? And if it does choose the improved borehole over the spring, what effect will this have upon its pattern of water use?

A closely related problem is how these responses may change when members of the family move from a farm to the city, and as they increase their incomes. How will their discrimination among sources, as well as their patterns of use, shift in those circumstances even though no improvements are made in the quantity and quality of supply?

Because the administrative capacity of most central and regional governments to aid in constructing improved supplies, let alone to maintain them, is very limited, it is important to find ways in which individuals and local communities can take effective part in the process. How can the greatest possible local participation be promoted and sustained? What is the significance of information, of financial incentives, of technical assistance, and of community organization in stimulating and supporting people in taking over local responsibility for improving supplies? How far can local groups go in maintaining satisfactory service? Ideally, the best government action would encourage the largest independent action by water users to assure themselves of supplies of desired quality at minimal social costs.

The answers to these and other practical problems would be easier if there were a generally valid model to describe the decisions made by individual water users. If we knew precisely how the housewife makes her choice, what factors influence her preference for one water source over another, and how these factors are related to personality and to conditions of social and physical environment, we could predict more accurately how individuals and groups would respond to alterations in those conditions or in opportunities to use them. Explanation of individual behavior must be sought in the mode of individual choice at the source or tap as it is shaped by shared values and by the inaction or action of government.

Models of Decision Making

A few simple models are, in fact, in common use in planning new water schemes. One is the economic optimization view of man as seeking to obtain the greatest returns from time and energy spent in drawing water. It assumes that water users are well acquainted with the relevant information about water and its costs, are rational in their assessment of it, and seek to select the optimum. This shows in the argument that the nearer consumers live to a standpipe the more water they will use, or that people use the acceptable source involving least cost. It shows, too, in the assumption that volume of water use will be profoundly affected by the rates charged. The evidence presented in chapters 4 and 5 casts some doubt on the East African water user's sensitivity to cost and net returns, while confirming its importance for certain classes of urban consumers and for setting limits of willingness to pay for capital improvements. Response to cost is not sufficient to explain differences in use of water sources.

Another more practical view of the water user often is taken by the engineer designing new works. The consumer is seen as acting in the future as he has in the past, but as welcoming any technological change which offers improved supply without changing costs radically. This shows in the estimation of future demand by linear projection of past trends and in the assumption that if a new borehole provides bacteriologically pure water without cost where unclean sources are common the water user will shift to it.

A variation of this model, man as a captive of custom, sometimes is advanced by those who are unable to explain responses to new wells or higher water charges in terms of economic optimization. They view man as bound by the hardened customs of his group. Under this view, he tends to conform to a group pattern of habitual behavior and to innovate only in special circumstances which provide special incentives for those few members peculiarly disposed to try something new or which disrupt the group's culture. It is assumed that water users have little information about their opportunities and are governed chiefly by the traditional group valuation rather than by rational analysis. Taking this view, the reluctance of users to patronize a new borehole sometimes is explained as superstition; the use of small quantities from abundant sources is explained as custom. Undoubtedly these are significant, but here, too, the East African experience raises doubts as to the adequacy of the model.

In an attempt to identify other aspects of the decision to draw water, we have employed a more complex and embracing model. This views the water user as a person who perceives the choices open to her with varying degrees of accuracy and who judges according to her own perception of the quality of the source, the technical means available to her in drawing on the source, the expected returns and costs, and the interaction with other people which such use involves. The emphasis is on the user's individual perception of the situation, as distinct from its definition by scientists or government officials. The decision is based upon awareness of the range of alternatives and upon the value assigned to the likely outcome of choosing one rather than another. Each valuation is seen as representing a personal preference which is conditioned by the customary behavior of the culture and encouraged or discouraged by whatever formal social action is taken by the society.

The problem arose of how to record and handle the user's views on alternative sources. There seemed to be many ways of analyzing them, and we considered two in detail.

One method was to use a binary notation. The possible alternative sources were considered by each interviewer and compared with the alternatives offered by the user. This scheme is illustrated in table 20.1. If the user was unaware of a source this was recorded as 0 on a binary scale. If there was awareness on her part the potential source had a score of 1 and she was asked why she did or did not use it, and the response was recorded in one of five classes of factors. She might consider the water quality as good or bad and the technical means for getting at the water as either available or not. She might regard the cost of drawing water as bearable or as prohibitive. And she might accept it or reject it according to other people's use of that source. When no response was elicited no score was recorded. If a value of 0 is given for unfavorable and 1 for favorable comment on each factor, a set of six numbers is obtained. It was assumed that

Table 20.1 Schematic matrix of potential water sources as perceived by users

Theoretical Alternatives	Considered as a Source	Perception of Alternatives				
		Source Quality	Technical Feasibility	Economic Efficiency	Effect of Other People	"Source Rating" Summary Valuation
Binary						
Possible scores	0,1	0,1	0,1	0,1	0,1	0,1
Examples						
A_1	1	1	1	1	1	1
A_2	1	0	x	x	x	0
Five-part notation						
Possible scores	0,1	0,1,2	0,1,2	0,1,2	$-2, -1,$ $+1, +2$	0–10
Examples						
A_3	1	0	x	x	x	0
A_4	1	2	1	2	x	4
A_5	1	2	1	2	-2	2
A_6	1	2	2	2	$+2$	10

a score of 0 for any one factor would mean the source was regarded as unusable. They therefore may be multiplied to give a summary valuation of either 0 or 1. Examples A_1 and A_2 show the two possible outcomes.

It was decided to use a rather more discriminatory system in the study of households without piped connections in order to register degrees of preference and to describe situations where a source was used in spite of perceived disadvantages. For the user's view of water quality, technology, and economic efficiency, a score of 0, 1, or 2 was given, with 0 representing unfavorable, 1 favorable, and 2 highly favorable. The first item—perception of a given alternative—clearly could only be positive or negative and is recorded as 1 or 0. On the other hand, the effect of interpersonal linkages on the decision whether to use a potential source could vary from definite discouragement, through neutrality, to active cooperation and was therefore given a value from -2 to $+2$. When no comment was given by the user on a given factor it was not scored and was marked as x. The critical part of the scoring was assigning a value to each factor as perceived by the user. This served to identify her reasons for rejecting or preferring sources. The scores were combined in a summary valuation, or source rating, by multiplying the first five factors and then adding or subtracting the score for effects of other people.

The rationale for making a source rating was that it served to sum up the user's expressed judgment and to give appropriate weight to each component part. It specified why a source was judged good or bad. Where a user dismissed a potential source as unsuitable because the water quality was considered unsafe, it often was impractical to elicit judgment about other factors (see example A_3). A woman would say "Why should I go there? The water is bad." It would be possible to find out why she thought the water bad, but not what she thought of expending energy to go there by comparison with other sources. However, if two or more sources were considered acceptable, her response to a question about why one was preferred over another would evoke comments on one or all of the factors (see example A_4). Feelings about the effects of interpersonal linkages on choice almost never led to dismissal of the source from consideration on other grounds. Therefore, the score for highly unfavorable interpersonal effects would be subtracted rather than multiplied (see example A_5). A maximum score occurred when the source was judged highly favorable on all counts (see example A_6). The result of the summary evaluation was a kind of branching analysis that identified the current judgment of each source and the relative importance of each critical factor. We did not anticipate an order in the branching of decisions, but it emerged in the field interviews.[1]

Choice by a Sample Household

This model analyzing the way people perceive their choices of water sources is now applied to one household in a humid section of Ganda territory north of Lake Victoria to illustrate the method and some of its limitations (see fig. 20.1 and table 20.2). The family living at the house has within easy walking distance five sources where water

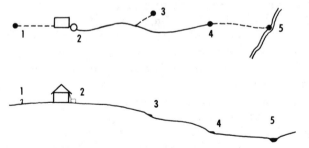

Figure 20.1 Diagram of available choices for perception matrix for a sample household.

Table 20.2 Perception matrix for a sample household

Theoretical Alternatives	Considered as a Source	Resource Quality	Technology	Economic Efficiency	Effect on Other People	"Source Rating" Summary Valuation
A$_1$ Borehole	1	0	x	x	x	0
A$_2$ Roof	1	1	0	x	x	0
A$_3$ Spring	1	2	2	2	−2	6
A$_4$ Spring	1	2	2	2	x	8
A$_5$ Stream	1	2	2	0	x	0

might be drawn. Source 1 is a borehole where pure, mineralized water may be had for the pumping. Source 2 is the iron roof of the mud and wattle house. In the opposite direction a path leads to a spring at 4 and a small stream entering a swamp at 5. Somewhat nearer than 4 and along a side path is another spring at 3.

An interview with the woman of the household shows that she is aware of the borehole but does not go to it because she considers that the water, judging from taste and the color it gives tea, would be injurious to her children. We score this zero under resource quality, and are unable to assign scores for other factors. She also realized she could take water from the roof if there were a suitable gutter, drain pipe, and barrel, but she doesn't know how to arrange this equipment and has not pressed her husband to try to buy it. The water would be acceptable although less satisfactory than spring-water, yet it seems inaccessible and we score source 2 as zero under technology. She regards a protected bit of the stream as almost as good as source 4. Since 4 is much nearer, she rates it higher on economic grounds and eliminates 5 on the same grounds. We score 5 a zero for economic efficiency. Source 3 rates high on quality, ease of obtaining water, and distance. It, however, requires going across the land of an irritable neighbor, and rather than risk that friction she goes to 4. We now have total scores of zero for three sources, a score of 8 for spring 4, and a score of 6 for spring 3. The former is the source which she currently uses and we end with a rather helpful description of why she prefers it over others.

What the User Says

Such inquiry runs the usual heavy hazard of the interviewer's failing to elicit a complete reply or the companion risk of his biasing the response through mode of inquiry. Moreover, some interviewers were more effective than others and their personal biases differed, although they came from similar educational backgrounds and shared a common training for the work. Some of the lessons were learned too late to change the field methods. Our field experience nevertheless encouraged us to accept many of the opinions obtained. To a remarkable degree all respondents in both rural and urban areas regarded questions about water sources and use as pertinent. They may encounter aspects which they feel are so complicated that they must defer to an expert or to superstition, as when a site must be found to dig a well, but generally they regard themselves as qualified to judge.

Water is a vital topic, and a large number of the respondents were interested in getting opinions as to the quality of their supplies and ways of improving them. Indeed, the keenness of curiosity and the hopes for improvement led us to wonder at times whether we might be doing a disservice by arousing questions about supply. We were careful to avoid suggesting that the study would result in any direct action by public or private agencies and guarded against raising false hopes, but we did not avoid stirring up local discussion of sources and their quality. Such discussion could hardly lead to anything but improvement, for it widened the range of choice in many instances and caused people to think about what could be done locally.

The framework within which we have structured users' opinions on possible sources may be questioned, because it assumes at the outset the factors which have some significance. However, the actual discussions with users were not hurried and followed an open pattern. Although at some stage the specific questions were asked, the user was not approached checklist in hand, and comments on choice that did not fit the structured questions were recorded in full. In fact, few cases were encountered in which any considerations other than those noted above appeared to influence the decision. The descriptive framework rarely failed to include the factors noted directly or by implication. Indeed, objections to the framework are of the opposite sort: it is so broad and complex that it does not lend itself to generalization or analysis. The conclusions drawn here must be regarded as partial and in no sense exhaustive of the process they seek to describe.

At three urban sites the data on choice were poorly and incompletely recorded and had to be discarded, although the other data on use were completely satisfactory.

Complexity of Decision Situations

The facilities at present available to a user of water may not only affect the volume she uses but may also influence her response to changes in supply. The reactions of an urban housewife enjoying a piped supply to alterations in service will differ from the response of a rural user, accustomed to several practical choices. Although in the city ethnic groupings have less effect upon the volume used, they are related to attitudes toward cost and adequacy of service.

One of the problems in understanding decisions concerning water is that although some of them may be made by the woman of the home, they are often constrained by the husband's independent action. His pattern of water use and his location of the house may or may not take into consideration how far she has to go for water—very often it does not appear to do so. In the rural setting, the wife usually has little or no control over any cash the family makes from the sale of crops or cattle. As with the gathering of firewood (Wills 1968), she may not have the option of trading cash for energy by getting a metal roof or gutter and a rain barrel. Or she cannot buy water from a carrier. In the urban situation where the household is run on cash, she may have greater latitude in deciding whether her household money goes for food, water, clothing, or something else.

The simplest decision situation in an urban setting would be one like high income housing in Dodoma, where the water user is supplied with the full complement of plumbing accessories, enjoys satisfactory service, and has no theoretically useful options: wells would not be necessary or feasible, and the precipitation is too sparse to warrant cisterns. In these circumstances the only choice is how much is used from the taps. The effects of such a limited decision have been explored in chapter 5. More complexity is introduced where there are readily available alternative sources, such as cisterns to catch roof water or shallow wells which can be pumped at low cost by electrical devices.

The simplest choice situation in a rural setting—really Hobson's choice—would be one where the household has only one source from which to draw, as in an arid area where the second nearest waterhole is beyond the reach of a day's round trip. No such situation was found in the East African sites studied, although they are reported in northwest Kenya in the Lake Rudolph region and in much of the northeast. In every household investigated there was a theoretical choice of at least two sources. Indeed, the theoretical range is so great in all but a few instances that it is difficult to describe them all.

Take, for example, the Hoey's Bridge rural area. Of thirty households studied, fourteen volunteered that they considered only one source in drawing water. These were rated from low to perfect in general character, but all were regarded as yielding potable water. In the other sixteen households, nine sources were flatly rejected as unsuitable: four because of water quality, three because of energy costs, and one each because of technical difficulties or irritating relationships with other users. Among these households twelve sources, although perceived as usable, were not used because of preferences for quality (2), technology (1), energy costs (3), and combinations of resources and economics (2) or technology and economics (2). In one case the user was willing to sacrifice energy for better quality; in another case, the opposite held.

At Alemi, out of thirty households in the northern part of the area only five fail to perceive more than one source. Of the others, twelve reject alternatives because of water quality and thirteen reject them because of energy cost. In expressing preferences among the remaining sources, five are favored because of cost, three because of quality, and three because of interpersonal linkages. One user may reject several sources, giving a different reason for each.

Range of Choice

Although it might be expected that households that have used one source for a long time and are satisfied with it would not perceive other sources as alternatives, more than 60 to 80 percent of users in areas lacking a piped supply volunteered awareness of one or more alternatives. These households were at least aware of that degree of choice and were articulate about it. The other households would suggest alternative sources when pressed to do so. It is a mistake to think that because a resource manager habitually uses a certain part of the landscape she is unaware of or unable to express her awareness of alternative sites. The awareness may be different where municipal supplies are readily available.

There is some limit to the number of alternatives rural water users are able to identify. When asked to enumerate all the other sources from which they might obtain water, women commonly named one, two, three, or four sources (table 20.3). The mean for all respondents was 2.3, and the minimum, of course, was 1. The number rarely

Table 20.3 Source perception and density, by sites

Site Number	Site Name	Perceived Number Mean	Range	Sources per Square Mile	Percentage Using Some Rainwater
11	Kiambaa	2.0	1–3	2.8	32
13	Mukaa	1.7	1–3	4.9	9
14	Masii	2.3	1–4	3.4	9
15	Manyata	2.9	2–4	5.2	12
16	Hoey's Bridge	1.7	1–4	3.3	7
17	Mutwot	1.3	1–2	13.5	7
19	Mkuu	2.5	1–8	1.5	67
20	Moshi	2.2	1–3	16.7	100
22	Kipanga	2.7	2–4	0.6[a]	n.a.
23	Alemi	2.4	1–5	1.9	6
24	Iganga	3.5	2–6	12.3	100
25	Iganga (urban)	2.7	2–4	6.7	94
26	Kamuli	3.6	3–4	6.3	81
27	Mwisi	2.0	1–3	6.7	23
28	Kasangati	2.3	1–4	12.3	67
29	Mulago	4.2	3–5	13.3	100

[a]At Kipanga almost every house has a nearby pond and all houses with tin roofs collect water, but these sources provide water for only a few months at most. The sources used in the calculation are the improved wells.

was higher than 5. This suggests that 5 represents the maximum number of options water users usually consider, a number lower than the 7 suggested by Miller from his laboratory study. It raises the question whether there is a finite and rather small capacity for considering alternatives just as there appears to be a finite capacity for making judgments of simple sensory attributes (Miller 1956).

Considering the diversity of the sites, a question arises whether the range of perceived alternatives varies significantly from place to place. One simple bit of evidence is that at several sites in the banana belt, where the same interviewer did all the field inquiry among the same ethnic group, the numbers perceived were substantially larger in a rural environment than in a nearby urban one. Thus Soga users in the countryside reported 3.5 sources whereas those in the nearby town of Iganga reported 2.7. The same kind of contrast appears for users in Gogo territory and for Chagga users. But the opposite is true for rural and urban dwellers in the vicinity of Kampala. Indeed, the largest recorded mean (4.2) is in the Mulago area, and the next highest in the periphery of Kamuli. As is shown in table 20.3, each site has a relatively narrow spread around a mean. Field studies in

Bolivia (Teller 1969) and Colorado (Schmoyer 1967) suggest similar groupings at study sites. The means, then, do mark a difference in perception that cannot easily be charged to the interviewer or the ethnic background.

An effort was made to compare the perceived alternatives with those theoretically possible. The latter must be defined by the investigator, because the water user already has defined her view of the possible by those she mentions. Yet the criteria for a theoretically possible source are difficult to apply from place to place. In lower dry reaches of Mukaa only three sources are within a radius of one mile from several of the households and some users walk as far as three miles. In the banana forest area of Iganga as many as twenty sources are within one square mile. If a three-mile radius were taken the number would be in the hundreds. A further complication is that the same stream, as in Mutwot, presents opportunity for one or more sources in every short reach. Similarly, wherever there is a shallow aquifer, as along the borders of Mwisi swamp, it is theoretically possible to dig a well, and, short of crowding on the shore, there can be as many new wells as there are interested and competent diggers. Personal judgment in this regard varies tremendously among investigators.

One measure of the magnitude of theoretical choice is the density of actively used sources within an area of one square mile. These were identified by adding to the number named by respondents any others found to be in recent use. Because most households studied carry water less than one-third of a mile (the mean being 1,400 feet) and only a small proportion go beyond one-half mile, this may be considered a reasonable unit for measuring relative number. The mean density is given for the various sites in table 20.3, along with data on the percentage of houses using some rainwater during the year. In the calculations no weight is given to rainwater as a source. Wherever a house has a tin roof, rain collection is a possibility, and the density is as great as the number of such houses; but the practicality of storing water, as is shown in chapter 3, is related to the length of the dry season.

The number of sources regarded as possible by at least one resident and the observer range from two to fourteen per square mile, exclusive of rainwater collection from the roof. These densities are roughly associated with water balance, but the larger ones are found on urban margins of wetter areas. The number of sources perceived does not vary similarly. The ratio of possible sources to perceived is related to habitat. In the dry lands of Mkuu, Masii, and Kipanga the ratio is two to three times as high as in the wet forest areas of Iganga and Kasangati. The rural-urban distinction does not apply uniformly, for the perceived range of choice is lowest in the town of Karuri where people look chiefly to the county council standpipe, and highest in Mulago where there is great awareness of the variety of improved and unimproved sources.

Many of the sources shift with the season. When the rains stop, a Kasangati household will go to a spring instead of depending upon the roof supply. When a stream runs dry in Mukaa, water will be drawn from a more distant spring or well. However, the areas where seasonal shifts are common do not show especially keen perception of alternative sources. Even where a household is obliged to change sources seasonally and is located among a large number of alternatives, the choice typically is made among two to five sources.

This tells us nothing about why other theoretical sources are not perceived. When probing questions were asked about these other alternatives, a manager often would admit their existence and perhaps add an unfavorable comment about them. But it was not practicable to pursue all the options through the interview.

Why Sources Are Rejected or Preferred

Here we ask why reasons are given for rejecting or preferring a perceived water source, how these explanations are related to each other, and how they seem to be related to the particular situation and the experience of the user.

The two most frequent reasons for rejecting a source were its cost and its quality. Economic considerations were mentioned as a reason for not using at least one source in all sites (table 20.4). Women said that the path was too long, or the climb too steep, or the charges by water carriers too high. Water quality was mentioned similarly in all sites but Masii. Quality was judged variously: oily appearance, turbidity, mineral taste, and color were mentioned often. Without being able to specify any of the health costs reviewed in chapter 6, the woman at the well is likely to have strong judgments about what appearance of water is good for her family. Sites were evenly divided between those where quality of resource was the dominant reason for rejecting sources and those where economic efficiency was dominant, but for the aggregate of sampled households rejections were highest for economic reasons. Technology was of secondary importance, but it occasionally showed in a woman's inability to build a

Table 20.4 Factors associated with choice of water source for households carrying water

Site Number	Site Name	Rejected because of: Q	T	E	L	Preferred because of: Q	T	E	L	Q and E	T and E
11	Kiambaa	4	2	6	–	–	–	2	1	5	1
13	Mukaa	1	2	1	–	2	1	4	1	2	–
14	Masii	–	3	6	–	3	–	5	1	6	1
15	Manyata	10	4	8	–	2	–	–	2	1	–
16	Hoey's Bridge	4	1	3	1	2	1	3	0	2	2
17	Mutwot	1	–	5	–	–	–	1	1	–	–
19	Mkuu	16	2	4	–	4	1	1	–	4	–
20	Moshi	6	1	7	–	1	–	1	1	1	–
22	Kipanga	16	1	1	–	–	–	3	–	–	–
23	Alemi	15	1	14	–	3	–	6	3	1	0
24	Iganga (rural)	3	2	31	–	2	–	5	–	26	–
25	Iganga	5	–	3	–	2	1	14	–	7	–
26	Kamuli	1	–	13	–	1	1	10	–	13	–
27	Mwisi	5	–	2	–	6	4	4	–	1	3
28	Kasangati	2	1	12	1	5	–	5	1	6	–
29	Mulago	2	1	24	–	1	–	8	4	–	2
	Total	91	21	140	2	34	9	72	15	75	9

Q = Quality   E = Economics
T = Technology   L = Social linkages

cistern or dig a shallow well. Avoidance of contact with other people figured clearly in only two instances.

When reasons were given for preferences among those sources not rejected, heavier weight was placed on economic efficiency than on resource quality, the two were given as combined reasons in the largest number of situations, technology had an even smaller role, and linkages took on greater importance. Users found it somewhat easier to say why they would not use a source than to explain why one was preferred over another. Fewer preferences than rejections were expressed, and more than 40 percent of the preferences combined two considerations, such as resource quality and economic efficiency. Once the field of choice is narrowed to those sources regarded as usable, the articulated reasons become more ambiguous.

To the extent that households perceived only one source and were unable to give reasons for excluding other sources, or could not articulate their preference, it was impossible to account for all choices in a study area, and the totals of rejections and preferences shown in table 20.4 do not add up to the number of sources mentioned. At Mutwot, for example, many users said that they considered only the nearby tap or well, and could not discriminate among other possibilities. The resulting statement thus shows the reasons given but ignores the unspoken judgments.

One way of measuring the value placed on water quality is to examine instances where a household goes to a more distant source to obtain what it regards as better water. A similar comparison is found where the household consciously uses an inferior but acceptable source because of the economy of carrying a shorter distance. What one user would consider acceptable another might reject. In five of the households studied a conscious preference was voiced for a source which was cheaper, although less satisfactory in quality. In seven other households the user preferred to go a longer distance for what she regarded as better water.

In no case did people use sources they declared to be clearly unsuitable in quality, although in numerous cases they lamented the distance to a suitable source. They can only select among sources that appear to them to be viable options: it is fruitless to rule out all the available sources as not potable. What is considered usable in Kipanga may not be in Mutwot. This is sometimes puzzling to the water supply expert. At the Alemi borehole several households go to the relatively pure pump supply only when they regard the shallow wells as polluted.

Within all the study sites there is deep and widespread respect for quality of water as observed in color, turbidity, taste, and odor. Nowhere did we find widespread casual or indifferent evaluations of water sources. Most users had evaluative judgments, and most were interested in ways of improving their supplies. If they appeared to act contrary to the judgment of an expert it was for reasons convincing to them. The gap between the two judgments does not seem to rise from lack of motivation to gain healthful supplies; it comes from differences in information and its assessment.

At Kipanga, with its large number of small ponds, the discrimination is between the quality of the nearest pond, many of them rejected, and a more distant pond or well; cost does not seem a major determinant in the choice where all sources are relatively near. At the other extreme, at Iganga with its high rainfall, many corrugated roofs, and springs of similar quality, the great number of rejections of perceived choices is on the basis of energy cost; springs are rejected where rain collection is practicable, but between potable springs the closer generally are selected.

Perceived water quality also seems relatively less important in Kasangati, Masii, and Mutwot and in the urban fringes of Kamuli and Mulago. At Mkuu, where sources of different quality are located at similar distances from many households, emphasis is upon quality. And at Mwisi, with only modest differences in quality among equidistant sources, it is heavily so. These judgments on water quality can be compared with the values expressed in group attitudes and practices.

Attitudes toward Water

The value users assign to water is revealed in a variety of ways. It is shown by the care they take with it in the home or at the source, which necessarily involves a judgment as to its proper control. How much they are willing to pay for it offers a second measure. The interview results suggest how these attitudes are related to the decision situation and to the personal experience of the user.

Care for Water

In the varied ethnic groups and natural environments afforded by East Africa it is illuminating to ask whether the actual regard in which water is held in the family circle or in the community differs significantly. Inevitably, concepts of ownership enter into attitudes toward community or individual responsibility for water sources.

The community norms generally are enforced by personal comment or group discussion. Opinions and norms are most confused in the urban peripheries, where a variety of ethnic groups have to cope with new sources and work patterns.

Users often express a preference for the quality of a source because it is well cared for, or they avoid sources where other users are careless or slovenly. The requirements for help in cleaning and improving water sources are stronger in rural territory than in the towns. With the possible exception of Chagga arrangements, work crews and periodic cleanup operations tend to be informal and in response to the initiative of concerned households.

Where there is sharing of sources, there is usually some feeling of responsibility for keeping the facilities clean and in working order. The Gogo do not improve the sources much, and so there is no strong organization for this purpose. One woman may, however, use the hole dug in a dry stream bed by someone else. Among the Lango there is a strong feeling of responsibility. A group of women together will dig and clean the small hole which constitutes a well and keep it clean. They will not then prevent other women from using the well, but they may make remarks about their laziness in failing to build their own or to maintain the common one.

The strongest tradition of cooperation in waterworks improvement lies with the Chagga and their long record of irrigation from the streams of Kilimanjaro (Stahl 1964). A strong social system grew up around the need for irrigation, and although water for domestic purposes is freely accessible to any Chagga, he or she has considerable responsibility in the system. In the better watered areas a little furrow of water may run through or near every household. The system of pipes with outlets, as in Mkuu, may have been a natural outgrowth of earlier organization.

Where sharing of water sources has been the custom, the introduction of piped supplies serving only part of the population may cause a problem. At Karuri, one resident complained that "the neighbors require water from people who own piped supplies." Some owners of piped supplies solve this social dilemma by selling water to their neighbors by the tin, or by collecting rainwater from the roof in drums and letting their neighbors use this freely.

How a Source Should Be Used

The tribal groups differ considerably in their prohibitions regarding the use of water sources. Among the Gogo the same pond may be used for drinking, washing, bathing, and watering the cattle. There are no restrictions at all. The other tribes prefer not to use a source for drinking that is used for washing or bathing or where the cattle come to drink. This does not seem to be a strict prohibition, for a Nandi woman will complain that her husband will not improve her spring and keep the cattle out, as his only interest is that the cattle get good water. A Kamba local chief may order that cattle not be allowed to drink from a section of a reservoir used by housewives. Judging from the hoofprints at the spot, this rule is not much honored in practice.

Except for areas such as Kipanga, where ponds are available for the exclusive use of a single household, it is understood that users will not throw refuse into a common source at the point where withdrawals are made. For example, a Kikuyu or Embu family expects others to do laundry or bathing below the outlet pool of a spring. Opinions are less clear about the effects farther downstream, and commonly there are no prejudices against upstream use, beyond the range of vision. Thus pollution is deplored, but in highly circumscribed conditions. There are often customary places for washing clothes at the source. The Chiga, Soga, Ganda, Kikuyu, Kamba, and Embu all feel that washing clothes should be done in a designated place, or at least downstream from a domestic source. The Ganda rather frown at any clothes washing at the source, although some people do it. There may also be a designated place for bathing, as at one delightful Nandi spring, screened by a thicket of bushes at one end, with a fine view of the plain at the other. Women rarely bathe at a source, especially among the Ganda and Soga. The older men often bathe at home, but among the Lango and Chiga teenage boys would not expect to have bath water carried for them.

The tribal groups on the whole agree that water should be accessible to everyone for his daily needs. However, this free access is in practice somewhat restricted, especially if a well is close to a house. The Nandi show a distinct preference for a spring or well that is used by only one or a few families; some have wells near the house which are strictly their own and may even be kept locked. At Hoey's Bridge, where Nandi and Luhya people are new neighbors and may use the same well, each group seems to feel it would rather not share in this fashion, and many of the recently arrived Nandi expressed a desire to dig their own wells.

In the Mutwot area of the Nandi, a strong preference for individual sources was expressed. The pipeline, with an outdoor tap for each family, was a financial failure, but those with access to the water are pleased to have it. The Nandi tradition of absence of a central authority or government organs may contribute to their strong feeling for individual sources (Oliver 1965).

Among the Soga, Ganda, Chiga, Chagga, Lango, and Gogo, there is general acceptance of freedom of access to water, although if a well is in someone's backyard, it may be considered private property. At Mwisi, in Chiga territory, one man was accused of filling in a well just beside his house when he got tired of neighbors' drawing from it. Where rainwater is used, as among the Ganda, the owner of a tin roof may let a neighbor share the runoff.

There may be uneven adaptation in any one community of culture to changes in social organization. For example, in Manyata, Embu District, water has traditionally been freely available to all people. Recent land consolidation has taken the form of individual holdings of strips of land running from the tops of the ridges down to the streams. Where springs or favorite sites on the stream fall into someone's private land, the owner may now begin to object not to access to the water, but to trespass on his land. This may definitely affect the choice of a source, as people there do not like to go where a neighbor is unfriendly or may actually forbid them entry.

Paying for Water

Each respondent in an unpiped household was asked whether he would be willing to pay for water. Each respondent in a piped household was asked whether he thought the current charges were fair. These responses give one indication of willingness to pay for water. Since people's expressed opinions and their actual water payments were associated it is likely that these inquiries have some validity.

Sixty percent of those carrying water and willing to voice an opinion were disposed to some kind of charge if water supply were improved. Any response of this kind to a theoretical question is of necessity conjectural and cannot predict reaction in the face of a concrete project. A large proportion of the users found it difficult to state an opinion. It may be significant that the only sites in which more than half the respondents expressed opposition to the idea of payments were in the highly rural areas of Mwisi, Iganga, Manyata, and Mkuu. In all but three sites there was some opposition, and the matter obviously is one for local debate.

East Africa bears the imprint of English riparian law, but it also shows the influence of Arab culture with a view of water as a free and precious good to be made available to all men. In households having piped supplies there was general willingness to pay for water. Only in high-density sections of Dar es Salaam and Dodoma was there any substantial expression of doubt. These were, of course, in areas most directly affected by recent abolition of payments for water at municipal standpipes. The responses of users in households having piped supplies do not show marked associations with other aspects of supply. Nor do they correlate with the users' previous experience with water shortage or with unpiped supplies. Urbanization seems to rapidly reduce the hesitation rural users of carried supplies show in considering possible charges for improved service.

The general picture emerging from the interviews is one of respect for water quality as the user happens to perceive it, and for increasing willingness to pay for better water as the user is exposed to its benefits. The consumers' capacity to obtain the supply for which they are willing to pay is affected by their disposition to community action and the government's readiness to foster it.

Water Management in Community Life

Is the drawing of water a basically integrative function in dealings among people? We began with the thought that an individual in low-income societies tends to use common water sources because of the benefits they yield in social communication and discourse, and that the individual water user would be more interested in using common sources than in independent sources. The shared well was believed to be an attraction. As the study went on this hypothesis was thrown into doubt.

Religious beliefs may strengthen or weaken the desire for a communal source. For some Muslim women the trip to the standpipe may be an occasion for meeting friends. For others, as for the Swahili women of Dar es Salaam, it is not. Many of them prefer not to appear on the street and therefore buy their water from a vendor, even though it is free at the standpipe. Here they clashed with official policy, for in 1967 the government rounded up the city water carriers as nonproductive laborers and tried to eliminate them by settling them in the country as farmers. The Swahili women were joined in their protest by clerical workers and other single men, as well as by employed couples, who had no time to spend waiting at the standpipe morning and evening. The outcry was sufficient to make the city council back down and allow the carriers to remain (*Drum* 1967).

The interviews in both rural and urban areas failed to show that women had any strong desire to go to a water source to meet others. Although they often visited with friends at the well or spring or standpipe, seeking a source where friends would be encountered seemed decisive only when the available choices were acceptable on other grounds. Perhaps more important, they avoided sources where irritation, unpleasant talk, or careless behavior might arise. Avoidance of trouble at the well seemed stronger than search for companionship.

When presented with the choice of a private source restricted to one or two families and a more public source with many users, the East African drawer of water seemed to prefer privacy. She would make the best of the line at the well, but she behaved as if she would rather have an individual well or tap. Thus, the tendency is for users to seek individual sources they consider to be of high quality.

We have found that both men and women at the sites studied are accustomed to join in modest and informal efforts to maintain supplies. However, there is little evidence of organized efforts to provide a major new improvement by any means other than government action supervised from the outside. Men of the community rarely join together to apply new technology to obtain the improvements which would satisfy their family's requirements at acceptable costs. This is reinforced by the views and practices which prevailed so

long among colonial officers and that were held until recently by
government administrators.

What the Government Officer Says

One district official's perception of the choice open to the Ganda
woman described in table 20.2 reveals some of the thinking that has
entered into government action to guide the course of domestic water
development in East Africa. He is European in origin, trained in
technology, and earnestly committed to helping improve the health
of the people of the area. Although he is aware of a much larger
number of sources than those reported by the sample household, he
regards only one as suitable. The springs and stream are dismissed
because he is confident they are impure. He concedes that rainwater
from the roof might be satisfactory in quality but considers the
technology of collection and storage too difficult and too costly for
the family. He knows that although a rural pipeline might be technically feasible it would be impractical because the consumers in his
opinion would be unwilling to pay. This leaves the borehole as the
sole desirable choice and the one which he expects that the women
of the area would prefer on grounds of purity and convenience. His
perception of the choice can be recorded as shown in table 20.5.
When reminded that the local people have not been using the borehole, he is sad and suggests that they do not care sufficiently for
water quality and health. He is oriented to drilling wells or building

Table 20.5 Perception matrix for one official's view of a sample household

Theoretical Alternatives	Considered as a Source	Resource Quality	Technology	Economic Efficiency	Effect of Other People	"Source Rating" Summary Valuation
A₁ Borehole	1	2	2	2	1	9
A₂ Roof	1	1	0	0	x	0
A₃ Spring	1	0	x	x	x	0
A₄ Spring	1	0	x	x	x	0
A₅ Stream	1	0	x	x	x	0
A₆ ... ₁₀ Other springs	1	0	x	x	x	0
A₁₁ Possible pipeline	1	1	2	0	x	0

Perception of Alternatives

pipelines, and he believes that the local community would be incapable of providing either type of improvement on its own.

Such an assessment of the choice open to one household should not be taken as representative of all officials responsible for water. In another area at the same time it would have been possible to find an engineer who, judging from his performance, would have rated the choices somewhat differently. Without downgrading the borehole, he would have scored rainwater as feasible and have offered technical assistance to the household in installing the necessary equipment. The springs would have been viewed as being sufficiently susceptible to type 1 or 2 improvement to provide acceptable water. He would have explored the possibility of a rural pipeline and this might have led to inquiry about willingness to pay for a private source.

Both officials reflect their experience in a European country. Both enjoy government housing providing free water and the facilities for extensive garden use.

The difference between these two responses illustrates the spread in official views of individual choice and desirable policy affecting water. Both personnel and policy were changing rapidly in the late 1960s, and our study came at the beginning of new administrations and programs. The effects of alternative policies will be examined in the concluding chapter. As preparation for that appraisal it will help to state the views reflected in government policy in the middle 1960s. These are views encountered in numerous developing countries and transferred from high-income countries where they often prevail.

Social Guides to Individual Choice

Public agencies intervene in individual choice of water source in three important ways: they alter the quality and availability of supplies, provide information about water quality and technology, and vary the cost of water from a particular source. The latter results either from bringing water closer to households obliged to carry their supply or from changing the price charged for water at a standpipe or at the tap. Information is provided in a wide variety of forms and does not necessarily affect the perception users have of the opportunities open to them. Improvement of supply is, of course, the most widespread kind of guidance.

With several notable exceptions such as the Ministry of Health program in Kenya, the prevailing domestic water policy in East Africa in the middle 1960s was to build improved supplies. Typically these were rural boreholes equipped with simple hand pumps, a few pipelines such as those at Mkuu and Mutwot, a few large reservoirs like that at Mukaa, and urban systems combining standpipes with multiple-tap installations. The range of choice was largely restricted to works which could be designed and constructed under supervision from professional offices.

Among those improvements—largely classes 6, 4, and 2—the responsible officials examined new projects with criteria quite different from those of the individuals they were helping. They rated water quality chiefly on the basis of bacteriological contamination and were unwilling to improve any source not meeting high standards. Most of their expenditures were for carefully designed works sturdy enough to be free from failure in ordinary circumstances. Officers were very reluctant to build any improvement that could be criticized by European standards as impure or weak. Although customarily no charge was made for borehole water, the other works were designed with the intent of recovering the full cost of construction and operation. This meant that major rural improvements were carried out by designers and builders from the central or district organization and that elaborate contracts for repayment were negotiated with local government authorities before construction was begun. In putting down boreholes the skilled geological and well-drilling personnel often selected sites after consultation with local officials, installed the hole and pump, and then departed after instructing a nearby administrative officer about proper maintenance.

Technicians familiar with the countryside were able to identify areas where domestic supplies were short or conspicuously polluted, and were sensitive to expressions of need voiced by local officials and their legislative representatives. The decision to allocate funds for a rural improvement was made in district or national headquarters, taking into account need, political equity, and technical and economic feasibility. Rarely did it involve any canvass of the users' perception of the situation. That would come only indirectly in seeking contracts for payment for a finished design.

The decision to make urban improvements was reached by a somewhat similar process, except that larger cities such as Kampala and Nairobi had their own competent engineering staffs. After assessment of needs and political pressure, funds were assigned to installation of either standpipes or multiple-tap systems. High standards of purity were sought, although not always achieved under the handicap of inexperienced local operators, and financial solvency was a prime consideration. Water supply officers were under special pressure to get adequate supplies to government installations, housing estates, and planned residential developments.

The social guides to individual choice of water supply are rigid in one direction and almost wholly lacking in another. Insofar as the drawer of water has access to a borehole supply, it is as a result of a remote government decision. The rural pipeline follows only after a scheme meets central engineering standards and promises financial repayment. Urban standpipes result from interaction of city administrators, engineers, and political leaders. Multiple-tap installations come after formal decision by government authorities requiring contracts with the users.

The household interested in improving its individual (type 1) supply receives almost no information or technical assistance. Likewise, informal groups of water users (type 2) receive little help unless they join in contracting for a major construction project. In terms of information, advice, financial assistance, or formal regulations they are largely on their own.

In these circumstances, a general model of individual choice may be hypothesized from the recorded response of water users. It is a model that views the decision as comprising but not dominated by economic considerations, that assumes that future use cannot be predicted by extrapolating past performance, and that recognizes customary practice as susceptible to rapid change in conformity with the users' perception of alternatives and their outcomes. Drawers of water perceive differences in water quality which lead them to reject some sources and to consider others acceptable. Among the latter they prefer the source with the lower energy cost subject to three constraints. First, if the sources have about the same costs, preference for quality may determine the final choice. Second, difficulties of coping with the technology of drawing water may rule out certain sources. Third, the possibility of meeting other people at or on the way to the source may influence choice among otherwise suitable sources. The choice may be thought of as a branching process. From the not more than five or six possibilities that are considered, those failing to meet some minimum limit of quality are eliminated. The remainder are reduced to those regarded as not too costly. Among those meeting both quality and economic criteria, refined perceptions of quality, technique, and relations to other people operate to govern the selection of source used. Users value the quality of water as they perceive it, and are willing to pay a good deal for what they consider improvements.

Customarily, rural water users in East Africa observe certain standards in using and caring for sources. They do so with relatively little guidance from their government, and the common types of public intervention tend more to restrict than to widen the choices open to them.

Evaluating the Consequences of Water Management Projects

One of the frequently mentioned devices for dealing with improvement in environmental protection on the global scale is to establish some sort of uniform standards. Here I think our lesson from water management is—caution. We must recognize that in some cases where we have established standards prematurely, we've had disastrous results.

. . . the Western world has perpetrated standards of quality for household water—imposed by engineers from Birmingham and New York and Ghent—on Africa, South America and Asia. The net effect of imposing these Western standards has been to set back the course of water supply for developing countries. It has led to gross neglect of small-scale improvements in rural areas that would be somewhat below Western standards, but still highly effective, and it has promoted paralysis in the development of facilities in growing urban areas of Asia, Africa and South America. Western standards have impeded the course of improvement of water supply.

Unpublished paper, Columbia University, 21 March 1971, 9

Notes

1. An interesting variation in this method was tried by Olinger among rural people in the mountains of northern New Mexico in 1969 (Olinger 1970). She scored the perceived alternatives by comparing them with whatever was the source in current use, thus emphasizing the advantages and disadvantages the user would see in

changing to another source. This method may be especially helpful where a public agency is anxious to forecast consumer response to proposed improvements.

References

Drum (Nigerian edition). June 1967.

Miller, George A. 1956. The magical number seven, plus or minus two: Some limits on our capacity for processing information. *Psychological Review* 63:81–97.

Olinger, Colleen E. 1970. Domestic water use in the Española Valley, New Mexico: A study in resource decision making. M.A. thesis, Department of Geography, University of Chicago.

Oliver, Symmes C. 1965. Individuality, freedom of choice, and cultural flexibility of the Kamba. *American Anthropologist* 67:421–28.

Schmoyer, R. D. 1967. Decision making in the development of domestic water systems in Prowers County, Colorado. Master's thesis, Department of Geography, University of Chicago.

Stahl, Kathleen M. 1964. *History of the Chagga people.* London: Mouton.

Teller, Charles, 1969. Domestic water use in Tarija Valley, Bolivia. Unpublished manuscript.

Wills, Jane. 1968. A study of the time allocation by rural women and their place in decision-making: Preliminary findings from Embu District. Faculty of Agriculture, Makerere University College, R.D.R. 44 (draft).

21 Geography and Public Policy

This paper, published in 1972, was written in 1969, amidst the heavy
and heady days of strife arising from the quest for peace in Vietnam
and the related questioning of societal and ecological justice at home.
This strong, straightforward statement gave voice to the frustrations
of younger professionals seeking to mesh their personal convictions
with their professional actions.[1] It troubled those peers already upset
by the turmoil within the university, and so deep were the divisions
that some colleagues urged White not to publish it. It was written
at a time when he had already decided to leave the University of
Chicago and was about to accept an offer to direct the Institute of
Behavioral Science at the University of Colorado.

Unfortunately, perhaps, the paper is still relevant today. We do
not "look back with amusement to those harassed days of 1970 when
we entertained serious doubts that man can avoid a nuclear holocaust
or genuinely prevent global disorganization or keep from fouling his
nest irreparably." While there is now a cohort of geographers who
ask, "Will it help?" "Can I do it?" rather than "Is it Geography?"
in selecting research to do or courses to teach, the profession as a
whole has not accepted any responsibility for public policy or facilitated the efforts of those who do. And sadly, there may be a new
generation emerging for whom neither the issues of peace and justice
nor the issues of freedom of research and teaching appear relevant
to its professional pursuits.

The concern for social applications of geographic thinking in teaching
and public action is commendable. We should respect that concern,
applaud its persistent and imaginative expression, and heed its lessons. But to what extent do we now do so? And are we addressing
ourselves to the truly urgent questions? These are, why the Association until recently has given so little attention to problems of

Reprinted from *The Professional Geographer* 24, no. 2 (May 1972): 101–4.

public policy in a period of crisis, and what it might do to help its members deal with such problems more effectively.

Let us first consider the opinion that the human race, long hardened to vicissitudes of place and circumstance, has indeed entered into an unprecedented period of crisis. The principal elements in that judgment are familiar to all of us.

On a global scale, wherever economic development is seriously considered, there is recognition of the world community's failure thus far to find effective ways of closing the gap between the rich and poor nations. The twin problems of resource use and population density continue to worsen. As members of the human family we know very little about how to speed development of the poorer countries, how to slow population expansion in relation to economic growth, or what would be an adequate expression of these illusive notions of optimal population and carrying capacity. We do not truly understand how serious are the threats to survival of accelerated changes in global stocks of soil, oxygen, carbon dioxide, and other vital resources.

At home, the tensions between racial groups, regional groups, and the intellectuals and the hardhats become deeply divisive and strong. Our cities decay physically while spaghetti-like expressways carry the upwardly mobile to suburban havens. We are fast becoming disillusioned with GNP as an index of quality of life.

It is no longer academic or fanciful to pose again and again the question of whether the world society, in which the people of the United States currently are the most powerful and richest segment, can survive. Is this a hysterical, exaggerated view of the world we share with nearly four billion others? I think not. I would be delighted to be persuaded that it is and that with a little bit of luck in a halcyon time twenty years from now we all can look back with amusement to those harassed days of 1970 when we entertained serious doubt that man could avoid a nuclear holocaust or genuinely prevent global disorganization or keep from fouling his nest irreparably. Humanity's capacity to do any of these things is new and undisputed.

That it is alarming is a matter of personal and corporate judgment. If the sober judgment is that these are imaginary or highly improbable pictures of the future, there is no need to change stance or pace. If it is more than a remote possibility, then we are impelled to ask why so little attention is paid to them by geographers, and what, if anything, our Association should do to enable us to respond more constructively.

It is my contention that geographers have some useful contributions to make to ways of coping with most of these problems and
that the lack of more explicit concern with them is due in considerable measure to lack of skill in doing so and, in a broader sense,
to commitment of the profession to the development of a respectable
discipline. Many of us act as though we lack both status in the
academic halls and a sense of efficacy in affecting currents of thought.
Being uncertain as to the relevance of what we are doing to the
immediate problems of the world, we are more secure in talking to
our peers and to future generations of graduate seminars than in
trying to influence the course of events. This is enhanced by the fact
that until the present year, young geographers have been produced
largely in a highly restricted seller's market, encouraging them to
knot together in protected departments.

There has been ferment within the profession in this regard. A
small number of geographers are taking vigorous action in a wide
range of fields including urban blight, regional planning, resources
management, and poverty. The report of the Survey of Behavioral
and Social Sciences refers to some of this. The Detroit Geographical
Expeditions express similar aspirations more concretely. The 1970
annual meeting of the National Council for Geographic Education,
focused wholly on metropolitan problems, illustrates a parallel trend.
The Association has embarked upon at least two projects which
consciously cultivate social change. The Committee on Geography
and Afro-America frankly seeks to remove racial constraints. The
High School Geography Project set out to alter the high school
student's image of the world and to change teaching methods at that
level.

We should ask, however, whether these and similar efforts are
sufficient. The question can be examined in terms of the desirability
of action along at least three lines.

First, it may be helpful if those who share these concerns were
to consider their criteria for selecting research problems in the light
of possible social implications. Beyond the conventional questions
of what might contribute to theory or method, what would be practicable and what would be interesting, are questions such as these:
What is the prospect that the results will help advance the aims of
the people affected? To what extent is it feasible for those affected
to join in or consent to the research? Is the design of the research
pointed to applying the results? It would be ridiculous to seek any
uniformity in these or other criteria, but it would seem important

for us to encourage explicit appraisal of them and their consequences, using concrete cases wherever appropriate. Speaking only as one individual, I feel strongly that I should not go into research unless it promises results that would advance the aims of the people affected and unless I am prepared to take all practicable steps to help translate the results into action. One of the common and commonly destructive questions about research runs "But is it geography?" I would like to see us substitute "Is it significant?" and "Are you competent to deal with it?"

The same questions that have been suggested for research can and should be put to teaching.

Second, to the extent that members feel a strong sense of urgency or of problem in looking to the world around them, we may ask what kinds of steps might best be taken by the Association to advance and support that concern.

An immediate and reasonable step would be for the Council and the regional divisions to provide systematically for discussion at Association meetings and in its publications of aspects of the national and world situation that appear to lend themselves to fruitful geographic analysis. The same meetings might examine the experience which geographers and others have had in designing such research and in translating findings into changed public policy. Examples of topics would be the methods of coping with urban decay, the implications of national goals for population distribution, means of responding to perceived environmental hazards, the consequences of projecting and attaining a stabilized population, the adequacy of environmental impact statements, and the reduction of racial tensions.

Another relatively simple step would be for the Association to create committees dealing with what it regards as urgent problems. These would sponsor the exchange of experience among Association members and seek support for those who wish to push further in either individual or cooperative investigations.

The Council and regional divisions should be prepared to distribute information to their members on these topics and to take public positions where geographers demonstrate competence and consensus. The moment the Association indicates its intent, in addition to other activities, to explore implications of geographic teaching and research to public policy it will face a continuing and troublesome process of sorting out what seem to be significant issues and valid modes of action. It should welcome the necessary debate on these matters. It should be alert to distinguishing the fatuous problems and the activities that are pedestrian fire fighting or flabby reform.

It should not let other work lag. And it should be willing to ponder dreams.

What has been stated first as a problem of individual choice in teaching and research and then as a problem of corporate responsibility by our professional association has still greater significance to us as members of academic communities. Each academic faces in his institution the issue of how to reconcile our jealous protection of freedom of inquiry in research and teaching with our conviction that education should be an instrument of social change toward peace and justice. In taking, at one extreme, the position that the university should refrain from any political stand on issues of social value we ignore the impossibility of its doing so—for many of its corporate actions such as admissions, property management, and governance inevitably involve value judgments—and risk the destruction of the institution through its insensitivity to the needs of students. At the other extreme, in taking the position that the university should be highly responsive to the anguish of human frustration and suffering, we risk its destruction through the subversion by local politics of the probity and imagination which must be protected there.

History offers no ready-made precedent for the solution of this dilemma. There has never been a time of such rapid change in the stock of knowledge and of analytical methods. Never have religious and ethical values been in such a worldwide state of flux. Never before have the technical possibilities of abundance for all and of destruction for all been so attainable. We must work with all our heart and mind to find ways of coping with these opportunities and threats, and this may mean radical renovation of our universities. This, too, should be among our agenda.

I put forward these ideas with no aim of gaining acceptance for them by the Association individually or collectively. The issues involved must be thoroughly discussed if we are to play our proper roles as teachers and scientists. The problems can be defined more accurately; the solutions can be analyzed more thoughtfully; the implications can be described more felicitously. But the issues should not be ignored.

Let it not be said that geographers have become so habituated to talking about the world that they are reluctant to make themselves a vital instrument for changing the world. This position will no longer do for research, for teaching at the college level, or for teaching at the high school level. It can survive only at the peril of the society which permits its comfortable and encapsulated existence. If we wish to direct geography's very modest contributions to the structuring of new social process and organization, we can act now in three ways. We can commit ourselves to a continuing and persistent questioning of our own teaching and research in relation to its definition and reduction of social problems. We can advocate the adoption by our Association of measures to sharpen and support such activity by groups of us here and on the international level. We can give our thoughts to the reshaping of the university as an educational institution. None of the urgent questions will be resolved within a decade or two. But the immediate issue of survival may be. What is important is where we stand in relation to the tasks of society. Little is to be gained by critically pointing fingers at white faces in textbooks, at vapid generalities about world power, or at observations about resources and man that are perfectly true, perfectly general and perfectly useless. Each of us should ask what in his teaching and research is helping our fellow men strengthen their capacity to survive in a peaceful world. That inquiry may go on in office or on mountain trail or ghetto street or in front of a wall map. No organization can do it for us. But a common organization can stimulate our thinking, help us examine information and insights, and support new and unpopular exploration.

This is the path on which we should be moving. What shall it profit a profession if it fabricate a nifty discipline about the world while that world and the human spirit are degraded?

Response to the Presentation of the Iben Award

There is no generally accepted definition of interdisciplinary or multidisciplinary research. We can distinguish at least four types of activity which do cross the traditional boundaries. There is the "renaissance man" type of research which resides in individuals, who, acting independently, manage with great illumination and insight to integrate the experience and concepts of several fields. We are favored with only a few such gifted people, although I am able to count one of them as an economist colleague at the University of Colorado. It is a rarity that, like other blessings, cannot be organized and can only be welcomed when conferred.

Much more common is the "umbrella" type of research. Two or more individuals from different fields gather together under the protection of a symposium, an edited volume, or a grant. The results of independent work bearing on a common topic are presented with little or no intellectual teamwork. This can describe aspects of many problems but rarely leads to problem solving.

More complicated is the "federated" type of research. Two or more individuals join together to work toward an agreed goal with understanding as to the contribution which each will make but reserving to each the responsibility for conduct of the research and the character and quality of findings. This lends itself to the solution of less complex problems and can deal with more difficult topics given sufficient time and patience.

The type of research which directs itself immediately to problem solving is what might be called "managed" research of the type described by our NASA colleague. From the outset there is an explicitly stated goal and an agreement that all participating in the analysis or synthesis will direct their activities to stated parts of the investigation.

(1972m), 1288–89

Notes

1. For an extended discussion of the conflict between conviction and professional action, see below, vol. II, chap. 7 (Feldman).

22 Natural Hazards Research

This summary statement (as of 1971) describes an important pathway of geographic research—studies of natural hazards. In describing the pathway, White uses the opportunity to consider the road not taken. Geography, at least in the United States, had as a whole opted to pursue one of its four traditions beyond the others by elevating spatial analysis, focused on cities, and ignoring its earth science, regional description, and man-land traditions.[1] The turning away from the latter was most pronounced, coming ironically at the moment in history when society became aware of the ecological crisis. Scientific leadership in the crisis passed primarily to biologists, and geography as the traditional science of the human environment waned. But there are a few exceptions, and a major one is the work on natural hazards.

The paper draws together the myriad strands of natural hazards research and describes its evolution from floodplain occupance to a modest but universal research paradigm applied to seventeen distinct natural hazards in studies in fifteen countries. By forming a research agenda around a policy problem and social issue, a contribution is made to social change or public policy, new links are forged to other disciplines, and new questions, methods, and theories are formulated. This arises naturally, as dimensions of the problem emerge. The problem, rather than discipline or available methodology, dictates the research.

Written at the end of an ambitious international collaboration undertaken by geographers through the International Geographical Union, White expresses quiet satisfaction with this long line of research activity. Such satisfaction was not universal, and in recent years several critiques of this work have emerged.[2] Stripped of their occasional acerbic tone, there are two major sets of criticisms. One set focuses on the use of extreme geophysical events as a starting point for social analysis, noting the implied rejection of alternative and perhaps more suitable units of analysis, such as places, livelihood systems, social groups, or societies. Concentration on extreme

Reprinted from *Directions in Geography,* ed. Richard J. Chorley (London: Methuen and Co., 1973), 193–216. The author is indebted to Ian Burton and Robert W. Kates for comments on an earlier draft.

events detracts from comprehending the human ecology of everyday existence, which is the central task of the man-land tradition of geography. A second critique addresses the wide use of choice and decision models and the focus on individual perception. Such a view, the critics hold, implies that humans are masters of their fate and ignores the profound constraints, sometimes natural but more often social and economic, that limit the actions of both people and governments.[3] Today, it is a measure of the strength of hazards research that critic and exponent alike have expanded the basic paradigm to relate human behavior to both extreme and everyday events and to identify elements of both choice and constraint.[4]

To a remarkable degree during the 1960s, geographers turned away from certain environment problems at the same time that colleagues in neighboring fields discovered those issues. This cluster of problems relates to the relationship between man and his natural environment, with particular reference to the kinds of transactions into which man enters with biological and physical systems, and to the capacity of the earth to support him in the face of growing population and of expanding technological alteration of landscape. In their self-conscious concern for developing the theoretical lineaments of a discipline, geographers tended to overlook those problems with which they, by tradition, had been concerned and which do not fall readily into allotted provinces of other scientific enterprises.

By neglecting the theory of man-environment relationships and its applications to public policy, the geographer loses an opportunity to apply his knowledge, skills, and insights to fundamental questions of the survival and quality of human life. He also fails to sharpen and advance theoretical thinking by testing it in a challenging arena of action. Any critical examination of man's activities as a dominant species in an ecosystem draws upon and invites refreshing appraisal by workers in other fields.

This argument is demonstrated by the line of natural hazard research as it has taken shape over the past fifteen years. It is presented here as an instance in which pursuit of a public policy issue led to a simple research paradigm and a model of decision making dealing with how man copes with risk and uncertainty in the occurrence of natural events. The approach was refined and extended in a variety of situations, served to stimulate new methods of analyzing other geographical problems, and fostered a few changes in methods of environmental management by national and international agencies.

The study and policy activities related in this direction of research represent an attempt to deepen understanding of the decision-making process accounting for particular human activities at particular places and times. The research seeks application of new techniques to one of the old and recurring traditions of geographical enterprise—the ecology of human choice. The results are slim yet promising. The experience may point more to errors to be avoided than to procedures to be emulated. However, the approach deserves appraisal as a possibly fruitful way of orienting new research and teaching of an old problem.

Application of this model and paradigm does not require any drastic changes in institutionalized teaching and research. Nor does it claim to establish a new sector of geographical inquiry. Rather, it offers one device for bridging some of the divergent lines of current investigation.

The Problem

How does man adjust to risk and uncertainty in natural systems, and what does understanding of that process imply for public policy? This problem, raised initially with respect to one uncertain and hazardous parameter of a geophysical system—floods in the United States—provides a central theme for investigating on a global scale the whole range of uncertain and risky events in nature.

Genesis of the Research

Definition of the problem had its genesis in observation of the results of a massive national effort in the United States to deal with the rising toll of flood losses. In 1927 the Corps of Engineers was authorized to conduct a series of comprehensive investigations to find means of managing the river basins of the United States for purposes of irrigation, navigation, flood control, and hydroelectric power. The legislative authorization called for the presentation to the Congress (the final decision-making body with respect to new construction projects on interstate streams and tributaries thereof) of plans specifying the needs of each area, the types of engineering construction work which could be undertaken, and projects proposed for federal or state investment, giving the estimated cost and benefit. In the years following 1933 the so-called "308 reports" submitted to the Congress contained explicit benefit-cost analysis of possible construction projects.

In theory, to present a benefit-cost appraisal of a proposed project for a river basin required an analysis of the possible actions which man could take in managing the water and associated land resources of the area, and it also called for a systematic canvass of what, from the standpoint of society, would be the flows of social gains and losses to whomsoever they might accrue arising from any one of those interventions in the ecosystem. This was a monumental and presumptuous task.

Even in his most naive periods of technological mastery, man could not expect to understand the full set of consequences of any major interventions such as the channelization of the lower Mississippi River or the construction of a dam on the Upper Ohio or the building of a system of levees along the Sacramento. The investigator could make educated and hopefully intelligent guesses as to certain outcomes, for example, alteration of stream regimen. He could not hope to identify all possible consequences. Measuring them would be still more difficult. Moreover, to complete a genuinely competent appraisal of possible lines of action would require canvass of the full range of possible activities which might be undertaken. A proposed dam then could be compared with other steps such as a levee, upstream management of vegetation, or downstream management of the floodplain. Yet, the practical engineering and administrative imperative was to go ahead with such investigations, using the best knowledge then available and applying an elementary kind of economic analysis in order to show for those items which could be readily quantified an estimate of prospective benefits and costs.

Thus, a program of planning took shape which was to have major consequences for resources planning and scientific work in other parts of the world as well as in the United States. Benefit-cost analysis of water projects in the United States became the most sophisticated piece of social impact investigation for several decades. There were more careful and detailed methodologies for computing water benefits and costs than for any other type of public investment. The procedures as first developed by the Corps of Engineers were later revised and embodied in rules and regulations issued by the Bureau of the Budget and approved by the Congress in two separate stages and were the basis for extensive literature of economic analysis. The analysis was of a normative sort: it was designed to suggest ways by which estimates could be made as to the most effective investments to achieve specified public aims. Almost no time was given to finding out what in fact resulted from such investment. It was assumed that what was proposed—for example, the reduction of flood losses or the increase in waterway traffic—would in fact be realized if only the proper combination of technical means, discount rates, and time horizons could be found.

The 308 reports found their way into concrete action in a remarkably short period of time because they first appeared in the midst of the great economic depression of the 1930s and provided individual projects which could be used in mounting public works programs intended to relieve unemployment and stir economic recovery in the nation. The Tennessee Valley Authority was established with the intent, soon discarded, of using part of the Corps of Engineers 308 plan for that area. Large projects such as Grand Coulee Dam and the reservoirs in the Upper Ohio were authorized in the interest of revising a depressed economy.

Geographers took an early interest in this new line of planning but their more lively efforts either proved abortive or dwindled over a long period of time. They were active in the National Resources Planning Board—the first federal agency in the United States to attempt to draw together the plans of independent state and national agencies into single, comprehensive river basin plans—and they joined in analysis of area economic and employment problems.

This interest stirred an investigation of the range of alternatives with respect to flood loss reduction (White 1942).[5] It also stimulated the first comprehensive attempt to anticipate the full social impacts of a large impoundment. The impact study was carried out by the Bureau of Reclamation and associated agencies on the effects of the Grand Coulee Dam on the Columbia Basin (United States 1941). The latter work under the leadership of Harlan H. Barrows was not only a pioneer piece of interdisciplinary research, but defined in broad outlines and with notable gaps the problem of ecological impact which, while studied with considerable care for Grand Coulee, was not to be investigated again with similar energy or breadth until the late 1960s.

In 1936, following a series of disastrous floods affecting urban areas in the Mississippi system, the Congress authorized a national flood control policy which declared it to be the intent of the federal government to contribute to the cost of flood control works wherever the anticipated benefits from such works would exceed the anticipated cost. In 1938 a supplemental act provided that where reservoirs were selected as a means of flood control no local contribution should be required to the cost of projects inasmuch as the allocation of benefits among the several state beneficiaries was so complicated that it seemed best to charge it all on the federal account.

Twenty years passed, more than five billion dollars were expended on new federal flood control works, and in 1956 a geographic investigation was begun of what had happened in the urban floodplains of the nation as a result of the investments during the two intervening decades. That investigation was to be followed by more thoughtful and searching studies through which ran the common thread of a relatively consistent research paradigm.

Research Paradigm

In carrying out the 1956 appraisal of changes in land use in selected floodplains following the Flood Control Act of 1936, the geographic research group asked the following questions:

1. What is the nature of the physical hazard involved in extreme fluctuations in stream flow?
2. What types of adjustments has man made to those fluctuations?
3. What is the total range of possible adjustments which man theoretically could make to those fluctuations?
4. What accounts for the differences in adoption of adjustments from place to place and time to time?
5. What would be the effect of changing the public policy insofar as it constitutes a social guide to the conditions in which individuals or groups choose among the possible adjustments?

These questions were addressed to seven sites, chosen to give a diversity of conditions of floods, urban land use, and flood loss abatement measure (White, et al. 1958). A review also was made of the record for flood-control expenditures and flood damages for the nation as a whole.

Adjustments were classified in three groups as shown in table 22.1. From that view any human response to an extreme event in a natural system had the effect of (a) modifying the cause, (b) modifying the losses, or (c) distributing the losses.

Table 22.1 Types of adjustments to floods

Modifying the Cause	Modifying the Loss	Distributing the Loss
Upstream land treatment	Flood protection works	Bearing the loss
	Dams	Public relief
	Levees	
	Channelization	Insurance
	Emergency measures	
	Flood warning	
Flood	Evacuation	
Proofing	Structural changes in buildings	
	Land elevation	

A number of conclusions emerging from the field studies had an unsettling effect upon those who were responsible for federal flood-control programs, and triggered new investigations to probe unresolved questions. In brief, it was found that while flood-control expenditures had multiplied, the level of flood damages had risen, and that the national purpose of reducing the toll of flood losses by building flood-control projects had not been realized. Parts of valley bottoms were protected from floods, but increasing encroachment on the floodplain increased the damage potential from a smaller flood. One part of a city was protected by a levee, but new urban growth took place outside the levee. Works which controlled floods with a recurrence interval of 500 years were certain to fail with catastrophic consequences when the 1,000 year flood took place. The findings also indicated that because of the federal government's concentration upon flood-control works and upstream water-management activities to the exclusion of other obvious but relatively unpracticed types of adjustments, the situation was becoming progressively worse and showed no promise of being improved by a continuation of the prevailing policies. It was recognized that a rising flood toll might be beneficial if accompanied by larger benefits from floodplain use. However, the increased losses were contrary to the public expectation.

At that stage the study had (1) demonstrated that geographic research could have a direct bearing upon the formation of public policy in one country; and (2) posed a set of problems requiring further investigation if satisfactory policy readjustment was to be obtained. These problems centered on how to account for the differential behavior of individuals and groups in dealing with flood problems from one place to another. It had been shown that people did not behave as it had been expected they would when the benefit-cost ratios for several thousand flood-control projects had been drawn up. It was not equally clear why people had chosen the particular solutions they did and, therefore, what sorts of changes in public action would lead to genuine improvement in the character of their choices over a period of time. The effort to deal with this problem satisfactorily demanded further inquiry.

Models of Decision Making

In all of the benefit-cost analysis and in the earlier work on changes in floodplain use, it was postulated that the choice made by people living on floodplains was essentially economic optimization. This was in the tradition of economic analysis and conformed to the normative judgments on which the projects had been initiated. In essence, it assumed that individuals living in places of hazard would have relatively complete knowledge of the hazard and its occurrence, would be aware in some degree of the consequences, and would seek to make those adjustments which would represent an optimal resolution of the costs and benefits from each of the adjustments open to them. The ideal of the completely optimizing man was viewed as one rarely achieved in action, but as the framework within which a modified model, namely a model of subjective expected utility, might be explanatory. The subjective utility model held that man would seek to optimize but that his judgment would be based on incomplete knowledge and upon his subjective view of the possible consequences. It would be expected that if the view people had of the expected effects of using a particular piece of floodplain could be ascertained, it would be possible to judge their probable response by selecting those solutions which would give them the maximum net utility.

Neither the optimizing model nor the subjective utility model seemed to explain much of the behavior observed in the study areas. For example, it was found that although people seemed to recognize distinct differences in hazard from one part of the floodplain to another, they did not readily translate that recognition into differentials in assigned valuation of the property. People often-times returned to the use of land which had been severely damaged by floods being aware of the consequences of a recurrence and facing probable disaster of either a personal or financial character

from such recurrence. Adequately to describe behavior for purposes of predicting responses to changes in public policy required
the use of some other kind of model. Experimentation began with
other possibilities. The obvious direction in which to move was
the model of bounded rationality as described by Simon and others
(Simon 1959). It was proposed in a general sense for a variety of
resource management decisions and was developed in a more rigorous fashion by Kates in his study of Lafollette, Tennessee (Kates
1962). In examining the behavior and expressed perceptions of residents of a floodplain in the Tennessee Valley, Kates attempted to
find out how people perceive the hazard, how they perceive the
range of adjustments open to them, and what factors accounted
for differences in their perceptions. This required measurement of
clearly economic gains and losses as perceived by them but also
consideration of a number of other factors such as the information
available to the individual, his personal experience, and the physical nature of the event.

In the following years additional efforts were made to refine a
model of bounded rationality, the most recent being that developed
by Kates in connection with the collaborative research on natural
hazards (Kates 1971). It will be noted from a simplified version as
presented in figure 22.1 that a resource-management decision may
be hypothesized to involve the interaction of human systems and

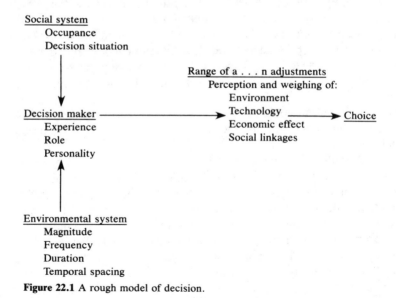

Figure 22.1 A rough model of decision.

physical systems in terms of adjustment to a particular hazard. The interaction is represented as a choice-searching process as affected by personality, information, decision situation, and managerial role. The result is a much more complicated model of how people make their choices in dealing with uncertainty and risk in the environment, one that did not lend itself as readily to careful field investigation but that promised more revealing explanations of individual and group behavior.

The Role of Perception

As it was recognized that the judgment of the resource manager could be more important than the judgment of the scientific observer, other types of investigations were stimulated. It was clear in the most elementary way that the definition of a flood hydrograph as developed by the meteorologists and hydrologists would be important in the decisions by individuals only to the extent that the described flood parameters were meaningful to the user. This resulted in suggestions of different modes of describing floods and different modes of presenting the results of scientific analysis to floodplain occupants. For example, the question was raised as to what sort of graphic display of a flood hazard would be meaningful to a person considering building a house on an area in which he had a choice between land above and below the maximum flood. To investigate this subproblem required knowledge of the occupant's perception— that process by which individuals organize exterior stimuli in order to form some concept of an event or situation—and as to the relationship between such perceptions, verbalized expressions of attitudes, and behavior. Here was encountered one of the truly difficult problems of social science: the relationship, if any, between verbalized attitudes and actual behavior.

In dealing with perception, it was recognized that psychological studies had been largely within the laboratory and had dealt with very limited physical phenomena. There was doubt as to their applicability to observations of more complex and gross phenomena. Hewitt investigated the theoretical ground for expressing extreme events in probabilistic terms (Hewitt 1969). At this point the interest of geographers in problems of perception and attitude formation converged with those of psychologists, sociologists, city planners, and architects who also were trying to specify perception and its implications (Burton and Kates 1964). Out of the concern for perception of floods came the first AAG symposium on problems of perception (Lowenthal 1967), a series of investigations dealing with perception of differing facets of the environment such as drought, recreational water, reservoirs, water supply alternatives, water recycling, and the like (Saarinen 1966; Baumann 1969), and Saarinen's geographic review of the perception literature (Saarinen 1969).

Out of the concern for attitudes developed a joint seminar between sociologists and geographers on problems of attitude formation, a joint investigation of attitudes toward water (White 1966), and a number of investigations bearing upon decision making and public participation in such decisions (MacIver 1970; Johnson 1971). Much of the research could not have taken place without strong cooperation with workers in other disciplines. Engineers were essential to appraisal of the effects of physical structures, and they often were the key professional group in applying geographic findings. Part of the field investigations were supported by the Tennessee Valley Authority's division of Local Flood Relations, and its engineering personnel were a chronic source of critical encouragement. Members of the Corps of Engineers engineering staff participated in a few studies (Cook and White 1962), and the agency later invited and used the results of appraisals of operating experience in floodplain management. Hydrologists from the U.S. Geological Survey shared in the design and assessment of floodplain mapping and its presentation to public agencies.

Wherever an urban area was studied there usually was collaboration with responsible city planning officials. Their critiques were illuminating, and in some cases produced interesting new ventures such as the combination of geographic planning and engineering skills in devising an urban redevelopment scheme for Waterloo, Iowa. However, it was necessary for geographic investigators to resist the temptation to become heavily involved in consulting activities. The pressures were to give time to applying the meager research findings rather than to expanding them.

Economists were drawn into the investigation and contributed fresh insights into the process of optimal decisions. Unconventional views of flood losses as nature's rental for floodplains were developed by two of them (Renshaw 1961; Krutilla 1966). A refined method of assessing losses and benefits from land-use regulation was devised by Lind (1966). A more rigorous analysis of the economics of natural hazards was carried out by Russell (1970). An investigation of rural use of floodplains in the early 1960s was supported by the Agricultural Research Service of the Department of Agriculture and brought geographers into working relations with agricultural economists (Burton 1962). However, it did not yield the anticipated refinements in economic aspects of flood hazard, and that sector of study awaits more intensive investigation.

Psychologists were drawn into examination of the personality traits affecting resource decisions. A simple sentence-completion test was devised by Sims (1970) and he collaborated in using thematic apperception tests (Sims and Saarinen 1969). A psychologist joined the research staff at the Department of Geography of the University of Toronto (Schiff 1971), and Kates collaborated with a psychology colleague in editing a review of hazard experience (Kates and Wohlwill 1966).

Interdisciplinary research in which workers in several fields genuinely interact is far more difficult to carry out than is research which draws from other fields at the pleasure of the investigator. In the latter case it is a readily manageable, sometimes gayly ostentatious, sometimes humbling exercise but always in the command of the investigator. When it is required by common commitment to solution of a problem, collaboration is not easily abandoned without personal hurt as well as cost to the whole enterprise.

Applications to Public Policy

It may help to briefly note a number of applications of this research paradigm and these models of decision making to specific public policy issues. The interest in each case is two-fold: (1) what was its use in forming public policy, and (2) what feedbacks, if any, did it have upon geographic theory?

At the outset of the floodplain occupance studies in 1957 an attempt was made to enlist the collaboration of people who were directly concerned in drawing up and carrying out such plans, and it was recognized that a principal alternative to the construction of flood control works was the regulation of land use. A representative of the Corps of Engineers took a year for a study leave to work with the Chicago geographic group and to produce a critical appraisal of experience with floodplain regulations (Murphy 1958). This led to tentative hypotheses as to community response to regulations and it also stimulated a legal investigation of the constitutional and statutory grounds for such regulation (Dunham 1959). The latter became the standard legal work on the subject.

Discussions of how people perceive the range of adjustments contributed to the establishment by the Corps of Engineers of a system of "floodplain information reports" which since have become operating practice. Early appraisals of providing floodplain information to residents of such areas showed the importance of individual perception in contrast to that of the scientific observer (Roder 1961) and fostered detailed experiments with modes of mapping sponsored by the U.S. Geological Survey (Sheaffer 1964). The Chicago metropolitan area became the first metropolitan area in the world to be completely mapped in terms of flood hazard. In the course of promoting and carrying out floodplain mapping through the Northeastern Illinois Metropolitan Planning Commission, further inquiries were made into the decision process. It became apparent that merely publishing the maps would be unlikely to have any significant effect upon decisions made by individuals or public agencies whereas if specific and favorable situations could be found in which the maps would be made available, the decision making might be changed. Thus, Sheaffer arranged for the organized group of land appraisers to make systematic use of the maps so that they in turn could attach a judgment about flood hazard to each land value assessment submitted to financial and mortgage agencies in connection with the purchase of buildings or property.

In addition, Sheaffer carried the first academic study of the possibilities of flood proofing, working jointly with personnel of the Tennessee Valley Authority (Sheaffer 1960). In time, that experience was the groundwork for preparing for the Corps of Engineers the preliminary manual of procedures for flood proofing for use of engineers, architects, and other technicians concerned with those alternative adjustments to floods (Sheaffer 1967). Geographers joined in the studies of receptivity that lay the foundation for the first fully operative national program of flood insurance in the United States (Czamanske 1967). They also helped sharpen the method of estimating flood losses (Kates 1965).

The eventual upshot of these investigations and their application in sample areas was the formulation of a new federal flood policy under a task force, in which geographers had a hand, established by the Executive Office of the President (United States 1966). The new policy involved basic changes in approaches and collaboration among nineteen different agencies, outlined a comprehensive effort by all interested agencies to deal with flood loss management, and inspired new lines of research and of data collection on their part (see table 22.2). An Executive Order (Number 11296) at the same time required all government agencies responsible, directly or indirectly, for locating new buildings on floodplains to take account of flood hazard in the location decision.

It would be a mistake to suggest that the resulting policy has been fully or effectively translated into action in all responsible agencies. Any basic change in bureaucratic outlook is slow at best. Yet, part of the geographic view of floodplain adjustments had been adopted within four years. A geographer had been appointed to head the floodplain study section of the Corps of Engineers. While reasonable progress was made by most units of government, several agencies dragged their heels against revisions in their procedures, as when the Department of Agriculture committed itself to building flow regulation and land treatment structures to the virtual exclusion of other types of adjustments (White 1970a).

In a number of states such as Iowa, Nebraska, and Ohio, geographers played a part in instigating and carrying out state efforts to apply the same comprehensive approach to flood problems in their respective areas. Under the leadership of geographers the Center for Urban Studies at the University of Chicago initiated several appraisals of experience with floodplain management which assisted in revision of federal operating policies.

Certain of the activities recommended by the Task Force found interested response in other countries. Thus, the preparation of maps of flood-prone areas was undertaken in France under the sponsorship of the Ministère de l'Équipement et du Logement (France n.d.; 1968). Studies of flood problems were sponsored by government agencies in Canada (Sewell 1965).

At the international level, the Department of Economic and Social Affairs of the United Nations joined with the Ministry of Reclamation and Water Management and other agencies in the USSR in sponsoring in 1969 a Seminar on Methods of Flood Loss Management. The Seminar brought together specialists, primarily engineers, from twenty-eight developing countries and a number of consultants, including one geographer each from Canada, Japan, the United Kingdom, and the United States (White 1970b). They gave careful thought to the approaches initiated in the United States, and in some countries the effects are now observable in national study activity. The United Nations report on the Seminar gives the Seminar findings, points out the implications of geographic research, and suggests new flood loss reduction policies for informal guidance of officials coping with flood losses in developing countries.

Table 22.2   Action recommended by the Task Force on Federal Flood Control Policy

To improve basic knowledge about flood hazard

The immediate listing of all urban areas with flood problems to alert the responsible agencies.

Preparation on maps and aerial photographs by the U.S. Geological Survey of reconnaissance delimitation of hazard areas.

More floodplain information reports from the Corps of Engineers and Tennessee Valley Authority.

Agreement by federal agencies on a set of techniques to be used in determining flood frequencies.

A national program by the Corps of Engineers and Department of Agriculture for collecting more useful data on flood damages, using decennial appraisals, continuing records on sample reaches, and special surveys after unusual floods.

Research by Department of Housing and Urban Development and USDA to gain greater knowledge on problems of floodplain occupance and on urban hydrology under the U.S. Geological Survey and HUD.

To coordinate and plan new developments on the floodplain

Specification by the Water Resources Council of criteria for regulation of floodplains and for treatment of floodplain problems.

Steps to assure that state and local planning would take proper and consistent account of flood hazard in:

    Federal mortgage insurance (Federal Housing Authority and Veterans Administration)

    Comprehensive local planning (HUD)

    Urban transport planning (Bureau of Public Roads)

    Recreational open space and development planning (Bureau of Outdoor Recreation)

    Urban open space acquisition (HUD)

    Urban renewal (Urban Renewal Administration and Corps of Engineers)

    Sewer and water facilities (HUD, USDA, Department of Health, Education and Welfare, and Economic Development Administration)

Consideration by Office of Emergency Planning, Small Business Administration, and Treasury Department of relocation and floodproofing in building flooded areas.

A directive to all federal agencies to consider flood hazard in locating new facilities.

The approach which seemed to be yielding results in the realm of flood losses was given application in several other sectors. The Office of Science and Technology established a special Task Force on Earthquake Problems which was patterned after the experience of the Task Force on Federal Flood Control Policy and which benefitted from the geographic contributions to the National Academy of Sciences report on the Alaska earthquake. Russell, Kates, and Arey (Russell, Kates, Arey 1970) pursued the problem of optimization in dealing with drought hazard as related to municipal water supply following the New England drought of 1965 (Russell 1970).

Table 22.2   (Contd.)

To provide technical services to managers of floodplain property

Collection and dissemination of information by Corps of Engineers in collaboration with USDA and HUD on alternative methods of reducing flood losses.

An improved system of flood forecasting under Environmental Sciences Service Administration.

To move toward a practical national program for flood insurance

A brief study by HUD on the feasibility of insurance.

To adjust federal flood control policy to sound criteria and changing needs

Broadened survey authorizations for Corps of Engineers and USDA.

Provision by the Congress for more suitable cost sharing by state and local groups.

Reporting of flood-control benefits to distinguish protection of existing improvements from development of new property.

Authorization by the Congress to include land acquisition as part of flood control plans.

Authorization by the Congress of broadened authority to make loans to local interests for their contributions.

Collaborative Studies on Natural Hazard

The various threads of inquiry were drawn together again in 1968 by a collaborative investigation of natural hazards supported by the National Science Foundation. Burton of Toronto and Kates of Clark joined with the author in examining the experience with a large array of hazards—drought, earthquake, flood, frost, landslide, hurricane, snow, tornado, and volcano—in a variety of settings (Burton et al. 1968).

Scientific as well as public policy response to these activities was sufficiently promising so that the International Geographical Union Commission on Man and Environment decided in 1969 to adopt as one of its two principal thrusts in the succeeding three-year period a program for international collaboration in the study of problems of environmental hazards. These joint investigations now comprise a number of comparative field observations and national studies as outlined in table 22.3.

The selection of study areas and collaborators was in many instances fortuitous: areas were chosen in terms of the inherent interest of the occupance and environmental problems but with a practical eye to the availability of competent personnel to carry out the work. It is hoped that from them will come a more rigorous and searching

Table 22.3 Field studies as part of collaborative research on natural hazards, 1971

Comparative Observations of Small Areas

Coastal Erosion	Scotland
	United States
	Wales
Drought	Australia
	Brazil
	Mexico
	Tanzania
	United States
Earthquake	Peru
	Sicily
	United States
Flood	Canada
	Ceylon
	France
	India
	Japan
	United Kingdom
	United States
Frost	United States
Hurricane	Pakistan
	United States
Landslide	Japan
	United States
Snow	United States

National Studies of One Hazard

Drought	Australia
	Tanzania
Hurricane	East Pakistan
	United States
Flood	Ceylon
Air Pollution	United Kingdom

testing of a number of the hypotheses that slowly have emerged over the years since the first office analysis was made of the range of adjustments to floods. Some of the early findings no doubt will be reversed or discarded. The principal hypotheses are now under examination by the collaborators at Clark, Colorado, and Toronto.

Perhaps the most fundamental of those hypotheses is that rational explanations can be found for the persistence of human occupance in areas of high hazard by examining the perception of the occupants of such areas and searching out their views of the alternative adjustments and the likely consequences of adopting any one of those opportunities.

In general, we suspect that there are three major types of response to natural hazards. Tentatively, we characterize these as follows: (1) Folk, or preindustrial response, involving a wide range of adjustments requiring more modifications in behavior and harmony with nature than control of nature and being essentially flexible, easily abandoned, and low in capital requirements. (2) Modern technological, or industrial response, involving a much more limited range of technological actions which tend to be inflexible, difficult to change, high in capital requirement, and tend to require interdependent social organization. (3) Comprehensive, or postindustrial response, combining features of both of the other types, and involving a larger range of adjustments, greater flexibility, and greater variety of capital and organizational requirements. We hypothesize that the United States currently is passing the peak of the modern technological type and is beginning to catch glimpses of the comprehensive type as it emerges here and elsewhere, but we do not suggest that there is a necessary sequence in the types of response.

It is also hypothesized that variations from place to place in hazard perception and estimation can be accounted for in considerable measure by a combination of factors embracing (1) certain physical characteristics of the hazard, (2) the recency and severity of personal experience with the hazard, (3) the situational characteristics of decisions regarding adjustments to the hazard, and (4) personality traits.

We have been inclined to try to describe choice of adjustment in terms of a perception model dealing with the individual manager's subjective recognition of the hazard, of the range of choice open to him, the availability of technology, the relative economic efficiency of the alternatives, and the likely linkages of his action with other people.

We further hypothesize that there are significant differences in the way in which these factors interact in relation to community action in contrast to individual action.

The Alternatives Approach

Another spinoff from the early floodplain investigations was application of the idea of range of alternative adjustment to other aspects of natural resources management. In elementary terms the alternatives approach in flood losses could be adopted to any other purposeful intervention in the environment. For example, in combatting stream pollution, the building of waste treatment works or of storage for diluting stream flow are only two of a much larger range of adjustments (Davis 1968). Alternatives would include such measures as controlling waste at the source, use of waste in agriculture, oxygenating streams, constructing special channels for waste transport, and the like.

This view was expressed in two reports from the National Academy of Sciences Committee on Water, with geographic participation (NAS 1966, 1968), that had a significant effect on national water policy in the late sixties. Attention to the full range of practicable adjustments converged with concern for systems analysis to produce water planning methods found in the North Atlantic Regional Water Study. In the North Atlantic study all interested federal agencies and twelve states joined in preparing river basin reports showing for each of three alternative aims (national economic efficiency, regional development, and environmental preservation) the range of possible activities, including nonstructural devices, which might be undertaken to meet perceived needs.

The same approach was embodied in part of the High School Geography Project. Its unit on environmental study introduces the student to analysis of alternative ways of dealing with flood losses in an industrial area. The treatment there coincides with increasing emphasis in the social scene on consideration of the range of possible social action in contrast to dependence upon simple technological solutions.

Appraisal

It is too early to venture an appraisal of how influential this direction of natural hazards research has been upon either public policy or geographic thinking, nor are we well equipped to try. In the short run it clearly has been linked to changes in methods of managing water and associated land resources in one nation and to a smaller degree in several others. What effect those changes will have in the long run is impossible to predict. As of 1971 they pointed to more searching examination of the range of choice available to man in coming to terms with his environment. Although the research has made only modest contributions to a theory of man-environment relations, it supported new efforts to specify the nature of environmental perception, to recognize the process of decision making for resource management, and to identify the landscape consequences of alternative public policies.

In essence, the activity was problem oriented and interdisciplinary. Such work is often tiresome and sometimes exhilarating. It requires research findings in a form highly intelligible to workers in other fields. It ignores conventional divisions of an academic field.

One lesson emerges from this history of investigation of a single environmental problem, using a rather unsophisticated research paradigm. It is that if environmental problems are pursued rigorously enough and with sufficient attention to likely contributions from other disciplines, they may foster constructive alterations in public policy but at the same time may stimulate new research and refinement of research methodology to the benefit of geographic discipline. Both may serve to advance man's painful, faltering, and crucial struggle to find his harmonious place in the global systems of which he is a part.

Environmental Perception and Its Uses: A Commentary

The evidence is slim for fields other than natural hazards. . . . So far as natural hazards are concerned it appears that the introduction of a perceptual dimension has led to somewhat more satisfactory statements in a few sectors, such as how people respond to warnings and how they decide to buy or not buy insurance and the circumstances in which they take steps to mitigate the deleterious effects of extreme events. The simplified explanations prevailing in the early 1960s are no longer accepted. At the same time the notion of perceived risk has found wide application in the burgeoning field of technological risk analysis, and it is commonplace to hear experts in a wide range of fields distinguish between "real" and "perceived" risks.

(1984a), 95

Notes

1. See William Pattison, "The Four Traditions of Geography," *Journal of Geography* 63 (1964):211–16.
2. E. Waddell, "The Hazards of Scientism: A Review Article," *Human Ecology* 5 (1977):69–76; William I. Torry, "Hazards, Hazes and Holes: A Critique of *The Environment as Hazard* and General Reflections on Disaster Research," *Canadian Geographer* 23 (Winter, 1979):368–83.
3. For elaboration of these constraints see below, vol. II, chap. 6 (Kunreuther and Slovic).
4. For these developments, see below, vol. II, chap. 9 (Whyte) and chap. 10 (O'Riordan).
5. All parenthetical references in this selection refer to the reference list at the end of the selection, not to White's bibliography at the end of the book.

References

Baumann, D. D. (1969) The recreational use of domestic water supply reservoirs: Perception and choice. *University of Chicago Department of Geography Research Paper no. 121.*

Burton, I. (1962) Types of agricultural occupance of flood plains in the United States. *University of Chicago Department of Geography Research Paper no. 75.*

Burton, I. (1965) Flood damage reduction in Canada. *Geophysical Bulletin* 7:161–85.

Burton, I. and R. W. Kates. (1964) Perception of natural hazards in resources management. *Natural Resources Journal* 3:412–41.

Burton, I., R. W. Kates, and R. E. Snead. (1969) The human ecology of coastal flood hazard in megalopolis. *University of Chicago Department of Geography Research Paper no. 115.*

Burton, I., R. W. Kates, and G. F. White. (1968) The human ecology of extreme geophysical events. *Natural Hazard Research Working Paper no. 1. Department of Geography, University of Toronto.*

Cook, H. L. and Gilbert F. White. (1962) Making wise use of flood plains. In *United Nations Conference on Applications of Science and Technology* 1:343–59. Washington, D.C.: Government Printing Office.

Czamanske, D. V. (1967) Receptivity to Flood Insurance. Master's Dissertation, University of Chicago.

Davis, R. K. (1968) *The Range of Choice in Water Management: A Study of Dissolved Oxygen in the Potomac Basin.* Baltimore: Johns Hopkins Press.

Dunham, A. (1959) Flood control via the police power. *University of Pennsylvania Law Review* 107:1098–1132.

France, Ministère de l'Équipement et du Logement. (n.d.) *Inventaire des Zones Inondables.* Paris: BCEOM.

France, Ministère de l'Équipement et du Logement. (1968) *États-Unis: Recherches Méthodologiques sur la Rentabilité Économique des Mesures de la Contrôle des Crues a L'Étranger.* Paris: BCEOM.

Goddard, J. E. (1971) Flood plain management must be ecologically and economically sound. *Civil Engineering,* September:81–85.

Hewitt, K. (1969) Probabilistic approaches to discrete natural events: A review and theoretical discussion. *Natural Hazard Research Working Paper no. 8. Department of Geography, University of Toronto.*

Johnson, J. F. (1971) Renovated waste water. *University of Chicago Department of Geography Research Paper no. 135.*

Kates, R. W. (1962) Hazard and choice perception in flood plain management. *University of Chicago Department of Geography Research Paper no. 78.*

Kates, R. W. (1964) Variation in flood hazard perception: Implications for rational flood plain use. In *Spatial Organization of Land Uses: The Willamette Valley.* Corvallis: Oregon State University.

Kates, R. W. (1965) Industrial flood losses: Damage estimation in the Lehigh Valley. *University of Chicago Department of Geography Research Paper no. 98.*

Kates, R. W. (1971) Natural hazard in human ecological perspective: Hypotheses and models. *Economic Geography* 47:438–51.

Kates, R. W. and J. F. Wohlwill, eds. (1966) Man's response to the physical environment. *Journal of Social Issues* 22:1–140.

Krutilla, J. V. (1966) An economic approach to coping with flood damage. *Water Resources Research* 2:183–90.

Lind, R. C. (1966) *The Nature of Flood Control Benefits and the Economics of Flood Protection.* Stanford University Institute for Mathematical Studies in the Social Sciences.

Lowenthal, D., ed. (1967) Environmental perception and behavior. *University of Chicago Department of Geography Research Paper no. 109.*

MacIver, I. (1970) Urban water supply alternatives: Perception and choice in the Grand Basin, Ontario. *University of Chicago Department of Geography Research Paper no. 126.*

Miller, D. H. (1966) Cultural hydrology: A review. *Economic Geography* 42:85–89.

Murphy, F. C. (1958) Regulating flood plain development. *University of Chicago Department of Geography Research Paper no. 56.*

National Academy of Sciences Committee on Water. (1966) *Alternatives in Water Management,* Publication 1408. Washington, D.C.

National Academy of Sciences Committee on Water. (1968) *Water and Choice in the Colorado Basin: An Example of Alternatives in Water Management,* Publication 1689. Washington, D.C.

Renshaw, E. F. (1961) The relationship between flood losses and flood control benefits. In "Papers on Flood Problems." *University of Chicago Department of Geography Research Paper no. 70.*

Roder, W. (1961) Attitudes and knowledge in the Topeka flood plain. In "Papers on Flood Problems." *University of Chicago Department of Geography Research Paper no. 70.*

Russell, C. S. (1970) Losses from natural hazards. *Journal of Land Economics* 46:38.

Russell, C. S., D. Arey, and R. W. Kates. (1970) *Drought and Water Supply.* Baltimore: Johns Hopkins Press.

Saarinen, T. F. (1966) Perception of the drought hazard on the Great Plains. *Association of American Geographers, Commission on College Geography.* Washington, D.C.

Saarinen, T. F. (1969) Perception of environment. *Association of American Geographers, Commission on College Geography.* Washington, D.C.

Schiff, M. R. (1971) Psychological factors relating to the adoption of adjustments for natural hazards in London, Ontario. Paper presented to Association of American Geographers, Boston, April 1971.

Sewell, W. R. D. (1965) Water management and floods in the Fraser River Basin. *University of Chicago Department of Geography Research Paper no. 100.*

Sheaffer, J. R. (1960) Flood proofing: An element in a flood damage reduction program. *University of Chicago Department of Geography Research Paper no. 65.*

Sheaffer, J. R. (1964) Economic feasibility and use of flood maps. *Highway Research Record* 58:44–46.

Sheaffer, J. R. (1967) *Introduction to Flood Proofing: An Outline of Principles and Methods.* University of Chicago Center for Urban Studies.

Simon, H. A. (1959) Theories of decision making in economic and behavioral science. *American Economic Review* 49:253–83.

Sims, J. (1970) Suggestions for comparative field observations of natural hazards. *Natural Hazard Research Working Paper no. 16. Department of Geography, University of Toronto.*

Sims, J. and T. F. Saarinen. (1969) Coping with environmental threat: Great Plains farmers and the sudden storm. *Annals of the Association of American Geographers* 59:677–86.

United States, Bureau of Reclamation. (1941) *Columbia Basin Joint Investigations: Character and Scope.* Washington, D.C.: Government Printing Office.

United States, 89th Cong., 2d sess. (1966) *A Unified National Program for Managing Flood Losses.* House Document 465. Washington, D.C.

White, Gilbert F. (1942) Human adjustment to floods. *University of Chicago Department of Geography Research Paper no. 29.*

White, Gilbert F. (1964) Choice of adjustment to floods. *University of Chicago Department of Geography Research Paper no. 93.*

White, Gilbert F. (1966) Formation and role of public attitudes. In Jarrett (ed.), *Environmental Quality in a Growing Economy.* Baltimore: Johns Hopkins Press.

White, Gilbert F. (1966) Optimal flood damage management: Retrospect and prospect. In Kneese and Smith (eds.), *Water Research.* Baltimore: Johns Hopkins Press.

White, Gilbert F. (1969) *Strategies of American Water Management.* Ann Arbor: University of Michigan Press.

White, Gilbert F. (1970a) Flood loss reduction: The integrated approach. *Journal of Soil and Water Conservation* 25:172–76.

White, Gilbert F. (1970b) Recent developments in flood plain research. *Geographical Review* 60:440–43.

White, Gilbert F., W. C. Calef, J. W. Hudson, H. M. Mayer, J. R. Sheaffer, and D. J. Volk. (1958) Changes in urban occupance of flood plains in the United States. *University of Chicago Department of Geography Research Paper no. 57.*

23 The Last Settler's Syndrome

White is always traveling. He is away from home overnight on the average of three weeks out of four and away from the United States at least six times a year. For this reason, perhaps, home has been extremely important to him, both his place of residence and its attached community. Washington, Haverford, Chicago (specifically the Hyde Park neighborhood) and Boulder are all places of deep attachment. In his middle years he even undertook a sentimental journey to Baden-Baden, the place where he was interned during World War II and reported it little changed.

In each place he makes time for some quasiprofessional community activity—a rehabilitation project in Hyde Park, planning the use of the Boulder floodplain, helping plan the scope and membership of the Boulder Area Growth Study Commission. It is for the latter that this public address was written.

In recounting Western history, White recounts a bit of his own history. He was born in Chicago in 1911. His father, a railroad man who had started work after grade school, selected Hyde Park in Chicago as a place to settle his young family, anticipating that the children could attend the new university being formed there. This anticipation had been nourished by his mother's visit to the World's Columbian Exposition of 1893 where, by chance, she took lodgings with a neighbor of the university's new president and learned of his intentions while chatting across the back fence. White worked summers on his father's Wyoming ranch, only to see it lost in the depression of the 1930s. As a last settler himself, he and his family have a large tract of land in Sunshine Canyon just above Boulder; they have tried to maintain it in its natural condition despite a suburban boom in mountain tracts and residence. His major recreations are taking care of the land and animals.

But he is amused and bemused as well by his own and his neighbors' values, sharing with first settlers the roots of a ranching youth, with late settlers a desire to keep things as they were, and with

Reprinted from *Social and Humanistic Aspects*, vol. 8 of *Exploring Options for the Future: A Study of Growth in Boulder County* (Boulder, CO: Boulder Area Growth Study Commission, November 1973), 80–85.

professionals a knowledge of the dynamics of change in place and changing places.[1]

Great waves of population movement accounted for the settlement of the west in the United States. Some thrust across the Great Plains, some pushed on across the Rocky Mountains barrier, some wasted on the plains and some receded. They left a series of impacts in that sector of the Front Range of the Rocky Mountains which we occupy here in Boulder.

We can see the remnants of these successive waves like old cultural beach ridges on the terrain. If one looks at the Rocky Mountains in the foothills section of the Front Range one finds traces of the physical remnants of these waves in remnants of old driving walls for the hunting of elk, in prospectors' pits, in the trailing bits of drift fences, and in the stumps of trees larger than any we see on the hillsides today.

In these traces we find one kind of pattern for the foothills and the front part of the Front Range and another pattern for the plains. Boulder sits on the hingeline between the plains and the foothills. In the foothills and mountains we can trace the Indians as the first settlers, coming sometime between 12,000 to 14,000 years before the present, and leaving very few imprints other than their effect on game and vegetation and to some extent the terrain. There was a long period of occupation by these first settlers of the continent. Next came the period of the explorers and the trappers, who left only small imprint. To be sure the trappers removed virtually all the beavers in a matter of a few years, but they left little evidence of their activities behind them. Then came the gold rush in the fifties, an invasion of people that radically changed the landscape and the cultural terrain of the mountains. As the miners pushed up Four Mile Creek and the ore road to Gold Hill, they established Ward and the other mining settlements. This was a time of vigorous movement in the development of the foothills, where the driving force was gold. In the tracks of the miners came the timber cutters who, using the ore roads and the railroads that had been pushed up toward the Continental Divide, began to cut out the trees. By the turn of the century, if one looks at the photographs that are now available as to the views from Boulder of the mountains, one sees that the whole mountain front was virtually denuded of trees. There were young Ponderosa pines growing, but the number of Douglas Fir that were to be seen anywhere were minimal. The whole vegetation of the Front Range was radically changed in that period.

In the tracks of the timber cutters came the ranchers, who wintered their cattle on the plain or in the foothills and then drifted them into the high country in June to graze on the nutritious summer mountain grasses. They constituted the fifth wave of settlers, along with a few farmers who tried and failed at farming in cleared foothills areas.

In the early 1950s the last of these waves hit as people moved into the mountains for residential purposes. In the beginning they were regarded as somewhat odd and eccentric and certainly impractical people to consider living in these areas. They certainly could not wrest a livelihood from the land. But they were the forerunners of what is now a major wave of suburbanites into this area which still goes on.

Down on the plains the events were somewhat more simple. After the Indian settlement had been disrupted by the military occupance of the area, agriculture began. The first form it took was livestock ranching, which was then relatively rapidly followed by irrigation farming. This spread and has sustained itself with accompanying commercial development until the recent wave of residential use spread out in the eastern part of the county and began to occupy farmlands for residential purposes. Thus we have in both plains and mountains today, a current wave of residential movement supported by commercial and governmental developments in Boulder and the growth of the university. All of these have brought increasing demands on service industries.

The character of this movement and particularly the character of the first settlers in the area has been, on the whole, well and beautifully documented in our American literature. We have broad sweeping interpretations, now the subject of much controversy, by people like Frederick Jackson Turner on the role of the frontier in American history. We have a whole series of literary interpretations of the life of the first settlers. The struggle of the early miners to find gold and to move on when they didn't was recorded for the western area by writers like Bret Harte and Mark Twain. The slow, gruelling efforts of the farmers to establish a permanent agriculture on the plains came alive in the works of writers like Rölvaag and Cather. For life in the mountains there is a great account by Isabella Bird, the English gentlewoman who traveled in the 1870s in the tracks of the mining rush and wrote about farmers on the plains and miners up in the hills.

She describes the first outpost in Estes Park and gives a beautiful interpretation of the strivings and way of life of these first settlers.

We have also a remarkable account of one of the last of the first settlers, those who took up public land in eastern Colorado in the early part of the twentieth century. Hal Borland went with his family to Badger Creek, due east of Boulder and south of Bush, and homesteaded public land. He gives a moving account, appropriately named "High, Wide and Lonesome," of the way in which that little family tried to make its place in permanent farming on the plains. One of the principal problems outside of drought, pests, severe cold and winds, were their predecessors in the preceding wave of settlement; the ranchers who cut their fences, ran over their fields with herds of cattle, and in every other legal and illegal way tried to discourage them from settling in that place. The family stuck it out, although Hal Borland didn't in the end. He came to the University of Colorado, then moved east, and became one of the eloquent voices of the environmental movement in the United States. For a long time he wrote the New York Times editorials on environmental matters, and he produced a series of books which are very sensitive in their interpretation of nature.

Most of this literary production has to do with the first settlers. But there is another end of the historical sequence, and that is the last settler in the area. There is as yet almost no literature of the last settlers, that is, the people who were last to come and are here now. I would like to suggest that many of these last settlers have certain characteristics, and that one can identify them by their beliefs and very often by their behavior. I would like to suggest what I think is a last settler's syndrome, some of its characteristics, and what might account for it as a unique feature of this whole set of waves across the plains. Then I will ask what its implications are for the communal life of a community like Boulder.

What I suggest derives from my observation, strictly unscientific, and I will present you with no detailed survey results, chi square comparisons, regression coefficients or other statistics. From my mountain-side observation, the principal symptom of the last settler is that he or she wants the area in which they live to remain exactly as it was when they arrived. It does not make any difference when they arrived. The important thing is, it must remain the way it was when they arrived, not a day earlier. This concern to keep things as they were when "I came," generally is accompained by at least three other secondary characteristics. One is that the person is militantly active in trying to keep things the way they were when he or she arrived. This can change; this is a positive symptom of the syndrome to which there is a negative. A second characteristic is that the individual is suspicious of any kind of change involving growth in the community or any change that would seem to threaten a departure from the situation that prevailed when he came. A third characteristic of the last settler's syndrome is a preference for environmental quality over economic well-being. The individual has already arrived and presumably has a job.

The first of these secondary characteristics has the possibility of shifting. The shift can be described as a move from militant action to disillusioned passivity. I am referring now to the individual who arrived in the year X and during the years of X + 1, X + 2, X + 3, and X + 4 was vigorously opposed to anybody else coming into the same area. In the year X + 5 he gives up, and while still displaying the basic preference, he feels that there is nothing he can do about it. He tends to withdraw from the tension of controversy about change and sits back and bemoans the tide of events. The people who exhibit the last settler's syndrome then fall into two groups, the militant activist and the disillusioned passivist. At the present time most of them are in the former group.

One can ask how we account for this kind of unique development. One of the characteristics of the first settlers was that they moved on. There were many, as recorded by Turner and others, who in crossing the plains moved three, four, five, six, seven times. Most of the miners moved many times and generally did not put their heels down as a group.

However, those who did put their heels in and stayed had very high aspirations for the development and growth of the area in which they remained. One finds this beautifully recorded in Willa Cather's "My Antonia." It is the story of the immigrant girl who is determined to make a family for her daughter, who sticks it out through all vicissitudes, and who has a deep belief in the capacity of a little community on the plains to develop a new and braver kind of life. Fussell, in his comments on the history of the Great Plains, places emphasis on the search, carried all the way to the Pacific Coast, for a kind of transcendental ethic which could be fashioned in some new method out of the new environment with new mixes of people. There was great concern for development but also movement.

If one looks at the last settlers in contrast to the first settlers, one finds that there is also movement, but a different kind of emphasis on development. One of the characteristics of the last settlers is that they are highly transient; out of the people in Boulder County in

1970, 60 percent had come in the preceding ten years. In the survey of opinion that the City of Boulder carried out within Boulder this past year, 40 percent had been here five years or less, including students who had a small share in the sample. Students, when one examines the total array of people in Boulder County, should not be considered as particularly transient members of the population. The average student undergraduate seems to stay about four years. No one seems to have accurate statistics on how long graduate students stay, but my hunch is it's a long time and probably longer than the average residence of the employees of one industrial employer in Boulder where half of the employees have been there less than five years. Probably the transience is no greater in the university than in one of the other large employers who has reported there is as much as a 24 percent turnover in a year. The average length of tenure for a faculty member at the university now is about eight years. We're talking then about a population in which there is a large amount of circulation and movement in and out. It would be a mistake to think those who are avid last settlers have been here a long time, or that they necessarily will stay here a long time.

Another characteristic of last settlers that I have observed is that they tend to be cooperators. They cooperate in stopping things and they cooperate in starting things. I am not quite sure what the balance is but they cooperate in both ways. Like some of the first settlers, they are very high in aspiration. Some of the studies made by the Boulder Area Growth Study Commission indicate that the kinds of considerations most often taken into account when people are asked how they feel about growth can be marshaled under a few headings: developing high environmental quality and decreasing pollution, increasing diversity within the city, seeking development of community cultural facilities, and improving well-being of members of the community. These are high aspirations and account for the special kind of concern for the well-being of the community that is exemplified by the efforts of the City and the County of Boulder at this present time. There is an attempt to assess where the community is going, if it ought to manage growth, and if so, in what fashion.

If you accept the argument that there is some sort of last settler's syndrome, one implication is that we have an unprecedented situation in our cultural history. For the first time we have a very substantial group of people who are not primarily concerned with changes in the direction of economic development but are concerned with maintaining settlement as they knew it when they came and in enriching it in that form rather than extending it. This is different, and we have little precedent in the political or educational processes for dealing with people with this set of mind.

A second implication is that although the emphasis is on keeping things the way they were, the population is constantly changing, so that the averge definition of what things were readjusts every year. One might find some people who will say "I can remember South Boulder when the Bureau of Standards was at the edge of the country and that is the way Boulder ought to be." Or some might say, "I can remember when Table Mesa was really out at the edge of things and that is the way Boulder ought to be." Then some who live even further south will say, "Well, the way it is now is the way we ought to keep it." You have then a succession of different views of what constitutes an optimal size or an optimal arrangement of the community which will not remain stable but will change as the people change.

A third implication is that rates of change and consideration of the ideal not only differ within the community but are constantly changing. Several questions grow out of this. One question is: is there any need for some special kind of education and information to help in the assimilation of this constantly renewed sector of the population which is seeking to preserve the place as it was when it came? In earlier years when the emphasis was on development, the stage could be set in terms of increases in population or extension of municipal facilities and other manifestations of growth. The future course of action was much more clearly defined.

Today, we have a kind of reversal of the situation. I suspect that much of our information, of our teaching and of our literature, is adjusted to the old rather than to the new. This presents a special problem for the schools. It also must be taken into account by the university and other large employers in the city and county who are responsible for the large influx and outflux of people. Is it possible and desirable to devise an education of a new sort that can prepare people to live with some kind of growth without falling into the disillusioned "passive" category? We know, as a result of some of the work of the Growth Study Commission, that considerable growth is inevitable in terms of the occupation of lands already blocked out for development. Is it possible to help people find ways in which, even with some degree of change, they can direct their efforts in part to trying to keep things the way they were but also towards the enrichment of the community? Can they put their militant inclination into action so as not only to preserve qualities of the environment which might otherwise be degraded but also to deepen and enhance the flavor of the life of the community at this size or at some larger size? There is a danger that if the issue is put as an issue of keeping things as they were or giving up, we lose a great asset of concern, energy, and imagination that might be directed toward enrichment. Can we somehow find means of continuing to cultivate among the last settlers the same kind of high aspiration for an overriding ethic of community well-being which was so deeply engrained in the first settlers?

Organizing Scientific Investigations to Deal with Environmental Impacts

Why is it that comprehensive studies of alternatives and the possible significance and character of impacts from each alternative have never been carried out? A multiplicity of very human elements accounts for much of this lack of planning. The persons who make a decision sometimes are actually unaware of the consequences and study them only after it is too late. This lack of foresight is due only partly to stupidity or ignorance; part of it is caused by lack of orientation and training concerning such problems or a lack of scientific knowledge about certain ecological relationships. Sometimes there is awareness of the problems, and even preliminary planning to investigate them, but the very complexity and uncertainty of the problems create perceptual distortion, and the hazards are minimized. Often, however, it is the technicians below the official decision makers who informally treat an ecological problem as unimportant; this is often in line with their professional traditions, the limitations of their expertise, and the failure to learn about or utilize past experience in other projects. Many economic considerations enter in; governments do not like to "waste time" in preplanning once a project is under way; the rewards of the primary planned-for impacts seem to outweigh the costs of secondary-impact damages or problems, so these are not seriously evaluated at the time; and budgets and construction costs create crises during actual implementation of a project which cause the so-called peripheral programs to be curtailed or cut out (with much higher costs often descending on the country after the damage has been done). Administrative prob-

lems, agency rivalries, the difficulties of interagency and interdisciplinary cooperation also constitute barriers to effective planning.

(1972e), 914–15

Notes

1. The population dynamics of Boulder are part of a worldwide
transition described in vol. II, chap. 13 (Burton and Kates).

24 Domestic Water Supply: Right or Good?

The following paper attempts to do for health and engineering professionals what the earlier paper on floodplain occupance (see selection no. 6) did for planning and engineering professionals—it seeks to raise their goals for improving water and sanitation beyond their current stagnant state and to do so by suggesting a broader approach than that in use. This paper on water supply draws from a major detailed field study of water use in East Africa (see selection no. 20) and a broad international survey,[1] as did the paper on floodplain occupance, which used detailed studies of seventeen floodplains and a broad survey of 1,020 urban places. Reviewing water supply provision since 1962, White finds a snail's pace of progress, with which three decades of further effort would find 1.5 billion rural people still with hazardous water supplies.

As with floodplains, he proposes a basic shift in orientation akin to a shift from controlling rivers to living in harmony with them—a shift from considering clean water as an economic good to that of a human right. This shift in orientation, if combined with a heightened technological flexibility and a culturally determined view about the steps that would lead to improved water supplies, could bring safer water to 95 percent of the world's people by the end of the century.

White was not alone in calling for such a global goal.[2] Many individuals, indeed several nations, have long placed universal access to clean water high or highest on their development priorities. Whatever the source, the idea caught on. In the U.N. Habitat Conference of 1976, the goal of bringing safe water to all the world's people by 1990 was established, and in the U.N. Water Conference of 1977, the major outcome was a proposal for a ten-year International Drink-

Reprinted from *Human Rights in Health*, Ciba Foundation Symposium 23, London, 4–6 July 1973 (Amsterdam, London, New York: Associated Scientific Publishers, 1974), 35–59. © 1974, Excerpta Medica. The author is grateful to Anne U. White, Bernd H. Dieterich, David Donaldson, and Miriam Orleans for reviewing the draft paper and to David J. Bradley, Ian Burton, and Robert W. Kates for comments on an early outline. A number of the observations have been influenced by discussions in seminars on rural water supply and sanitation organized by the International Development Research Centre in August 1972, and May–June 1973.

ing Water and Sanitation Decade. But as the time frame shortens, it becomes obvious that the call for new approaches and techniques has not been fully heard or accepted. Nonetheless, there is now, in the third year of the Decade, evidence of the acceptance of improved water supply as a long overdue development, of some greater sensitivity to cultural diversity, and of a flexibility in technology not seen before. The effort comes largely from national resources, even in these years of recession when increased international funds have not been widely accessible. Sanitation lags far behind. Even though the goal of complete coverage could be attained at one-twentieth the cost of annual expenditure for armaments,[3] it is unlikely to be reached.

Dimensions of the Problem

The most nearly accurate and helpful set of data on the status of domestic water supply is that compiled by the World Health Organization. Based on preliminary data from questionnaires addressed to member developing countries, the statistics for 1970 from 90 selected countries account for 45 percent of the total world population.[4] The results are compared in figure 24.1 with a similar

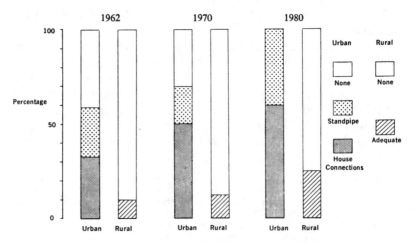

Figure 24.1 WHO estimates of populations of developing countries provided with adequate water for 1962 and 1970, and target for 1980. (See note 1.)

canvass of 75 developing countries in 1962 and with targets suggested by WHO for 1980, at the end of the United Nations Second Development Decade.

For statistical purposes the population surveyed is divided into rural and urban categories according to whatever classification is used in each country. Urban services are grouped according to whether the population (1) has a water connection within the house, (2) must carry water from a public standpipe, or (3) has neither service. Rural dwellers are classified according to whether or not the housewife has 'reasonable access' (not requiring 'a disproportionate part of the day fetching water') to a source that is uncontaminated.

The statistics are subject to several sources of error. They underestimate the population served insofar as the reporting agency is uninformed about local conditions or about the activities of other government agencies. An agency also may consider that a user lacks reasonable access to water because the agency did not provide the source. The reports often overestimate the population served in that they assume that once a community system is installed it is used or operated by the consumers in the way intended by the builders. Many a traveller in a developing country has encountered an unused well or a distribution system where the treatment plant no longer functions. On balance, the statistics probably overreport the community services and underreport the achievements of individuals and informal local groups. For any one country the figures may not be accurate, but the aggregate probably is moderately near the present situation.

On the basis of these data WHO concludes that for the 90 countries about 70 percent of the urban population, and 12 percent of the rural population, accounting for one-quarter of their total population, are served by improved supplies (fig. 24.1). The situation in 1970 marks a slight increase in the proportion of the rural population supplied between 1962 and 1970.

Change was far more rapid in Latin America than in Africa and Asia. For the total of 27 Latin American nations the proportion of urban users served from house connections and public standpipes increased from 60 to 78 percent while the proportion of rural people with improved supplies rose from 7 to 24 percent.[5]

This picture of status and change can be supplemented in several ways to give a more nearly complete outline of the world problem. To do so requires some long leaps of estimation in order to fill in four major gaps.

First, to account for the other 55 percent of the world's population, one must attempt an estimate for developed countries as well as for developing countries not reporting to WHO. Of the latter, the most conspicuous is the People's Republic of China with its 800 million citizens, give or take 50 million.

Second, it is desirable to classify populations according to spatial pattern. A helpful division in terms of problems of water supply and sanitation is as follows:

> *Cities:* Organized urban areas and their satellites.
> *Peripheries:* Disorganized shantytowns, bidonvilles, barrios, and other (one hopes) temporary living areas on the immediate fringes of cities.
> *Rural—clustered:* Settlements, primarily for agricultural purposes, of households grouped together.
> *Rural—dispersed:* Widely scattered households lacking grouping and nuclei.

These distinctions are important because the differences in density and arrangement of habitation are related to the types of management techniques used, the cost of providing water and waste services and the health hazards associated with them, as summarized in table 24.1 . The relationships vary according to density of settlement, the type of physical environment, and the volume of water used; a more detailed classification, for example, would distinguish between nucleated villages in arid regions and nucleated villages in wet lowlands.

Economies of scale and concentration make the water supply of a city relatively low in cost for low use per head, whereas the difficulties of waste disposal are relatively high as use increases, and the health hazards from both water and waste are high. Dispersed settlements may require high costs per head for water supply but they carry low health risks.

Table 24.1 Characteristics of settlement patterns for water supply and waste disposal

Spatial pattern	Cost of water supply	Health hazard from water	Cost of waste disposal	Health hazard from waste
City	Low	High	High	High
Peripheries	Medium	High	Medium	High
Rural				
Clustered	Low	Medium	Medium	Medium
Dispersed	High	Low	Low	Low

Third, while the prevailing standards of water quality constitute desirable targets for planning, they are not a basis for a binary classification of water service as a contribution to well-being. The definition of a 'safe' supply reflects the local judgment of acceptable risk. As Bradley shows,[6] the quantity as well as the quality of the water used makes a diference to the health of the user. The *hazard* to health from water should be thought of as ranked along a continuum from very hazardous to insignificant. In parallel to that continuum runs a second dimension of *amenity*, in which supplies can be graded from those having negative value to those yielding high degrees of convenience and pleasure. We may think of any individual or community supply as falling somewhere along the scales outlined in figure 24.2.

The two scales of water hazards and water amenities do not always harmonize: a Lango housewife may withdraw abundant supplies of Nile water to use for bathing, but the water may be so contaminated that her household chronically suffers intestinal disease; and a family in dry areas of Rajasthan may lament the small volume it can draw from the village standpipe during the dry season but enjoy the security of a fully protected supply. In general, the greater amenities are associated with large usage of supplies having insignificant hazard to health.

The volume used daily varies roughly among the four major types of population distribution and organization as shown in figure 24.3.

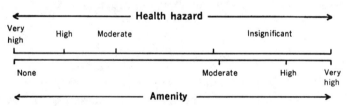

Figure 24.2 Scales of the health hazards and amenities resulting from the water supply.

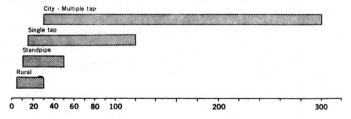

Figure 24.3 Daily consumption per head in liters for major classes of water improvements.

For city dwellers the mean daily consumption per person is 10 to 50 liters for those dependent upon standpipes, 15 to 120 liters for those having only a single tap in the household, and 30 to 300 liters for those with several house connections.[7] Rural consumers without tap connections or standpipes use 4 to 25 liters: their use of single tap and standpipe supplies is similar to but generally lower than that of urban consumers.[8] The picture is complicated by the many arrangements made to combine domestic with livestock, irrigation, and other uses. The actual measured consumption is somewhat less than that commonly used for design purposes.

Fourth, any evaluation of water supply as an environmental component of well-being implies some safe mode of disposal of waste, waste water, and human excreta. The more concentrated the population and the larger the volume of water withdrawn per person, the more difficult is the sanitary disposal of excreta (see table 24.1). In cities it involves either carriage and disposal by water or the collection of nightsoil and its disposal on land. In many urban peripheries disposal is casual and unorganized. In clustered rural settlements the technical problems are less acute because of the sparser population at risk and the greater space for disposal facilities. In dispersed settlements the cost of providing suitable water supplies may be large for any one household, depending upon local terrain and hydrology, but the difficulties and hazards of waste disposal are minimized. A program for water supply necessarily implies provision for waste. Sewerage lags far behind water supply in developing countries.

An Alternative Description of the Problem

Taking these four considerations into account, I shall present the problem of the world water supply in an alternative way. By dealing with the total population and by first classifying it according to spatial pattern, we can estimate water service in 1970 as shown in figure 24.4. In doing so, we are making rough judgments about the adequacy of water services in rural areas where WHO reports are not summarized. These rough estimates are presented here on a preliminary basis, in the hope that more nearly accurate estimates will be provoked or forthcoming.

For the higher-income countries it is estimated that 1 to 10 percent of the rural population is not now provided with adequate water and sanitation where required. Among the developing countries, it is estimated (on the basis of personal interviews while travelling in the

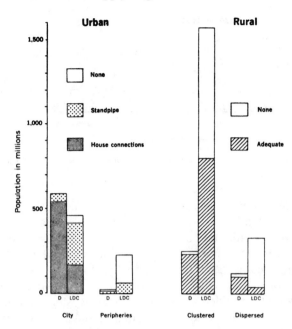

Figure 24.4 Estimated distribution of water services for the world population in 1970. D, developed and LDC, less developed countries.

country) that the People's Republic of China has provided standpipe service for 90 percent of its city dwellers, that the peripheral populations are relatively small, and that about 70 percent of the rural population, virtually all clustered, have adequate supplies and sanitation, if we take into account the facts that the Chinese tend to boil their drinking water and that they remove nightsoil for use as fertilizer on agricultural land. A major and rapidly growing component of the population in other developing countries is the peripheral group of shantytowns around tropical cities which are largely neglected and may rely on standpipes and water vendors.

On the basis of these rough estimates we obtain a picture which appears to be one of relatively adequate water for city people in developed nations; deficient supplies for more than half the cities in developing nations; generally inadequate supplies in the peripheries of tropical urban areas; and thoroughly adequate supplies for only about half of the rural dwellers—chiefly those in clustered settlements.

Figure 24.5 shows for the 1970 WHO statistics how the same data for 90 developing countries might be presented on a scale of health hazard. The important difference between the approaches in figures

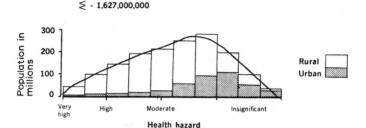

Figure 24.5 Distribution of population for 90 developing countries on a scale of health hazards.

24.1 and 24.5 is that the latter shows a continuum rather than a hard and fast dividing line. In that framework, the primary challenge is to move the population as far as practicable within financial constraints towards a sector in the distribution curve where health hazard is insignificant, rather than to transfer blocks of population from unsatisfactory to satisfactory categories. For example, in many tropical cities now provided with central water systems the most productive investment is to assure that standards of quality are maintained.

Applying the same approach and a great deal of guesswork to the total world population, we arrive at the estimate of the 1970 situation shown in figure 24.6. This global picture recognizes the very great progress already made by high-income countries, by builders of city water supplies, and by the People's Republic of China. It also suggests that the immediate activity on the world-scale which would offer the larger benefits would be the movement of the mass of rural and peripheral urban dwellers along the spectrum towards a situation of insignificant health hazard (fig. 24.7)

Obstacles to Improvement

This brings us to the most puzzling of the problems relating to water supply. Why is it that in almost all low-income countries the concern of the common people for improving their water supply has been insufficiently strong and persistent to force programs to provide such minimal improvements for all?

A possible explanation is that people's concern is a function of income and the way of life providing it, and that only as nations increase their income per head to some minimum will the basic right

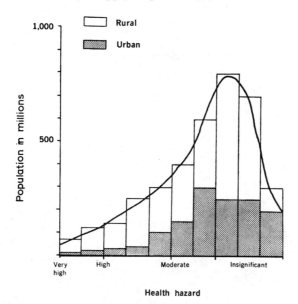

Figure 24.6 Estimated distribution of world population on a scale of health hazards for 1970.

to clean water be met generally. Such an explanation overlooks at least two important considerations. Some countries with a much larger Gross National Product still fall short of the goal. Argentina reports 16 percent of its rural population served, and among the higher figures reported for a large nation in Latin America is Venezuela's 46 percent. Cuba reached 60 percent in 1966 and Costa Rica 56 percent. At the same time, the People's Republic of China with an average GNP of $150 appears to have met the basic needs of a very large proportion of its rural population.

The argument that concern is related to income is not convincing, and we must look for other obstacles. They include the volume of supply, technology, finances, administration, training, and a catchall often loosely described as 'motivation'.

Enough Water

At the outset it is clear that in virtually all parts of the world with any significant population density, the amount of water available is adequate to meet reasonable domestic needs. The mean daily flow of the River Nile is enough to supply all of the domestic needs of

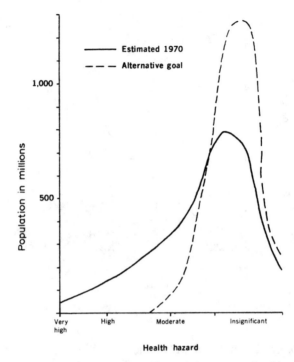

Figure 24.7 Estimated distribution of world population on a scale of health hazards and amenities for 1970, and for an alternative goal.

the world's population. This excludes industrial and public service requirements and the amenities of lawn-watering and swimming pools.

Even though there are places where there is less water readily available than the population would like for domestic purposes, the practicability of extracting water and transporting it over long distances is such that many (although not all) clustered desert populations can receive a supply. We can make only rough estimates of the financial hardship involved in problem areas.

Technology

There are many ways in which the techniques for finding, extracting, conveying, storing and treating water could be enhanced. There are a few ways in which a breakthrough in technology, such as a fixed-bed disinfectant without taste or odor, could bring about a sweeping change in the availability of adequate water and sanitation. It is painfully evident that the available technology is often not applied effectively: for example, disinfection in numerous tropical urban supplies.

Burton outlines the opportunities and the constraints elsewhere,[9] but for our purposes two generalizations seem warranted. The first is that technology *is* available to provide water to all inhabited areas. The other is that the costs of such techniques are so high in some places, or the skill of applying them is so demanding, that many dispersed rural areas are unable to use them.

Finances

The construction cost of water improvements runs from about $1 to $300 per head, depending upon environmental conditions and the desired quality of service. As outlined in table 24.2, multiple-tap supplies in cities fall in the range $10–$300; single taps cost $5–$40, and standpipes about $3–$20. Services to rural dispersed areas range from $1 to $30 in cost.

If we accept the view that all people should have an adequate supply, in order to reduce the health hazard, then the financial problem of providing it becomes one of finding the lowest effective cost and of paying the costs efficiently. The record for design of low-cost, multiple-tap supplies in cities in Latin America is on the whole excellent.[10] This suggests that consumers in cities and in many clustered rural settlements can be expected to repay all or a large part of the costs through volume or tax charges. In numerous cities where consumers using standpipes pay on the basis of the water withdrawn,

Table 24.2 Range of capital costs for domestic water supply improvements

Type of improvement	Cost per head in dollars (1970)
City	
Multiple taps	10–300
Single taps	5– 40
Standpipes	3– 20
Peripheral	
Standpipes	3– 20
Rural	
Clustered	3– 10
Dispersed	1– 30

Source: G. F. White, D. J. Bradley, and A. U. White, *Drawers of Water: Domestic Water Use in East Africa* (Chicago: University of Chicago Press, 1972).

users spend more per unit of volume than those with house connections, but the operating costs are so high and the collection of revenues is so cumbersome that system designers often tend to avoid standpipes and to disregard consumers who lack house connections. Slum dwellers may pay water vendors 20 times the metered rate. Figure 24.8 illustrates the range of changes paid by rural and urban consumers in East Africa.

It seems likely that for many nucleated population centers the capital investments in water improvements can be returned over amortization periods of about twenty-five years at low interest rates if there is strong public support, often in the form of grants, for

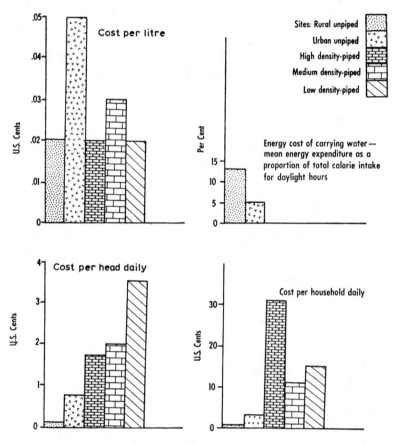

Figure 24.8 Direct costs of water in selected East African sites. From G. F. White, D. J. Bradley, and A. U. White, *Drawers of Water: Domestic Water Use in East Africa* (Chicago: University of Chicago Press, 1972).

design and for the training and supervision of operational and administrative services for the poorer communities. A difficult question of policy arises in the design standards in cities of low-income countries. The tendency in some of these areas is to design for more multiple-tap connections and higher daily consumption than probably would be required to bring the health hazard down to a truly insignificant range.

For the urban peripheries and dispersed and smaller clusters of rural population the financial problem is more acute. The costs per head are often high in relation to income, the difficulty of preparing adequate designs adjusted to local conditions is large and sometimes insurmountable when handled through a central organization, and operation and maintenance is impeded or destroyed by improperly trained but well-motivated personnel. The suggested solutions run along several lines including self-help, centralized management, or a combination of these.

Financial agencies argue that self-help projects, in which consumers share in the design and provide labor and materials so far as practicable, rarely work out, and construction agencies say that such works fall short of professional standards, but the number of systematic experiments with self-help is small. So long as the lending agency tries to assure full repayment and the construction agency is committed to completing works meeting the critical approval of its peers for water safety and durability of works, the process is slow, likely to be very costly, and unlikely to arouse strong local participation.

The works initiated and completed by central agencies incur higher costs in the long run, do not necessarily command use by the intended beneficiaries, and may fail to function when placed in local hands. Often it is asserted that the central agency must continue to operate the works if they are to attain their purpose.

Were the national government to regard the improvements as a means of redistributing income and of helping people to claim a right to clean water, the financial arrangements might be quite different and would recognize the full social costs, in contrast to financial costs, of inadequate supplies. In cities the charges would be adjusted to provide single-tap or standpipe service to the peripheries free or at nominal charges. The rural users would be encouraged to make their own improvements, with national tax income being used primarily for technical advice and sophisticated materials when required. In especially hazardous environments national subsidies to the users would be warranted, but the basic stance would be one of facilitating self-help by those individuals or groups that wish to improve their supplies and lend themselves to assistance of the type described above.

Administration

Wherever water is managed for the public good, the administration of the planning, construction, and operation tends to be complex because of the frequent need to serve several purposes—domestic, livestock, irrigation, electric power and the like—by multiple means. Domestic water is no exception. In cities it usually is a straightforward municipal service, but water supply is often under a different management from waste collection and disposal. Many peripheries are an administrative no-man's land. In rural areas domestic water may be handled in a variety of ways: as a principal aim; as a subsidiary to irrigation; as a sector of a national water agency; as a sector of public health ministries; or it may be divided among geological well drillers, engineering builders of surface works, and inspectors of sanitary conditions.

In most countries, the activities of national water agencies are focused chiefly on providing water whose quality meets the government's standards of purity with works that meet engineering standards of safety at costs that—it is hoped—will be repaid by the consumers. With notable exceptions, whether the costs are to be repaid or not, the works tend to be designed and built by experts from the central government organization. It is practical for central agencies to stimulate local initiative and contributions. Only in relatively few instances, such as the Chinese programs, does the agency serve primarily as a consultant to local groups or focus on basic education about the consequences and means of improved supply. In Guatemala and Peru, where the rate of improvement has been rapid, the central agency designs the works and promotes community contributions to the cost of construction and operation. The Peruvian communities pay for all local operation and maintenance; in Guatemala, the proportion ranges from nothing to the entire cost.

Training

Much of the training for work with water supplies and waste disposal has centered on the production of professional sanitary and civil engineers who can handle a wide assortment of environmental problems. In many countries short-term training enables intermediate personnel to operate simple works and keep them in repair. As with administration, the emphasis is on central responsibility for planning and construction with the expectation that the local groups will use, operate, and pay for the works. In only a few instances are workers trained to provide education in water needs and opportunities, or to encourage local groups to carry out their own improvements with whatever levels of health hazard they are willing to bear.

Motivation, Preference, and Value Systems

When known techniques are not applied, or peasants are not willing to pay improvement costs within their reach, or a highly organized national program fails to expand its projects, or consumers refuse to drink the clean supply, it is common to say that the people are ignorant or lack motivation. We know only a little about why many water and waste disposal schemes have failed or who so few are started, but several observations seem warranted by the studies made so far.

In virtually all settings where household water use has been studied with care the consumers have been found to be concerned with their own health and that of their families and to be willing to go to considerable lengths to safeguard their health. The steps they are willing to take do not necessarily include the improvement of the water supply. The housewife may be unconvinced that a bacteriologically pure, mineralized supply from a government well is more healthy than a pleasant-tasting supply from a polluted spring. Her husband may mourn the death of a son but fail to recognize that a day's work contributed to protecting the spring might have helped to prevent the death. It is more than a nicety of phrasing to avoid speaking of 'man and water supply'. For two-thirds of the world it is the woman who daily draws the household's water from rural source or urban standpipe, and who, if she has the option, often chooses the source that she regards as more healthy.

Rural consumers generally have clear preferences about water in a community value system in which the use and improvement of water is only one aspect.[11] Improvement programs, however sound technically, which ignore these perceptions of water, preferences, and values are likely to run into severe trouble.

In city multiple-tap or standpipe systems where the consumer has few or no real alternatives and where the system must be constructed by a central, professional agency, the assessment of such preferences and values may be unimportant. People either are willing to be connected up to the system or they are not. In rural situations where there usually are numerous choices and where individual contributions of labor or money may be essential to construction, such assessment may be crucial. The two essential and related ways of taking preferences and perception into account are to canvass people before improvements are designed and to involve the beneficiaries as deeply as is practicable in defining their needs and in designing, constructing, and operating whatever improvements they may choose to undertake. Their willingness to help themselves then becomes a criterion for external technical assistance, as it has done is Argentina.

The Underlying Orientation

The strengths of the organization required to meet the needs of city populations and high-income rural dwellers may become barriers when the same approach is applied to the poor of the urban periphery and rural areas. The first approach is oriented towards higher and relatively rigid standards of risk and depends on professional engineering skills, cash repayment by the consumers, and administration and training aimed at fostering such modes of action. Meeting the needs of the mushrooming urban peripheries will probably require a financial and design policy of getting clean water to all the people in minimum quantities necessary for sustaining health without elaborate repayment. To reach the more than one thousand million rural folk who need improvements will probably require a more flexible approach which first finds out local perceptions, preferences, and value systems, then initiates education on the health consequences of improvement, and later offers technical assistance in planning, constructing, and operating works to meet the aims of the user. Both approaches would hinge on a shift from regarding water primarily as an economic good to regarding water primarily as a human right.

Is a 25-year Goal Practicable?

In a period of twenty-five years would it be practicable to improve the water supply for those populations now inadequately served, so as to provide water supply and sanitation with low or insignificant health hazard to 95 percent of the human family? I am taking the time period of a generation as the maximum that a major world effort might be expected to accept in dealing with a matter of vital and universal concern. Five percent is a minimum estimate of the number of people who live in such rigorous and remote habitats that improvement will not be feasible for a long time to come. It is assumed that sewerage would be provided only where the health hazard would otherwise be very high, high, or moderate. If we extrapolate the trends of the past decade for ninety developing countries the answer is 'no': over three decades at least half of the rural population, or more than 1.5 billion people, still would not enjoy the benefits of improvement. If we assume that the level and tempo of technical development, administrative management, and training and education are stepped up as part of an international initiative based on helping governments to enable people to claim their right to clean water, the answer is a hopeful 'yes'. These are heroic assumptions that need to be examined critically.

Indeed, some who are familiar with both the exigencies and the remarkable advances in community water supplies over the past century would ask why one should launch a larger world program. Why not let work go on as in the recent past, simply speeding it up? According to this view, all that is needed is more money and more training of people to spend it. But the record gives no confidence that international funding could be increased or that the pace of national expenditure will accelerate beyond present rates, or even be maintained in some circumstances. Contrariwise, it might be argued that a time horizon of 1998 is far too long—that it is callous and complacent to expect events to march at such a slow pace.

With most ambitious undertakings the enthusiasm and energy harnessed by the initially high aspirations are accompanied by the danger that scattered or endemic failures in the early stages throw the whole effort into jeopardy. This might well be so with a 25-year, worldwide water supply and sanitation program undertaken either independently or as a part of larger health program. A preliminary planning component of about five years would seem essential. The preparation would include the refinement and popularization of possible technological choices for particular urban and rural habitats, the careful appraisal of the recent successes and failures of different management techniques in providing improvements in peripheral and rural areas, the financing of national training and educational programs, and the cultivation of an awareness among responsible national leaders of the opportunities and problems presented by accelerated programs. Basic to the new research, administrative studies, and tranining and education, is the fundamental change in approach noted here, with its emphasis on flexible standards and self-help responsibility.

In moving from a view of water as an economic good marketed where the comsumer can afford it to a view of water as a right all are entitled to claim by their initiative, national governments, co-operating through the World Health Organization, would commit themselves in three important ways. They would publicly accept the obligation to help consumers to achieve adequate supplies within some target period, recognizing that rights are to be claimed and merited by contributions of time and money, rather than given. They would train people to assist local and regional groups in making technically sound choices. They would allocate sufficient funds to pay for environmental health components in education, continuing technical assistance, and in-service training and would promote manufacture within the country of essential equipment which would otherwise make heavy demands on foreign currency.

Massive transfers of capital for construction projects would be neither necessary nor desirable. For the type and level of construction involved, foreign loans would be required chiefly for planning large city projects where revenues would repay the construction costs. The speedy provision of peripheral and rural facilities in many countries would call for international assistance in research on key technological problems, in establishing institutional competence to carry out education, training, and technical assistance programs, and in developing the capacity, either nationally or regionally, to produce the needed materials. International financial agencies would probably have the chief role in setting up the necessary institutions, but nongovernment international agencies might have influential parts to play.

Is this aim and this assessment of the practicability of reaching it within twenty-five years wholly visionary? I think not. More effective tactics and operating methods than those suggested here might well be devised and almost certainly would emerge if a major international effort were launched. Even so it might fall far short of the mark. Perhaps more certain is that unless there is a worldwide marshaling of technical knowledge, administrative wisdom, and political enthusiasm, based upon the simple principle of a healthy environment as a human right, the end of the century may see few advances in the lot of the poor on farms and in urban shantytowns in terms of water for domestic well-being.

Natural Hazards and the Third World: A Reply

Only one of the studies touched upon the Sahelian region, but the experience reported from there since 1973 bears out much of the observation which can be drawn from the natural hazard studies under review: hazard always arises from the interplay of social and biological and physical systems; disasters are generated as much or more by human actions as by physical events; the present forms of government intervention in both traditional and industrial societies often exacerbate the social disruptions from extreme events; if we go on with the present public policy emphasis in many regions upon technical and narrow adjustments, society will become still less resilient and still more susceptible to catastrophes like the Sahelian drought. It has been much easier for governments to turn to cloud seeding, well drilling, and relief operations than to move toward fundamentally harmonious patterns of land use. There are already signs that prevailing policies are being challenged and in a few instances reversed. To assist in both diagnosis and prescription of constructive remedies the social scientist needs to do more than bemoan the spread of short-sighted development measures or attempts at cross-cultural comparisons. There is urgent need to help in the design of alternative policies which will be sensitive to indigenous values, perceptions and creativity, and will stimulate rather than constrain local initiative.

(1978m), 230–31

Notes

1. C.S. Pineo and D.V. Subrahmanyan, *Community Water and Excreta Disposal Situation in the Developing Countries*, Offset Publication no. 15 (Geneva: World Health Organization, 1975).

2. See below, vol. II, chap. 1 (Wolman and Wolman).

3. R. Leger Sivard, *World Military and Social Expenditures, 1982* (New York: Institute for World Order, 1982).

4. Pineo and Subrahmanyan, *Community Water and Excreta Disposal.*

5. D. Donaldson, *Progress in the Rural Water Programs of Latin America (1961–1971)* (Washington, D.C.: Pan American Health Organization, 1973).

6. D. Bradley, "Water Supplies: The Consequences of Change," in *Human Rights in Health*, Ciba Foundation Symposium 23, London, 4–6 July 1973 (Amsterdam–London–New York: Associated Scientific Publishers, 1974), 81–91.

7. G.F. White, D.J. Bradley, and A.U. White, *Drawers of Water: Domestic Water Use in East Africa* (Chicago: University of Chicago Press, 1972).

8. White et al., *Drawers of Water*; D. Donaldson, *Progress and Programs: Rural Water Supply in Latin America*, mimeograph (Washington, D.C.: Pan American Health Organization, 1973).

9. I. Burton, "Domestic Water Supplies for Rural Peoples in the Developing Countries: The Hope of Technology," in *Human Rights in Health,* 61–70.

10. A. Wolman, "Technical, Financial, and Administrative Aspects of Water Supply in the Urban Environment in the Americas," *Ingenieria Sanitaria* (November 1959):1–13.

11. R. P. Morfitt Associates, *A Non-Conventional Mass Approach to Rural Village Water Projects* (Corvallis, OR: R. P. Morfitt Associates, 1969).

25 The Role of Scientific Information in Anticipation and Prevention of Environmental Disputes

Three lines of activities come together in this paper—White's Quaker orientation toward conflict resolution and avoidance, his growing participation in interdisciplinary "blue-ribbon" scientific panels, and the emergence of new interdisciplinary efforts in environmental risk and technology assessment. After joining the Society of Friends in 1940, White spent the war years as a conscientious objector in relief work in France during 1942–43 and then worked with the American Friends Service Committee (AFSC) in Philadelphia on relief for India and China during 1944–45. Later, as a member and then chairman of the AFSC Board, he took part in high-level meetings of diplomats searching for ways of conciliation and reduction of world conflict. His first major attempt to mesh his values and professional knowledge took place in the context of his Mekong proposals (see selection no. 12). This paper represents another attempt.

In 1973, White was elected to the National Academy of Sciences and proceeded to work within the framework of the Academy's National Research Council, a major source of governmental and public advice on scientific matters. He led the Research Council's Environmental Studies Board and eventually its Commission on Natural Resources. Playing a parallel international role was the Scientific Committee on Problems of the Environment (SCOPE) of the International Council of Scientific Unions; White served first as the representative of the International Geographical Union to SCOPE and then as its president for six years.

In 1974, he was invited to a symposium on benefit-risk decision making that encouraged him to link his experience in natural hazards risk perception and decision making to the newly emergent field of risk assessment and its predecessors, environmental impact and technology assessment.[1] All three strands come together in this pa-

Paper presented at the Conference on Adjustment and Avoidance of Environmental Disputes, Bellagio, Italy, 19–23 July 1974.

per, in which he attempts to sensitize scientists to the social characteristics that affect their ability to assist in reducing environmental conflict.

Environmental disputes ordinarily spring from an actual or prospective human intervention in the environment which provokes changes in natural and social systems. The dispute arises when the activity is seen by one or more of the groups affected as disturbing the complex interaction of physical, biological, and social processes of the environment in a fashion believed to be deleterious. When the chain of events from intervention to impact crosses a national boundary it becomes international: factory smoke in one country causes a change in acidity of lake water in another country; pesticides draining from cotton fields in one nation lodge in the fish of another nation's coastal waters; irrigation withdrawals upstream reduce power generation downstream.

There are two other important grounds for international disputes over environmental matters that do not stem from human intervention. The more general one is disagreement over the management of supply of a natural resource in one state as it affects the use of resources elsewhere: this is inherent in manipulation of supply and prices for scarce materials. A special and thus far unimportant circumstance is where climatic change independent of any human activity places new stresses on environments and generates fresh adaptations to available resources; a new ice age is not in any nation's five-year plan, but the possibility of widespread change is always present, and some observers think it may be under way.

Considering only those disputes triggered by human intervention, scientists until recently have tended to be drawn into them at the stage when the issue is drawn and fact-finding becomes an essential part of adjustment. The volume of water in a stream is established, the concentration of nitrates in a lake is measured, changes in rainfall are estimated. With massive expansion of technology during the past four decades, the capacity to alter natural systems has grown immensely, the number of possible interventions has multiplied, and the scientist increasingly identifies issues before the administrator, legislator, or judge calls for help in adjusting them. This is partly because some of the biological impacts of new chemicals are not apparent to untrained observers, in contrast to water shortages, salted land, coal smoke, and disappearing species. It is also because national governments, smarting from criticism of certain technological measures symbolized by the large dam, have taken steps to appraise more thoroughly the environmental impacts of new development projects.

The insights and competence of the scientist now are applied in a somewhat haphazard fashion on the international scene. Ways of marshaling those assets earlier and more effectively could be organized as part of a broader effort to anticipate and prevent the numerous disputes which now appear to be taking place.

Interactive Systems

It is helpful to remind ourselves that use of natural resources always involves an interaction between human users and the soil, water, air, and organisms of the place. Each is modified, sometimes slightly, sometimes so as to destroy the resource or the people. Two or more systems interact.

The major components in an environmental dispute are outlined in figure 25.1. A human action disturbs either a biological-physical system or a social system so that their interaction is altered. This provokes some impact in the related systems. Usually those affected make an explicit or implicit judgment, after trading off all the impacts, so that the net effect is either advantageous or disadvantageous to them. If those having rights in the area feel they may be damaged or at a disadvantage the dispute takes shape.

It may or may not lead to still further human intervention to extend, correct, or retract the earlier action. The perturbation in a

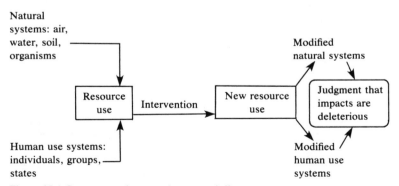

Figure 25.1 Components in an environmental dispute.

social system need not involve a change in commodities or resource use. It may only change perception of the significance of the resource and the value in which it is held by a society as, for example, when new information about the presence of a new substance in drinking water provokes objection to upstream waste discharge even though there are no verified changes in the health of the consumer.

This may be illustrated in a more concrete form by the case of weather modification as outlined in figure 25.2. A cloud seeding project is financed for wheat farmers in a semiarid region. They perceive that in their new farming practice, in which cloud seeding is an integral part, the seeding increases rainfall and crop yields mount. The costs are small by comparison with the gains in wheat revenue. A group of farmers in another jurisdiction downwind does not practice cloud seeding. These farmers perceive weather modification in the upwind area as decreasing rainfall for them and thereby depressing wheat yields. They suspect soil blowing will increase. Their judgment is that the intervention is deleterious to them. When they take the case to an administrative review group neither side is able to provide indisputable evidence that cloud seeding in the area is related to precipitation in a causal way. The lines showing the effect of cloud seeding upon rainfall are dashed to indicate doubt.

This case is chosen to suggest a situation that often prevails. A possible impact is believed to exist, but its precise character cannot

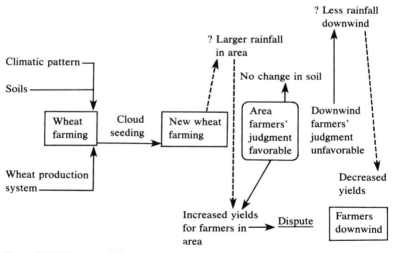

Figure 25.2 Diagram of dispute over cloud seeding in a wheat region.

be established, and failing such validation there is doubt as to who is impacted.

Local, Regional, and Global Impacts

The spatial extent of the impacts of any intervention is of basic significance on the international scene. The important distinction is among impacts (1) wholly local in significance within a nation, (2) affecting parts of two or more nations, and (3) those which have the potential of altering entire global systems. The types of interactions disturbed in the first two classes are the same. The distinction is a matter of boundaries. They include fresh water pollution, air pollution, eutrophication of lakes, soil erosion and stream silting, local weather modification, grassland deterioration, forest destruction, and elimination of species habitat.

In the global category are changes in atmospheric conditions which have the potentiality of triggering changes in world climatic patterns, toxic substances which find their way into oceanic and atmospheric circulation so that they attain virtually global circulation, and alterations in habitat or welfare of species carrying the threat of global extinction. The third class of impacts probably requires special treatment and is not amenable to all the kinds of preventive and adjustive measures suitable to the first two.

Scientific Understanding of Systems

Science has at least three roles in illuminating the grounds on which conflict resolution and avoidance may proceed.

Scientific investigation defines the systems which are affected, indicating where there is an established or presumed relationship among the various components. Definition of the system components may be fundamental because often disputes arise where it has been assumed that the impacts were less far-reaching than later demonstrated in rude practice. As in the experience with large man-made lakes, a whole sector of impact or of possible causal action may be neglected by the designers of projects.[2]

Second, science may describe the characteristics of the various components, including the ecosystems, the physical systems, such as atmospheric circulation patterns, affected social groups and organizations with their preferences and conventional modes of action. To identify the components is to deal with their interactions insofar as they are established. To assert that PCBs in one industrial harbor are related to the health of an aquatic species elsewhere is to presume knowledge of the function of each organism or water current in the chain of transport.

Third, science offers means of estimating the significance of impacts not only in terms of physical or biological quantities but in terms of the way in which they are perceived by the people and organizations affected. As noted above, it is important to distinguish between perceived relationships and impacts and those which can be established through laboratory or field observation. The first figures in the short run, but the second may lay the foundations for dealing later with more difficult problems and solutions. The experience of Mexico and the United States in jointly managing Rio Grande was one of progressively satisfying the international and domestic interests of the two countries while continually exacerbating the physical and social problems of channel aggradation, ground water exhaustion, and irrigation agriculture with which they were concerned.[3] Ribbons were cut with diplomatic agreeability while the resources of the river valley deteriorated.

It should be noted that none of the scientific analysis described above falls under the explicit heading of what often is called "systems analysis." Systems analysis has the high goal of desribing the relevant components of systems and their interactions, and it constitutes a major thrust of modern scientific investigation. So far, however, it has not been able to develop sufficiently accurate and comprehensive descriptions of systems to permit confident prediction of the consequences of particular interventions. This would be desirable and in time it will be achieved in many sectors of environmental studies, but for the present it is aspiration rather than a practical tool.

Ways of Using Scientific Information

Scientific input may be of use at any one of the three points noted above, but it rarely is received in any rational order for purposes of avoiding conflict. Investigations may identify an impact before knowledge of the interactions producing it has been revealed, as in the case of the acid lake water in Sweden. The volume of water available in the Colorado Basin was studied with care several decades after international and interstate treaties divided those waters. The prospective impacts of a dam on the Tigris are studied after construction is authorized.

What it would be desirable to do if scientists were drawn in from the start has been specified in a good many instances. For example, there are relatively detailed suggestions for agricultural development projects[4] and for international river basins.[5] Science may define the components of a system in which it is hypothesized that there may be significant interactions but be unable to demonstrate over a long period that the threat is real and imminent. There is a conviction among numerous atmospheric scientists that changes in particulate matter in the atmosphere, generated by grass fires, overgrazing, and fossil fuel may have a measurable effect upon world climate. They thereby define a possible problem without specifying its magnitude.

It would be comforting to think that scientific investigation could identify those environmental impacts that are likely to be significant in future before they arise and then to chart the means to either prevent their emerging or to cope with them before they exercise their full effect. This, in essence, is the thrust of technology assessment as it now is practiced in a number of developed countries. If technology assessment were to do its chore thoroughly and comprehensively it would help tremendously in anticipating the consequences of any technological action, including changes in social systems, before the perturbation takes place. The results generally are published, and sometimes are the target of legislative hearings.

Examples of technology assessment currently under way include the appraisal of supersonic transport and its effects upon climate and the analysis of new solar and geothermal energy sources and their consequences for social and biological systems. One of the more elaborate efforts is directed at the impacts of drilling for gas and petroleum in deeper zones of the outer continental shelf.

In the same vein and less detailed in method is the attempt by some national and international agencies to require environmental impact statements for specified kinds of resource management programs. With the leadership of the World Bank and the United Nations Development Programme, more stress is placed on making the preparation of estimates a condition of financial support in the light of available knowledge of likely environmental consequences. National applicants are obliged to conform in some fashion. All too often the statements are perfunctory or obfuscating. Administrators may resist full canvass of impacts on the grounds that the statement will cause undue delay or encourage opposition to the program. Both ambiguity and delay may result from lack of scientific information. That deficiency may be especially acute in estimating consequences of new chemical compounds: there is no common bank of test findings, and almost none of the negative findings by manufacturers are readily available.

But scientific inquiry rarely takes place in as formal and policy-oriented a setting as technology assessment and environmental impact analysis. Often it is incidental to the basic scientific effort to understand the interaction of systems and their processes. All too often a scientific contribution is invited after an environmental dispute has arisen and the contending parties either seek information to support their respective position or, taking a longer view, explore peaceful modes of adjustment.

What Affects Its Significance to the User?

When the International Council of Scientific Unions began its current attempt to assess what is known about major changes in the global environment, one of the basic sectors in the entire range of investigations was that of appraising problems of communication of environmental information and the societal response to it. Scientists recognized that there would be little point in elaborating the knowledge which they were producing unless they understood the circumstances in which they could communicate their information and its possible uses to responsible people and agencies in a fashion which would avoid denial, distortion, or hysteria. To gain that understanding requires a thorough canvass of the factors affecting the interpretations and applications that are made of scientific information, the process of risk assessment by individuals and groups, and the societal response to scientific evidence as presented in the rough framework of risk assessment.

At least six major factors are known to have some influence upon the way in which inputs of environmental information are received and used.

Timing is of obvious importance but often neglected. If the information can be made available before the intervention is taken it is possible not only to plan for dealing with the impacts but to avoid sometimes hysterical and sometimes evasive tactics which take shape when new information enters into what was thought to be a smooth-running program for resource management. What would be welcome at early planning stages is anathema when the program is in operation.

The origin of the information has a powerful influence upon its credibility and the rate at which it is circulated. In international disputes it may be initially irritating to have information come from one source rather than another, and it may be helpful to alleviate early stresses and strains by bringing the information from relatively impartial sources.

The message format may influence the way in which people respond to information. For example, it is known that the same set of scientific findings with respect to a possible threat to community health may be received with grave concern or with complacency according to the way in which the statistics are presented or in which threats to health are stated. The tactics of fear arousal have severe constraints in gaining a hearing for the message: a farmer may turn a deaf ear to threats of land exhaustion unless he feels efficacious in dealing with the cause. The authoritarian note may gain acceptance for messages in other circumstances.

The channels and networks of communication are likewise of import. It is known that just as media vary in their credibility, certain groups in society can command sober responses when they serve as the channels for information, as in the case of family physicians. Some citizen groups may be viewed with distrust by government agencies and, in turn, the government agency in some instances may be the least credible. A common response to threatening messages is to seek confirmation; provision must be made for confirmation devices if acceptance is to be sought. It is possible to investigate the channels through which information flows, the credibility of each, and the audiences with which they have respectable reception.

Last, and in a whimsical sense sometimes of less importance, is the accuracy and confidence of the scientific information which is transmitted. Here we must face up to what may be the most important single consideration with respect to scientific information on environmental disputes. In most instances the information relating to definition of system, description of its components, description of the interactive processes, and measures of their significance cannot be provided in a simple causal framework. Ordinarily, the sum of the descriptions and definitions are larded with large doses of uncertainty. It is not clear whether a particular factor is significant or whether it is linked in a precise manner with another factor. Even more difficult, the relationships where they are known to exist often must be expressed as probabilities rather than as firm linkages. The customary expression of findings is not, "if action A is taken, process B will be generated and effect C will occur in group D." It is more likely to be "if action A is taken, process B probably will affect some parts of system C with results which may or may not be grave for group D."

The most troublesome of this kind of probabilistic conclusion asserts that "if action A is taken, we are not confident it will not gravely affect group D." In those circumstances the chronic social issue is whether to take the action in the absence of assurance that there will be no deleterious effect. This leads us on to the problem of risk assessment.

Risk Assessment

What is meant by risk assessment is that process by which a group judges the risks to life, safety, health and environmental options in the light of the values of the society.[6] As a special form of cost-benefit analysis, benefit-risk decision making requires description of the prospect that a given consequence will result and an estimate of its probability and of the benefits and costs for the society. Such weighting of benefits and costs always requires an explicit or implicit statement of social aims and a measurement of the extent to which the anticipated impacts will affect those social aims.

As applied to resources management, it is an appraisal of how much risk society is willing to bear in gaining net benefits from a particular environmental intervention. Every action from crossing a street to building a nuclear power plant involves some risk (the supreme material risk would be the possibility that a program such as generating a thermal nuclear reaction might lead to destruction of the planet Earth). Perhaps the most sophisticated methods are used in assessing the risk of an accident in a nuclear plant. However, risk assessment is widely practiced in environmental decisions and there is much to be learned from examining the techniques and their implications. The United Nations Environment Programme (UNEP) has initiated a comparative study of these topics.

Journalists sometimes are inclined to assume that environmental decisions are taken without systematic canvass of the elements in risk assessment, and the intitation of the High Aswan dam often is cited as a prime example. In fact, the government of the United Arab Republic reviewed almost all of the prospective side effects from the dam, including relocation, fish, silting, schistosomiasis, and soil fertility, before starting the project. Some relevant information was used, numerous gaps were identified. A judgment then was made that, taking all available evidence into account along with the primary aims of agricultural and industrial production, the nation was warranted in assuming the risk. If there was deficiency, it was not one of neglect as charged but of judgment as to weights placed on scanty evidence. The context then was largely national, and ten years later a similar project would encounter questions as to its international implications.

Societal Response

The means by which a government arrives at and responds to an assessment of risk are affected by its scientific and technological capacity, by the value systems inherent in its culture, and by its distinctive modes of action. We are only beginning to understand a bit of why it is that the same scientific information about the effects of mercury on aquatic environments led to radically different societal response in Japan, Sweden, Canada, the United States, and the United Kingdom. We have only an inkling of why one country banned DDT within a few weeks of receiving a single scientific report while another country spent years debating at various levels of government the issue of control, and another country decided to neglect the threat for the present.

In attempting to anticipate environmental disputes it is important to understand the processes of risk assessment and work toward either an agreement on these processes among countries which are similarly affected or toward a frank recognition of the differences in response and the grounds they present for variation in modes of formal adjustment.

The current public concern for environmental quality grew and is differentiated among countries in a way that so far has defied careful explanation. The question of whether the wave of activity in the late 1960s and early 1970s will subside as did two earlier waves is unresolved. However, the values expressed now and in the years immediately ahead set the social limits within which administrative and policy developments may be expected to unfold.

Anticipatory Measures

The scientific community is drawn into anticipation of environmental disputes in a wide spectrum of activities. The individual investigator may run across a track which points to impending difficulties as when a biologist observes that unless more is learned about what is happening to chemical A there may be trouble ahead. Commonly these insights are shared in periodicals and symposia.

Occasionally scientists participate in the consulting-board type of review that appraises a large program proposal. They may join in World Bank and United Nations Development Programme missions to examine either country programs or specific projects. They figure prominently in preparation of environmental impact statements and in technology assessment. The sorts of normative efforts to suggest standards for new resource development programs drawn up under both governmental and intergovernmental auspices enlist solid scientific experience. Thus, the World Health Organization canvass of health hazards called upon a large array of environmental scientists.

At the base of these activities is the undramatic, tedious, demanding job of monitoring changes in environmental quality. The Global Environmental Monitoring System being established by the United Nations Environment Programme prepares to do this for the first time under international auspices on a worldwide basis. The opportunities are large, the difficulties are equal.[7] In countries where elaborate and discerning monitoring of certain resources such as water has gone on for a long time the interpretation of results is extremely complicated: multiple factors are at work, significant factors are found to have been neglected in earlier times, the processes measured are often in doubt. In other areas it will take a long time to catch up. Huge undertakings such as the WMO–ICSU Global Atmospheric Research Program are required to fill in major gaps in data, theory, and computation.

In the long run the monitoring activities are fundamental, but it would be sanguine to expect that they will point quickly to many of the emerging environmental issues. Perhaps more fruitful in the short run will be the judgments, hunches, and insights of the people who design and carry on the monitoring work and who address their research (as at Chelsea College, London) to the issues raised by it.

Are these measures enough in view of the pace of environmental deterioration, enlarged technological interventions, and heightened awareness of international obligations?

International Action for Prevention

In a strict sense all of the international collaboration in pursuit of science helps advance the definition and anticipation of the conditions in which disputes may arise. It should and will move forward in a deliberate, individually innovative fashion. There are, however, at least four concrete types of action which might be taken to speed up the process and to focus upon findings at times and in circumstances which would enhance the prospect that administrative and legal solutions for emerging problems can be found in time to avert undue conflict and delay.

One promising line of action would be to accompany the global environmental monitoring system with an explicit effort by associated scientists to identify those points at which there is likely to be controversy in the future. Some combination of participants in the intergovernmental Earthwatch and in the SCOPE assessment would be in order. This would harvest judgment which undoubtedly would neglect some significant future changes and overemphasize others. Nevertheless it might, at a minimum, help flag certain issues which will claim contentious attention in the future. For example, in the Global Environmental Monitoring System (GEMS) no provision is made for the monitoring of viruses in drinking water. At the present time there is no standard method for identifying viruses, and no major city has a standard program for measuring on a regular basis. The very fact that no provision is made in the first run of GEMS for virus measurements may indicate that special scientific efforts should be made to indicate those international waters in which it might be expected that future debate about the role of viruses would be urgent and controversial.

A second avenue of entry into the movement to anticipate environmental disputes is the proliferating exercise of technology assessment. At present this is chiefly national in sponsorship and execution. Yet, many of the questions are of concern to a large number of nations, and a few of them embark upon global analysis. Taking advantage of the informal network of communication already used by government officers, national academy groups, the International Council of Scientific Unions, the International Union for Conservation of Nature, and individual scientists, it should be practicable for the United Nations Environment Programme to invite the interested groups to share in an annual review asking which technology assessments are under way, which questions seem more urgent, and what questions are neglected. Without setting up any new institutional machinery this might speed up information on what is in prospect, and it might stimulate greater cooperation among investigators than now prevails. It is noteworthy that the largest governmental venture in technology assessment—the U.S. Congress'

Office of Technology Assessment—has no formal provision for exchange of experience with other nations.

A third possible initiative would be to provide for the organization under United Nations auspices of teams of scientists, technologists, and administrators to work together in canvassing regional environmental issues which are known to be emerging and for which there has not yet arisen a formal administrative negotiation or legal proceeding. Such an issue might be the prospective development of an international basin where plans have not advanced to the point at which commitment is made to construction and inspection schemes. There is precedent for that kind of sustained international effort in such basins as the Lower Mekong, the Lower Parana, and the Senegal. By joining over a period of years in collection of data according to mutually agreed standards with verified calibrations and with confidence in the veracity of the observers, it is possible to establish a base from which later negotiations can take place.

Fourth, the critical appraisal of modes of communicating significant information, methods of risk assessment, and factors affecting societal response should proceed on a comparative basis. A beginning has been made by UNEP and SCOPE. To the extent that this applies lessons which can be learned about communication of new information on environmental assets and hazards, it will speed up that process and reduce the possibility of distortions in public interpretation of new and sometimes disturbing findings. To the extent that it can identify the differences in national means of assessing risk and the roots of those differences, it will specify points for which future negotiators can be on the lookout in understanding contrasting judgments and negotiating positions.

There may well be other more productive means of accelerating scientific definition and preparation for emerging disputes. Whatever is undertaken needs to build upon the present network of government and nongovernment scientific enterprises.

Reservoir Systems—A Socioeconomic Perspective

When we start thinking about water quality I should make a comment about environmental impact statements (EISs), which I regard as one of the great hazards to the water management effort at present. That is not because I think EISs are undesirable but because of the way in which they are prepared and disseminated. White's Law (a sort of personal rule of mine) states that the more lengthy an EIS is, the less its writers know about the environmental impact. If they know nothing about it, they include complete lists of all the birds and insects that have been found in various ecosystems in the area under consideration. If they are not at all sure about the environmental impact, the last thing they want to say is, 'We know nothing about it; it is just a blank as far as we are concerned'. They very carefully state a lot of information that implies that they do know a great deal and that whatever they do know is nothing that would give cause for alarm.

I think we are going about EISs in the wrong way. We ought to have EISs that basically say, 'Here is what we know something about; here is what we know nothing about; and the only way we can answer the questions is to carry on further studies—and we must ask how much of that is worth doing'. In this way we can at least distinguish between established evidence and completely speculative or mythical evidence that is often marshaled. In a sense, the longer we follow the present procedures in preparing EISs, the longer it will take us to confront what it is we must find out. Unless we have some kind of revision in these procedures, we are going to be slowed down rather than speeded up in arriving at and coming to grips with understanding the effects of environmental perturbations.

(1981g), 556

Notes

1. For related discussions, see below, vol. II, chap. 6 (Kunreuther and Slovic); chap. 9 (Whyte); chap. 10 (O'Riordan); chap. 12 (Munn).

2. Scientific Committee on Problems of the Environment, *Manmade Lakes as Modified Ecosystems*, Report no. 2 (Paris: SCOPE, 1972).

3. John C. Day, *Managing the Lower Rio Grande* (Chicago: University of Chicago Department of Geography, 1970).

4. Raymond F. Dasmann et al., *Ecological Principles for Economic Development* (New York: John Wiley and Sons, 1973).

5. United Nations, *Integrated River Basin Development* (New York: UN Department of Economic and Social Affairs, 1970).

6. Committee on Public Engineering, National Academy of Engineering, *Perspectives on Benefit-Risk Decision Making* (Washington, D.C.: National Academy of Engineering, 1972).

7. Scientific Committee on Problems of the Environment, *Global Environmental Monitoring Systems* (GEMS): Action Plan for Phase 1, Report no. 3 (Paris: SCOPE, 1973).

26 Stewardship of the Earth

That the human race is a family, that some are deprived of the elementary needs of life, that all have a responsibility for the earth's future—this is the substance of one of the few personal expressions of credo in White's papers. How to meld such beliefs with citizenship in the richest and most technologically powerful country in the world and how to make personal choices that reflect these values while serving as a loyal member of this nation's elite has become a lifelong puzzle and task, engaging both Gilbert and Anne White as well as their circle of Friends.[1]

The people of the Western nations use about 60 percent of the foodstuffs, fibers, and metals produced each year from the earth. Were present trends in population growth and consumption to continue, within two decades 10 percent of the world would be consuming about 70 percent of the total resource production. This is an unlikely prospect for reasons suggested below, but it outlines a situation which is especially disturbing when five other facts about the globe are taken into account. The gap between the rich and poor is growing among and within most nations. The political and social effects of unequal location of energy and other mineral resources are acute. Population numbers continue to climb. The global environment shows signs of widespread deterioration. Both natural and social environments are increasingly vulnerable to catastrophic disturbances.

From the standpoint of resources allocation and social justice, the resulting problems are becoming more complex. There may, however, be a cheering challenge in the possibility that out of its struggle with these realities the human race may move a bit nearer to behaving as if it were indeed one family.

Paper submitted to the Friends World Committee for Consultation, June 1975.

The Rich-Poor Gap

The disproportionate sharing in material resources is far more extreme than might be suggested by reviewing the mean per capita income statistics for the nations of the world. These show for more than 120 countries a range of from about $100 to $4,000 per capita in a recent year. However, there are 20–30 countries, ranked as the very poor, in which the income statistics not only hover in the neighborhood of $100 per capita but reveal no significant improvement.

Among the whole range of countries the absolute gap in income between the very rich and the very poor continues to increase at an alarming rate. A World Bank economist estimates that at present rates of growth in income, India would require 100 years to gain as much an increment of material wealth for its citizens as does the United States in one year. Until recently many well-meaning people in the Western countries and many residents of poor countries entertained the dream that as a result of international aid and development the poor would come in time to attain the present status of working classes in developed countries. This has been borne out in only a handful of places. The gap is widening rather than closing.

The situation is made more acute by the fact that within most countries, and particularly the poor countries, the effect of increases in the mean income for the population as a whole is to make a substantial proportion of the population less well off than they were before. That is, while the nation reports an increase in net income per capita, the lower 30 or 40 percent of the population may actually decline in the share which it can claim of the national wealth. This is reflected most dramatically by the situation of millions of people who have moved from rural areas in tropical countries to the peripheries of growing cities. In shanty towns, barrios, or bidonvilles they live in conditions lacking most municipal services, with low wages, high unemployment, and high social instability.

The argument that growth once stimulated in a developing country will continue and that in time the poor will catch up and share in some greatly improved level of living has not been realized. Two decades of intensive international efforts at development have left both donors and recipients largely disillusioned. Many of the standard techniques for distributing "know-how," building up to a "take-off" point, or cultivating community development simply have not worked. Productivity for a nation as a whole has increased in most cases but not necessarily to the benefit of those in urgent need of improved conditions of food, water, clothing, housing, and employment.

The People's Republic of China is a dramatic exception to the experience in low-income countries. Since its revolution in the early 1950s it has made modest increases in per capita income, but apparently has come close to assuring a minimum level of food, clothing, housing, health care, and employment opportunities to every member of the society. A few countries such as Tanzania have moved in a similar direction with less conspicuous results so far.

Disparities in Natural Endowment

Perhaps the hardest fact to face in dealing with inequalities on the world scene is that countries do in fact differ tremendously in their endowment of climate, soil, minerals, vegetation, and water. The Western nations have grown up in situations which were especially propitious in terms of promoting modern industrial growth. They took advantage of special conditions of minerals and agricultural land in a temperate climate, and developed large stocks of capital and of technical proficiency. They drew heavily upon raw materials from other countries, and the colonial system contributed to the early dependence of those nations upon the Western nations for management of markets and manufacturing processes. By the twentieth century the Western nations had far outstripped the others in rates of growth, concentration of capital, and development of the human resources of trained and educated people. At the same time many of the low-income countries had become highly dependent upon shifting world markets for the raw materials which they produced as a major part of their income: rubber, coffee, hides, tin, and petroleum.

The early 1970s saw in the development of OPEC a turning point in the position of the high-income countries. Even before the Arab embargo had come close to paralyzing industrial activity in nations dependent upon oil imports, producers had begun to band together. This was the first attempt at a world organization of producers confronting the more wealthy consumers and manufacturers. Producers of other raw materials claim a larger share of the income from the processing of the materials, more control over the management of the extractive activities, and assurance of less fluctuation in price and demand, which affects their fiscal stability. Arab oil-producing countries leapt to the top of the list in income per capita and are struggling to find effective ways of using the new returns. Groups in the high-income countries are beginning to protest that they are "blackmailed" by the producers of oil and aluminum and tin. At the same time, factions in the low-income countries raise strident voices against their poor remaining poor or becoming poorer.

Concern about ownership of minerals from the deep ocean bed and of fisheries and minerals in coastal waters is symbolic of new views as to who should steward the unexploited resources of the globe. Developing countries are attracted by the vision of common ownership of the ocean bed but they also are alert to the immediate benefit from extending their territorial claims over the continental shelf from three to twelve to 200 miles beyond their shores.

Mounting Population

It is common to project total world population on the assumption that all of the factors which have applied in the past will apply in the future. While it is one way of examining the future, it may be greatly misleading. The factors affecting a nation's rate of growth are much more intricate than this suggests. The birth rate, for example, is not simply a matter of availability of contraceptive devices or even of the income level of the parents. The quality of family life, the sense of security the family has in government support versus children, the sense of responsibility the individual has to the community and nation, the availability of community services, the need for labor on the farm are among the factors contributing to family decisions as to how many children it will have. Their interaction is not well understood.

Thus, in Indonesia in recent years there has been a remarkable decrease in the birth rate in one of the very densely populated parts of the earth; rapid change had not been anticipated and is not well understood by students of population in other parts of Asia. Countries such as Ireland and Romania maintained stable or declining populations for substantial periods of time. No one knows exactly what has happened with regard to the total population of China but there are indications that its rate of growth was radically reduced in recent decades and that a period of stabilization may be in sight.

In any event, serious doubts are raised as to whether or not the world faces a period in which there will be great and repeated famines leading to reduction of population to a number which can be fed. It is argued by some observers that the globe already is overpopulated, and that nothing short of drastic decrease in population can be foreseen: international aid is regarded as a palliative that defers and exacerbates the lethal days of reckoning. Others see the limits as fast approaching. Still others argue that a much larger population can be supported at minimum standards, and that the task is to develop available resources rapidly, taking advantage of international food reserves and organization to help tide over a difficult transition.

An important aspect of this fluid situation is that the aspirations, value systems, and social relations of people may be more significant than the availability of contraceptive technology in shaping the number of members of the human family that may be struggling to inhabit the planet twenty or thirty years from now, and that the policies followed by nations in helping each other may determine the amount of misery they will undergo.

Resource Deterioration

While growth in total population is indisputable, and there is only question as to the course it will take, the stage of the global resource base is an object of serious searching and question.

There is evidence of enhanced productivity of some parts of the globe as a result of human ingenuity and organization. The productivity of soils in parts of Asia has been enhanced by careful molding of the terrain and management of water. The "green revolution" with its introduction of new seeds designed to yield much higher returns of rice and wheat and corn has increased production in conjunction with improved applications of water, fertilizers, and pesticides. The art of manufacturing and applying fertilizers has advanced along with techniques for soil treatment and for forest management. All of these technological advances contribute to enlarged production from available resources to feed the hungry mouths of the world, and technicians who are familiar with some of these devices claim that the capacity of the globe to feed the human race should be increased several fold by skillful application of present-day knowledge.

At the same time there is widespread evidence of deterioration of the resources, partly as a result of the misapplication of new technologies. The lands in some irrigation projects are going out of production because of unsuitable soil drainage or water, and in some regions the rate of destruction is almost as large as the rate of introduction of new lands through national and international projects. Grasslands are deteriorating in certain areas through overgrazing, sometimes in response to well-intentioned efforts to increase available water for the livestock. Soil erosion continues in many farming areas of the world. One of the senseless changes is the destruction of prime agricultural land by the growth of cities such as Cairo and Santiago.

There has been a kind of fascination with the suggestion that modern man is repeating on a world scale the "tragedy of the commons" in which at one stage in English history the individual livestock owners, seeking their own advancement, contributed to the deterioration of common grazing lands. In fact, where such destruction has developed in recent centuries there generally has been a countervailing social movement to correct it. In England the commons were abandoned and replaced by an enclosure movement in which a large part of the rural population was forced off to the city. It is a gross misreading of history to think that mankind inevitably uses up common resources in the pursuit of individual self-interest. There are notable exceptions, but societies generally have found ways of retarding the destruction of the resource base once the advent of such destruction is recognized. What is more in question is the mode and timing of corrective measures, and whether they will be soon enough and strong enough to avoid serious dislocations.

Destruction of nonrenewable minerals is another matter. Once the coal resources of Yorkshire or of Pennsylvania or of Bihar are exhausted they cannot be renewed, although mankind can seek other sources of energy.

There is also clear evidence of deterioration of parts of the environment through release of toxic substances in air and water, as in the case of particulate matter from coal fires or sulphuric acid from a manufacturing process.

Major controversy has to do with whether or not there are profound changes in the environment that have been brought about inadvertently. There are profound doubts about the effect of fossil fuel burning, grazing land burning, and forest destruction upon the climate of the globe. Some scientists argue that climate is changing in secular fashion, others that it is demonstrating greater variability, and others that human activity so far is masked in its effects by the magnitude of natural fluctuations. The consequences of enlarged radioactivity or of new chemicals, such as DDT or dieldin, are much debated.

As with much else concerning environmental systems, we know very little about the intricate processes of land, water, and organisms that are affected. We are unable to predict with great confidence the consequences of perturbing the system by a new chemical or water input or by soil destruction. Perhaps the most controversial environmental question facing Western nations today has to do with nuclear energy: proponents regard it as the only effective and cheap source of energy for the long run after fossil fuels are exhausted: opponents regard the dangers of waste disposal, accidents, blackmail, and low-level radiation as being so large as to warrant no further development until the full measure of the risk has been taken by society.

Attention should be called to two aspects of the present environmental situation affecting the role of the individual in husbanding the earth. One is that all of the processes are dynamic: we are not dealing with a static world which before the advent of human technology remained unchanged. It is very much a moving flux of forces in which climate change and appearance and disappearance of species is a part of the scenario.

Second, virtually all of the estimates of what is happening and what will be the consequences of future human intervention must be in terms of possibility or statistical chance. At best, many of the predictions are like the forecast that tells a housewife whose laundry is on the line that there will be a 40 percent chance of rain. It is rarely suitable to assert with certainty that a new pesticide or soil practice will have only a specified effect and no more. We are dealing with doubts, questions, earnest searching.

Enlarged Instability

In both natural and physical systems one major effect of modern human management has been to increase instability. Much of the earlier application of technology to food production and human health was designed to eliminate some great catastrophes of earlier history. Railroad development in India combined with irrigation projects for a time to prevent a repetition of great famines on the subcontinent. Improvement of water supply and waste disposal in London prevented the repetition of great epidemics of earlier centuries. These threats to the human family were greatly reduced. At the same time the proliferation of industrial organization and techniques led to greater disruption when a rare breakdown does occur in the much more complicated system. People have moved into more hazardous locations along the shores of hurricane lands in the Bay of Bengal or into dry margins of the Sahara or into the floodplains of great streams. The complexity of urban organization and its dependence upon refined technologies, as symbolized in the computerized management of power and transportation systems, means that the occasional accident has been virtually eliminated but that when something does go wrong with the system it can go very wrong. Societies are in for fewer but greater catastrophes.

Ethical Concerns

In the face of problems of this magnitude and complexity ordinary citizens may wonder what they can do. One attractive conclusion is that it is all too complicated and that individuals must leave it to higher authority or expert managers. A realistic assessment shows that the higher authorities and expert managers have not been conspicuously successful in coping with new conditions, that small rather than ponderous organization may yield greater benefits, and that many of the solutions seem to rest in considerable measure upon the actions of individuals in shaping new value systems and modes of life and livelihood.

At the outset, it is plain that there are no clear and simple answers. We should beware of simplistic suggestions that two hundred million people must die, or that food production can be doubled, or that matters will be settled by a new diet, a new technology, or a world food reserve. Surely a continuation of the past will not suffice: present modes of resource allocation, materials distribution, population growth, and environmental management must somehow be modified to provide for a more enduring mode of use of the limited resources of the world. The practicability of change is enforced by the record of momentous shifts in citizen preferences over recent decades: the widespread disenchantment with war machines, the interest in environmental quality, the shifts in birth rates. Recognition of this prospect generates at least three types of concerns growing out of religious conviction. Any individual can be expected to be concerned about ways in which some minimum level of living can be assured for the poor of the earth. It cannot be assumed to be cared for by simply addressing ourselves to increasing the average lot of the poor nations or by concentrating on emergency aid.

Individual concerns also must grapple with how people live with the growing disparity among the human family. Each of the devices for dealing with this disparity, ranging from the egalitarian approach of the Chinese to the sophisticated socialism of the Swedes, carries distinctive benefits and costs in terms of human dignity and spirit.

A third and elementary concern is in preserving the common environment and in holding open the options for its future use by later populations.

In all three of these directions the present time is one of turbulence and rapid change. There is large opportunity for individual innovation, but if the solidarity of the human family is to be preserved there must be a great deal of imaginative experimentation.

Friends has a long record of vigorous and sometimes constructive involvement in social experimentation. We need remind ourselves of only a few and note the record is studded with failures. Aside from the few large movements such as Penn's holy experiment in Pennsylvania, Quaker modes of action embrace a variety of organizational efforts and of individual testimony. Zambian laborers have been helped to create new housing for themselves, and Mexican peasants have been helped to increase both crop production and their own share in the returns. Quaker businessmen pioneered the institution of fixed prices in commercial transactions. Demonstration and educational programs cultivated public receptivity to new attitudes, as with work camps and the concept of shared labor or emergency relief and the concept of helping without discrimination as to race or religion.

Perhaps most widespread and influential is the individual witnessing to what she or he is led to regard as the right course of action. Both group effort and individual testimony flow from conviction as to the role of people on earth. In stewardship of the common heritage, a few simple beliefs recur: that all are indeed members of the same human family, that all share in responsibility for the others, that each is capable of responding directly to divine guidance. To seek to translate these into practical action with regard to soil or petroleum or the fish of the sea is not necessarily to do what is directly effective in changing society; it is to testify to a way that is harmonious with one's fellows and with a healthy earth.

Individual Choices

Individual consumption is the point of departure for most choices which people can make about their role in stewardship of the earth. While no one individual will alone affect the course of community decision about the environment or materials, it is clear that the aggregate of individuals can be profoundly influential.

The stance of the citizen of an affluent country is quite different from one in a poor country, but even there the relatively well-off members can begin by asking how their lifestyles affect their fellows. Savings of one-sixth of use of petroleum or of a cereal may have a profound influence on a nation's balance of trade or upon its capacity to assist others or to call for aid from others.

One argument that often is heard against changed individual consumption patterns is that technology is evolving so rapidly that savings now may have little significance for the future in any event: if substitutes are found for petroleum within twenty years it will make no difference whether we saved a gallon of petrol last week or not. Saving a gallon last week may indeed preserve a few more options for someone the next year or the next decade or the next century, but it also may have important symbolic effect in signaling changes in the way in which a family, community, or nation uses and shares its available resources. So, too, may be viewed the attempt of a group to recycle its waste materials or to reduce its use of meat requiring disproportionate cereal feeding.

There are numerous ways in which individuals can share in concrete programs for advancing the welfare of the poor within their own or other communities. The many failures of assistance in the international development field can be viewed as discouragement for further support of development efforts. They also can be seen as a challenge to find imaginative and sensitive ways of assisting the poor in gaining a minimum standard of living for themselves. Any such investments of time or money should be viewed not as ones that will guarantee early and large returns but as experiments in a great human laboratory in which many failures may precede a few successes.

In most parts of the earth individuals have some opportunity to take a stand with respect to actions which threaten the earth close to them or within their common purview with deterioration. One simple kind of a rule which may guide individual and collective action in those instances is to prevent irreversible changes in the environment, as when an area is completely destroyed by surface mining or a species is eliminated. This is not to suggest that every species can be preserved or that all areas must be preserved from any kind of surface occupation, but only that the options need to be assessed before such measures are taken and that, wherever practicable, means of preventing the irreversible be sought. This may be peculiarly the case in the controversy over nuclear energy that now promises to extend from developed countries such as Sweden, France, the United Kingdom, and the United States to developing countries, such as Iran, which are taking on massive new energy-generating facilities. There is no agreement among the scientific fraternity as to the long-term effects of low-level radiation upon the genetic characteristics of the human race. Nor is there agreement as to effective ways of storing the radioactive waste substances with half-lives of as much as 25,000 to 250,000 years. Local communities and nations are faced with decisions as to whether or not to go on producing these wastes without having solved disposal and storage problems, and citizen participation in the controversy may be determinative.

Finally, individuals in some places can influence the policies of their nations with respect to international trade, management of resources, international aid, and education. A new system of movement of materials and manufactured commodities is evolving in the wake of OPEC. The disenchantment with the modes of international assistance that developed during the 1950s and 1960s is moving toward a basic readjustment in those structures and modes of action. The kind of patronizing view which dominated so much of Western education and science over recent decades is being strongly challenged, and we can expect new efforts at the elementary, secondary, and university levels to help young people find more satisfying views of their role in relation to their fellows and to the common earth.

Drastic changes in world organization and practices regarding its resources undoubtedly are in prospect. Whatever they may be, it is likely that the lifestyles of a large proportion of the human race will be altered. These will spring from or require changes in individual values as to their relations to their fellow humans and to the earth for which they have common stewardship.

We could all view this prospect with less perplexity if we were confident that there would or would not be world famine or deep conflict over resources or disastrous environmental change. But no one can be sure. We shall have to live with that uncertainty for a long time, knowing that what we do might influence the outcome. We could also be comfortable in deciding what to do if we were confident that a particular course of action would be efficacious. But we are not. Eating a diet for a small planet, or promoting family planning, or backing a new community scheme, or stopping the destruction of a patch of green landscape, or a dozen other types of action might be useful or might turn out to be of no importance in terms of direct social change. However, we can be confident that

·action which is in accord with a few basic beliefs cannot be wrong and can at least testify to the values which we will need to cultivate. These are the beliefs that the human race is a family that has inherited a place on the earth in common, that its members have an obligation to work toward sharing it so none is deprived of the elementary needs for life, and that all have a responsibility to leave it undegraded for those who follow.

Report of the President

The gaps between scientific knowledge and its application to remedial programs are wide and growing. In case after case, such as the desertification of semiarid lands, the adoption of suitable measures to retard degradation or enhance environmental quality is impeded by the inability of social and political systems to use the information in hand. Economic development and environmental quality are no longer seen as necessarily in opposition, but the means of reconciling them in constructive fashion, even in relatively stable societies, are poorly understood and require intensified research bridging physical, biological, and social pressures.

It is becoming painfully apparent that an underlying reason for environmental degradation in many areas is to be found in instability of the social system. Without measures to reduce the growing reliance on violence and the sharp fluctuations in economic markets in many countries, the best-intentioned efforts to translate knowledge into action in environmental management are fruitless.

The mounting risk of nuclear warfare overshadows all other hazards to humanity and its habitat. Slowly it is recognized that a possible major exchange of nuclear weapons constitutes the greatest threat to survival of the planet as we cherish it.

Unpublished paper, V General Assembly SCOPE,
Ottawa, Canada, 16 April 1982, 1–2.

Notes

1. For explanations of this puzzle and task by three members of the circle, see below, vol. II, chap. 11 (Boulding) and chap. 13 (Burton and Kates).

27 National Perspective

Knowledge is power but also responsibility. White conscionably pursued the public implications of his private research. This selection is intended to provide a complete and relatively up-to-date sequence for his most sustained and successful endeavor, changing the pattern of human adjustment to the floodplains of the United States.[1] The audience is new: public interest groups join planners and engineers; there is new opportunity in the emergence of the key influence of the Federal Insurance Administration; and, if compared to previous analyses (see selections 2, 6, and 14), there is a major change of focus. The central issues at the water's edge are no longer issues of direction or orientation, but pragmatic issues of policy coordination and implementation. The strategy is in place, the tactics are still in doubt.

The nation is going through a period of thoughtful reevaluation of its use and abuse of floodplains. New steps are being taken in the direction of more economic and socially desirable management. The shift in emphasis from flood loss reduction to floodplain management is wide and strong. We are knee deep in new flows of activity. Whether we move into shifting sands or onto firmer ground will hinge upon how clearly we recognize the national scene on which we find ourselves and how forcefully we act to guide and respond to the next steps. What follows is a barebones outline of national scene, new developments, and persistent problems.

The directions which national thinking is taking are clearly reflected in a series of public discussions over the past two years. These have their roots in the Task Force report (House Document 465) a decade ago. During the summer of 1974 the National Association of Conservation Districts and the Wildlife Management Institute sponsored a major effort to examine citizen participation in floodplain problems and followed this up last February with a discussion of legislative implications. A panel of the National Conference on Water in April

Reprinted from *The Water's Edge, The National Forum on the Future of the Flood Plain*, (U.S. Department of the Interior, Minneapolis, MN, 17–19 September 1975).

1975 recognized forcefully the new emphasis and the difficulties of administering it. During the past summer the Engineering Foundation and the Office of Water Resources Research and Technology independently sponsored workshops on problems of floodplain management in the urban context. This meeting is the national culmination of a series of regional forums on the water's edge.

All of these efforts point toward managing floodplains so as to reduce flood losses wherever they are uneconomical by national standards, but also so as to promote stable use of the land and water from the standpoint of its distinctive qualities for social advancement.

National Scene

Consider a few indicators of what is happening on the national water front.

Per capita average annual property losses from floods remain about stable while population grows and the total climbs proportionately.

Deaths have been on the decline but show signs of increasing recently.

The benefits from floodplain use have been increasing and probably at a lower rate than damages: the basis for estimates is very weak.

The costs of controlling, modifying, and planning for floods are increasing. The Corps of Engineers construction program continues at about the same per capita level but is making innovative provisions for work on nonstructural alternatives. The Soil Conservation Service is moving less rapidly and is lagging in adopting new approaches. The National Weather Service is improving its flood forecasting capability and is enlarging its warning services. Local governments are expanding their investments in local storm drainage works. Private property owners are increasing their expenditures for flood proofing of structures.

Local land-use planning activity relating to flood hazard has experienced a quantum jump in response to the requirements of the federal insurance program. Technical assistance from state and federal agencies has failed to keep pace. Not more than twenty states have strong advisory service in support of the thousands of communities involved.

The federal insurance program has begun to reach a substantial number of communities and property owners. Like any promising effort, it has encountered difficulties.

National expenditures for research on flood problems has remained at a low level, and the overwhelming proportion is allocated to hydraulic and hydrologic questions rather than to understanding why people and communities respond as they do to flood hazard information.

One way of summarizing where we stand is to plot the types of possible adjustments to floods in terms of their likely effects upon the net benefits or losses from floodplain use and upon the vulnerability of the nation—catastrophic events causing severe social dislocations. This is suggested in figure 27.1 from a recent assessment of research on natural hazards.

Protection works alone would foster net benefits while increasing catastrophe potential. Relief alone and, to a lesser degree, insurance alone would promote both net losses and catastrophe. Land use, properly managed, has the larger potential to promote increased benefits while reducing the disrupting impact of the rare event.

New Federal Policies

The new federal policies which have developed over the past decade are of great potential significance. One need only to enumerate the principal ones to recognize their spread and complexity.

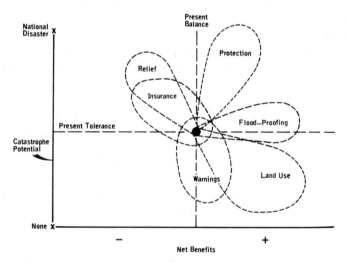

Figure 27.1 Trends and limits of adjustments to floods. From G. F. White and J. E. Haas, *Assessment of Research on Natural Hazards* (Cambridge: MIT Press, 1975). ©1975, University of Colorado. Reprinted by permission of publisher, MIT Press.

Executive Order 11296 of 1966 instructs federal agencies to take account of flood hazard in new siting decisions.

The Flood Insurance Act of 1968 provides coverage for floodplain dwellers, with 90 percent subsidy for existing structures.

The Flood Disaster Relief Act of 1973 prohibits financial assistance for noninsured properties.

The Disaster Relief Act of 1974: Section 201 supports states in preparing disaster response and prevention plans. Section 314 requires insurance on property to be restored. Section 406 sets minimum standards for assistance in disaster loans and grants. Section 501 outlines economic recovery planning for disaster areas.

The Community Development Act opens the opportunity to include floodplain management as a sector of community planning.

The Coastal Zone Management Act includes flood hazard as one feature to be taken into account in local and state planning for critical areas.

The Occupational Safety and Health Act has the authority of limiting work places in hazardous areas.

The Water Resources Development Act: Section 73 specifically authorizes nonstructural measures. Section 80 lays the groundwork for a new cost-sharing policy.

The Water Resources Council statement of standards and procedures for project evaluation explicitly provides for evaluation of a wide range of possible floodplain uses.

The pending report of the Water Resources Council on floodplain management has had for the last six years the potentiality of providing an influential framework within which floodplains may be managed. Whether or not it will realize this potential in the current year will depend upon the ability of the member federal agencies to reconcile their several views in a forthright and practical statement.

Other developments, such as the programs for assisting installation of new water supply and waste disposal facilities and for construction of highways, should be noted as affecting the course of floodplain use.

We ought to get on promptly and without further delay with a much more rapid delimitation of the floodplains, taking advantage of the most advanced methods but also taking advantage of the realities of local land-use planning and of the recognition that flood loss management is only one aspect of broader planning. There is substantial reason to think that the present mapping activities of the Federal Insurance Administration (FIA) can be carried out at one-fifth of the present cost, in which case the rate of mapping might be advanced four-fold.

Problems

The most immediate problem is how to translate federal words into action. Not all members of the federal establishment are working in the same direction or with diligence. The Department of Housing and Urban Development is blandly deficient in exercising its assigned responsibilities for community planning and disaster prevention and reconstruction. The recent General Accounting Office report shows how incomplete and flabby is the administration of some of the policies that are on paper, specifically Executive Order 11296. The Council on Environmental Quality takes a vigorous, positive stand with respect to floodplain management but has little influence on day-by-day management. The Office of Management and Budget, on fiscal grounds, solidly blocks new and imaginative use of non-structural measures. The Water Resources Council is dilatory in bringing forth a program for unified floodplain management.

Next in urgency is defending the current policies from erosion at both legislative and administrative levels. If legislation put forward by Senator Eagleton and Congressmen Casey and Molihan in the current Congress were to be enacted, the sanctions tying community land-use planning into flood insurance coverage would be removed. Further unhampered invasion of floodplains would be invited to the economic detriment of the United States.

Any discussion of floodplains on the national level should begin with recognition of the wide diversity of local situations. It patently is impossible for any one set of regulations and guidelines to apply in detail to all local situations. At the same time the new efforts should take advantage of the benefits deriving from dealing with local flood problems in a larger context. Not only do the origins and disposals of flood flows affect much larger areas than the local community, but the consequences of decisions as to allocations of land and transportation routes are far ranging. The new policies, and particularly the flood insurance program, are struggling with how best to do this.

A fundamental issue is whether or not future growth will be taken into account in delimitation of flood hazard areas under Federal

Insurance Administration procedures. FIA currently assumes present development in and above the floodplain. This makes it difficult for local groups to adopt more rigorous limits and guides, and it surely guarantees that whatever limits are set for the present will prove inadequate in the future in large sectors of the country. Not only does continuing urbanization change the frequency and configuration of flood hydrographs for the more frequent floods, but the construction of new buildings in the floodway and in its fringe storage areas alters the hydraulic characteristics of channels.

A recurrent issue is whether or not relatively rough delimitations of flood zones will stand up in courts. The argument is that unless the floodplain mapping is accurate and can be abundantly supported as neither arbitrary nor capricious it will be disallowed. There are doubts about this. The evidence is that even though more than 600 communities have had floodplains delimited for land-use planning purposes there have been few challenges of the constitutionality of the act and there have been no tough questions as to the quality of the floodplain delimitation. This is the case even though in some circumstances there are three separate delimitations, for example, the Federal Insurance Administration, the Corps of Engineers, and the Soil Conservation Service. There is reason to think that local communities are ready to accept relatively rough delimitations which can be changed as new evidence becomes available.

More important, most of the delimitations hinge upon the setting of the 100-year frequency line, a hard political compromise. People increasingly recognize that such a line is one of convenience and that a large amount of property above the line will be subject to floods of lesser frequency, thus making it incumbent upon the communities to at least consider what will happen to the use of that property. The question of precisely where the 100-year recurrence interval line goes may be of importance in some zoning, but where flooding is only one parameter taken into account in development plans it makes less difference.

The methods of floodway delimitation are of critical importance. Rough measures will be sufficient for most local purposes. Continued unrestricted construction of buildings or elevation of property in the fringe zones by landfill inevitably will affect the hydraulic characteristics of the floodways and the character of flooding both upstream and downstream.

A final problem among a potentially longer list is the policy of continuing 90 percent subsidy of insurance premiums on existing buildings. In the near future this has the great disadvantage of supporting disaster-prone conditions until new buildings replace them. In the longer run it encourages maintenance of present buildings in floodplains even though other forces of land management might work toward outward movement.

The opportunity at this forum is to take account of the basic shift in emphasis from flood loss reduction to making constructive and ecologically harmonious use of floodplains, critically examining the stream of federal developments of the past decade and recognizing the key role that can be played by state government. From this we should make explicit suggestions as to how: (1) present policies should be translated more effectively into action; (2) they should be defended against legislative and administrative erosion; and (3) they should be supplemented where the need is strong.

This is important for the future of floodplains. It is even more important as a demonstration, not yet achieved, of our nation's capacity to truly manage a complex natural resource for the long-term public good through integration of local, state, and federal activities.

A Water Policy for the American People

The nation is now at a unique stage in water development. For several reasons, it is a stage which will never again recur.

First, the nation is on the threshold of a tremendous increase in the volume of construction for federal water projects. The cost of projects now under construction or authorized is equal to the entire cost of all federal projects heretofore constructed. And the projects planned but not authorized account for costs at least four times larger.

Second, present mobilization plans impose heavy competing demands for construction materials, machinery, and men.

Third, accumulated experience with basinwide programs in such diverse areas as the Columbia, the Missouri, and the Tennessee basins offers guidance never before available as to the basic data essential to reaching sound decisions.

Fourth, technical information on water, land, forest, and mineral resources has accumulated rapidly in recent years. Much of this was not available to those who planned the authorized or proposed programs.

Fifth, most basins are relatively undeveloped. Only a few key projects have been built or started. There is still time to make the necessary changes if it is decided that radical alterations are required.

Once they are completed, major water control structures can be altered only with difficulty, or not at all. There are only a relatively few suitable dam sites, and once they are appropriated, the possibilities for economic multiple-purpose development are very limited. Once an irrigation project is developed, it cannot be moved because unfavorable soil or climate factors are discovered. There is a sobering finality in the construction of a river basin development; and it behooves us to be sure we are right before we go ahead.

(1950a), 18

Notes

1. For the entire sequence, see below, vol. II, chap. 2 (Platt).

28 Global Life Support Systems

by Mostafa K. Tolba and Gilbert F. White

This brief declaration marks the beginning of a UNEP-supported SCOPE effort to consider the joint interaction of the great life support systems of air, water, and minerals, using the biogeochemical cycles of carbon, nitrogen, sulphur, and phosphorus as the common ties of planetary life. It had been preceded by a major SCOPE effort (1977e) to identify major global environmental questions towards which international science might contribute, to develop improved methods for ecotoxicological impact and risk assessments, and to examine each of the major biogeochemical cycles.[1] Three years after this declaration it could be stated that the annual release of carbon dioxide into the atmosphere is equal to about 10 percent of that used by plants; nitric oxides and nitrate products produced by human sources are about 50 percent of what the biosphere produces naturally; and more sulphur dioxide enters the atmosphere from human activities than is exchanged naturally (1982b). Nonetheless, our understanding of the interactions between the major cycles is just beginning.[2]

The time is ripe to step up and expand current efforts to understand the great interlocking systems of air, water, and minerals nourishing the earth. This is essential for a reliable assessment of the opportunities for enhancing food production on land and sea. Moreover, without vigorous action toward that goal, nations will be seriously handicapped in trying to cope with proven and suspected threats to ecosystems and to human health and welfare resulting from alterations in the cycles of carbon, nitrogen, phosphorus, sulphur, and related materials.

Much has been learned over the years about these movements in and around the planet. However, only recently have scientists begun to see them in sufficient detail and perspective to permit establishing

Reprinted from *United Nations Environment Programme Information/47* (Nairobi: United Nations, 5 June 1979).

a base for prudent estimates of what may happen if people go on making major changes in energy and material flows.

The exchanges of these elements affect the productivity and quality of land, water, and air. They depend upon many physical, chemical, and biological processes, some of which are still unknown. It now is beginning to be possible to make preliminary estimates of global and regional budgets of the cycles. Scientists are striving to estimate the speed with which the exchanges and chemical transformations take place among the various components of the earth, ocean, and atmosphere. Sometimes these processes operate rather rapidly and are effective in transforming material over great distances. For instance, the atmospheric carbon dioxide content continues to increase even in Antarctica, far from major sources of fossil fuel emissions. In contrast, the processes sometimes operate slowly or over small distances, as exemplified by the centuries that it may take before polluted groundwater is purified.

Society depends upon this life-support system of the planet Earth for sustained and increased production of food, fodder, fiber, and fuel. Mankind also has a major influence on these cycles, as revealed by disturbances in the carbon, nitrogen, and phosphorus flows. Vast amounts of carbon are transferred from the earth to the atmosphere through burning of fossil fuel. The input of carbon dioxide to the atmosphere is intensified by the clearing of tropical forests and cultivation of land. This increase in atmospheric carbon dioxide might eventually affect climatic patterns.

Nitrogen and phosphorus fertilizers are applied to increase agricultural production as a part of the attempt to meet the world food needs. Better understanding of the function of the nitrogen and phosphorus cycles will help to optimize the use of limited land resources. In the near future the industrial production of nitrogen in fertilizers will exceed the quantities produced by natural biological nitrogen fixation. The effects of these additions on the global nitrogen cycle are the object of searching speculation. For example, nitrogen oxides, of which nitrogen fertilizers are one source, may be important factors affecting the stability of the stratospheric ozone layer. Yet the rates of production, transfer, and destruction of nitrogen oxides on the global scale are at present known only in broad outline.

Understanding the working of the biogeochemical cycles is made more difficult—and more necessary for policy purposes—by the evidence that each one of the cycles influences the others and may be related to trace elements and toxins. It is impossible to assess the capacity of biota to bind an atmospheric excess of carbon dioxide into biomass without considering other limiting elements such as nitrogen and phosphorus. The acidifying effect of sulphur in the atmosphere may affect the productivity of some land and aquatic ecosystems. These and other major dependencies must be appraised.

If in the next few years we achieve a better comprehension of these cycles we shall forge a powerful tool for identifying ways in which mankind can beneficially influence and utilize the resources of soil, vegetation, water, and atmosphere. We shall also possess a means of assessing many of the potentially serious risks to the global environment which so far have been studied without an adequate general frame of reference.

The work carried out so far is only a promising beginning. Based on results of such international scientific cooperation as the International Hydrological Decade (IHD), the Man and the Biosphere Program (MAB), the Global Atmospheric Research Program (GARP), and the Studies on Outer Limits to Biosphere Tolerance (UNEP), it is now possible to make crude estimates of the quantities and rates of exchange involved at the global level and reach preliminary indications of possible risks to the global environment. Nevertheless, these are insufficient to permit accurate prediction of how these cycles can be used to meet the increasing needs of mankind without causing undue deterioration of the environment. Presently, knowledge of the magnitude, rate, and in some cases even the direction of the exchanges of basic elements between oceans and atmosphere is severely limited by lack of reliable data.

Over the next decade it should be possible to establish the essential basis for such understanding of the global life-support system. To do so will demand expanded cooperation and greater commitment to basic studies involving many individual scientists and new ventures such as the World Climate Program and will call for an integrated effort of many disciplines, including scientists from all regions of the world. Continuation of efforts at the present level will be too little, too late.

Accordingly, we draw attention to the fundamental scientific importance of understanding the biogeochemical cycles which link and unify the major chemical and biological processes of the earth's surface and the atmosphere. The results will have practical significance for all of us who inhabit an earth with limited resources and who, by our actions, increasingly affect the quality of the human environment. Further, we invite members of the scientific community in the various disciplines to contribute to the design and execution of a collective endeavor.

Stronger support by governments and closer cooperation among them and with intergovernmental agencies and individual scientists are essential if the opportunities are to be fully realized. Steps should be taken by the responsible agencies to stimulate a comprehensive research program suited to the needs.

By 1982, ten years after the United Nations Conference on the Human Environment in Stockholm, we hope to see a major advance in understanding of the complex interactions among these cycles. If such advances are realized, it should be possible to present results promptly in a form useful to national policy makers and be a practical contribution to the requirements of environmental improvement in various parts of the world.

World Environmental Trends Between 1972 and 1982

Reviewing the decade, four dominant trends can be recognized. First, scientific and popular interest in environmental protection have come together to form a new kind of conservation movement. Second, there has been a data explosion in the environmental field, but much of the information is of limited value in assessing trends or as a foundation for decisions and actions. Third, new understanding of the structure and functioning of environmental systems offers a prospect of more reliable planning. Fourth and finally, it has become apparent that the lack of social organization, education, training, and political will, are commonly the limiting factors in environmental improvement rather than a shortage of scientific knowledge.

(1982e), 27–28

Notes

1. For an elaboration of these SCOPE efforts, see below, vol. II, chap. 12 (Munn).

2. B. Bolin and R. B. Cook, eds., *The Major Biogeochemical Cycles and Their Interactions,* SCOPE Report no. 21 (Chichester: John Wiley & Sons, 1983).

29 Environment

Science, the premier interdisciplinary scientific journal, celebrated its 100th anniversary in 1980, providing White the opportunity to assay the evolution of environmental science, policy, and values. The survey is strongly colored by his chairmanship of the Commission on Natural Resources of the National Research Council and his presidency of SCOPE, and in a sense the trends he perceives are trends he is actively trying to create. Thus while the language is descriptive—"now appearing prominent on the scene are at least six trends in scientific thinking and action"—the list is really a prescriptive agenda for crosscutting, collaborative, scientific efforts.

Readers of this and the companion volume will recognize the six trends in White's own work. The beginnings of a holistic framework are found in the broadened approach to the floodplain (see selections 2, 6, 14 and 27), arid lands (see selections 8 and 18) and the evolution of water management (see selections 4, 16, and 17). Similarly, the broadened range of choice appears in many of these same papers and explicitly in the "Choice of Use in Resource Management" (selection 9). The shift from studies of natural hazard perception (selections 15 and 22) to general issues of risk is evident (selection 25). White is also always monitoring his environments of interest: floodplains, arid lands, river basins, attitudes toward the environment (selections 15 and 19), resource use (selection 3), water spply (selections 1, 7, 16, 20, and 24), and the West (selection 23). His repeated retrospective looks encourage his interest in real-time "streamlined" monitoring networks. And the emphasis on basic life support systems and the global scale of problems rests only lightly on an evidentiary base that life support systems at a global level are at threat. These emphases are heavily based on White's moral eval-

Reprinted from *Science* 209 (4 July 1980): 183–90. © 1980, American Association for the Advancement of Science. The author thanks the following people who kindly offered comments on an earlier draft: W. C. Ackermann, W. D. Bowman, J. E. Cantlon, E. N. Castle, D. W. Crumpacker, T. A. Heberlein, R. W. Kates, E. B. Leopold, T. F. Malone, N. Nelson, R. Patrick, W. Robertson IV, R. C. Rooney, R. E. Sievers, L. M. Talbot, A. U. White, M. G. Wolman. I also drew heavily upon experience with the Commission on Natural Resources of the National Research Council and with the Scientific Committee on Problems of the Environment of ICSU, and upon discussions with environmental scientists while lecturing at the University of Pittsburgh.

uation expressed as stewardship of the earth (selection 26). In a review of developments that spans 100 years, half of which the reviewer participates in and gives leadership to, it is not surprising that trends detected and trends created intermingle.

The role of science and technology in shaping and solving environmental problems is changing radically in response to shifts in the goals of the environmental movement and in modes of scientific analysis. The emerging perspective on problems of resource management and preservation contrasts with the prevailing approach of the past decade and has much in common, albeit at a more refined level, with the concerns of investigators of the human environment in the 1880s. The new perspective applies sophisticated techniques to questions of the health of people and ecosystems in a fashion that lays less emphasis upon regulation alone and more upon a wide range of remedies and that examines positive as well as negative effects of anthropogenic alteration. In so doing, it stresses the maintenance of basic life support systems, methods of risk assessment, and the interrelation of environmental factors on local, regional, and global levels.

Background

The environmental students of the 1870s and 1880s shared a deep concern for appraising the full impacts of human occupation upon areas as a whole. In Powell's landmark *Report on the Lands of the Arid Region of the United States,* completed in 1878, the potentialities and severe natural limitations inherent in the Great Plains and the Great Basin were examined in a regional context and applied to a proposed new policy for public land management.[1] Building on the results of field investigations and the pioneering interdisciplinary scientific surveys of possible transcontinental railroad routes,[2] Powell went beyond the description of economic opportunities to point out those interrelated conditions of climate, water, soil, and vegetation that would restrict the sustained use of the regions. The time was one of massive human conversion of landscape in the western United States and Canada, in Australia, and in parts of Latin America but also one of critical inquiry into consequences in terms of the interrelationship of people and resources within areas. Monumental efforts were under way to describe the changing face of the earth in its entirety. Ritter and von Humboldt had led the way, and Reclus between 1876 and 1894 completed the great *Nouvelle Géographie Universelle* which synthesized available knowledge about the planet's surface in 13 volumes.[3] Marsh was deep in his final revised appraisal of the alterations in natural systems induced by past and contemporary civilizations, drawing heavily on the history of soil and forest degradation in Mediterranean countries.[4]

Although the world and regional perspectives found favor in literary and educational circles, cautions about environmental deterioration had almost no influence on public action for a long time. The principal exceptions were the establishment of the national forests and parks systems in the United States. Embodied in the White House Conference of 1908 was concern over the danger of resource exhaustion and wasteful use.[5] However, development was the predominant theme for a half a century. Technology for earth moving, dam building, crop growth, forest exploitation, and mineral extraction and refining contributed heavily to the expansion of resource use. National programs for resource development set the tone for science, engineering, and data collection in those fields either in supporting a program or an alternative, as in the case of forests and floods. What Hays calls the "Gospel of Efficiency" was powerful.[6]

During the 1930s a strong surge of responsibility for stewardship of deteriorating resources left its marks on the governmental shore. A program to conserve soils led to the Soil Conservation Service with its research effort centering on test plots and watersheds treated in the entirety of physical movements of soil and water. This was accompanied by watershed programs in the Forest Service and by the river basin plans of the Corps of Engineers, the Tennessee Valley Authority, and the Bureau of Reclamation (now Water and Power Resources Service). The 1950s and 1960s saw these initiatives continued but with relatively little attention devoted to verifying the consequences of development.

A powerful new aspect of human modification took the form of a massive, steady increase in the volume of chemical compounds released into the environment through waste emissions, pesticides, herbicides, fungicides, and solid waste disposal. In number of products and in volume of production the world chemical industry increased dramatically in the decade beginning with World War II. Bioassays of toxicity of waste accordingly entered a new period in the 1950s. It is variously estimated that the number of different chemicals currently marketed in some volume ranges between 60,000 and 70,000,[7] and this may be as low as 40,000. In the United States 105 of these chemicals are manufactured in quantities exceeding 25 million pounds annually; 80 of them exceed 100 million pounds annually. The volume of synthetic organics produced in 1980 will be more than 100 billion pounds, marking an increase from 10 million in the 1940s, as shown in figure 29.1. Perhaps as many as 1000 new chemicals are introduced annually.

Also beginning in the late 1940s fundamental questions were raised about the future capacity of the globe to support its rapidly growing population. These were expressed most frequently in popular or semipopular publications sometimes labeled as Malthusian.[8] They later were stated most dramatically in a systems framework in the Club of Rome report in 1972,[9] which, while subject to scientific criticism, extended an earlier point of view and provided a broad theme for many investigators concerned with resource degradation and the rising population.

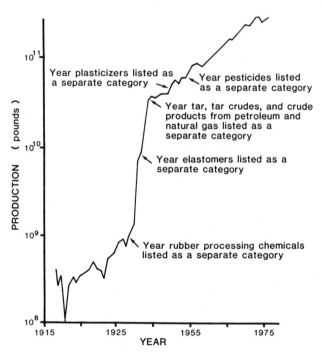

Figure 29.1 Production of synthetic organic chemicals. From National Academy of Sciences, *Science and Technology: A Five Year Outlook* (San Francisco: Freeman, 1979). ©1979 by the National Academy of Sciences.

The 1970s

Through a combination of forces, the last years of the 1960s and the decade of the 1970s saw a shift in emphasis to the maintenance of environmental quality. Stress was placed on air and water quality standards, regulations, subsidy of pollution-abatement works, preservation of endangered species, wilderness, and scenic areas, and critical review of the environmental effects of proposed new developments. Earth Day, the National Environmental Policy Act (NEPA), the Council on Environmental Quality, the Environmental Protection Agency, the Endangered Species Act, and clean air and clean water acts were landmarks. The United Nations Conference on the Human Environment, held in Stockholm in 1972, reflected and in turn stimulated similar changes in other countries, principally Western Europe and Japan.

The enthusiasm expressed in the 1970 Earth Day and the Stockholm Conference was not merely a response to the mounting scientific evidence concerning changes in environmental systems or the extent to which many of those changes in air and water involved external effects and the use of common resources. Rather, the environmental movement of the 1970s expressed frustration with the workings of big business, big government, and large universities; it apparently was in part a reaction to the material affluence of the time, to the moral and social impacts of the Vietnam war, and to other stresses in the social fabric that were widely publicized by the media. Whatever the precise climate and conjunction of the forces at work, they defied the observer seeking to explain the new emphasis. It was a reaction to more than the identification of resource destruction and hazards of the type reported in *Silent Spring*[10] or in scientific reviews.[11]

Many a scientist or engineer was alternately confounded and entranced by the speed with which regulations were adopted with incomplete supporting evidence, by the rejection by some public interest groups of what previously had been hailed as beneficial measures—like the Echo Park dam or the Alaskan pipeline—and by radically different values placed on various risks, such as nuclear power, automobile fatalities, and pesticides.[12] In scores of cases public concern was voiced over the alleged miscarriage of well-intentioned technological projects.[13] Anxiety grew over carcinogens and radiation hazard. Citizen groups became sensitive to hazards carried involuntarily by individuals.

The experience of the 1970s is significant for the future because it reminded the scientific community that, if it is to help the bodies politic deal intelligently with environmental problems, it must understand the factors that affect how different publics value nature, and the complexities of local community response to scientific evidence and to perceived difficulties in manipulating natural systems.[14] It also must understand that social response may be sluggish by comparison with technological change.

It is notable that although the programs of the 1970s provoked improvement in some sectors of the environment and critical opposition by some interest groups, they commanded larger expressions of public support in the United States in 1978 than in 1971. As shown in figure 29.2, this strengthening of concern was reported in all regional, ethnic, income, and educational groups in the face of economic challenge.

Events generated by the 1973 oil embargo had a profound effect on the U.S. definition of resource needs and of appropriate scientific research to help meet them. Within a few years the first approximation of a national energy program was under way, and this encompassed new inquiry into the environmental health effects of energy technology, the management of public lands, energy conservation, and alternative sources. This set of efforts has just begun to be reflected in scientific and technological advances, but the mounting perception of the severity of possible energy shortages lends urgency to appraisal of environmental policies.

Ten years after NEPA, measures to improve environmental quality are subject to attack from several quarters. It is claimed that impact assessment and pollution regulations hold back innovation, and that they are cumbersome, unduly costly to society by impeding economic growth, and in some instances, like the clean air standards, are based upon unsound or inadequate scientific evidence. Environmental activists are branded as hypocrites not prepared to sacrifice the comforts of modern society.[15] Each of these arguments is in debate and surely will be debated for a long time to come because the environment is regarded as important. Fundamental to them is the question of whether or not in each sector in which regulations and new control technologies have operated the quality of the environment has, in fact, improved. One direct result of these controversies has been to stimulate more searching examination of the full range of consequences of the various remedial measures as well as of other social forces affecting the environment.

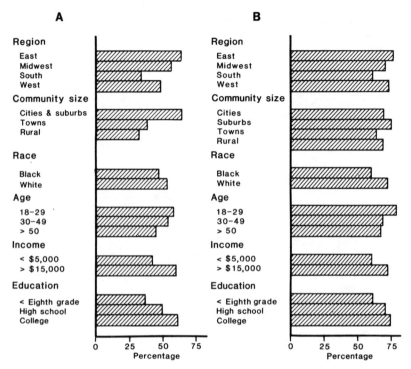

Figure 29.2 People feeling that water pollution is very serious as (A) a local problem in 1971 and (B) a national problem in 1978. Source: (A) *Harris Survey Yearbook of Public Opinion, 1971* (Louis Harris and Associates, New York, 1975), p. 260. Question wording: "Do you feel water pollution is a very serious problem around here, somewhat serious, or not a very serious problem?" (B) A Harris survey conducted in May 1978. Question wording: "Here is a list of different kinds of environmental problems. For each would you please tell me if it is a very serious problem, a somewhat serious problem, or only a small problem for this country—pollution of lakes and rivers."

Thus far, the analytical tools available for that purpose are at best modest, and in many instances the data required for adequate appraisal are deficient. Many environmental impact statements are written without a sound basis for estimating the impact. This was given sharp emphasis by the interaction of scientists and environmental lawyers in the litigation made possible by NEPA.[16] Much of the controversy between industrial and environmental groups hinges on effects that are either unverified or speculative. For example, the assertion that clean water regulations retard manufacturing growth and reduce profits is heard frequently, but there is little solid information on which to assess it.[17]

To deal effectively with the whole range of environmental problems that are evident or emerging would call, ideally, for perfect knowledge of the natural systems to be affected. The knowledge would be so nearly complete that for any proposed intervention in the environment it would be possible with a high degree of confidence to estimate the expected effects on people and ecosystems, and to specify the full costs and benefits of any measures that might be devised to enhance resource use. As illustrated by the program for reducing air pollution, this ideal never will be reached. At best, new technologies will present new solutions and new complications, and further investigations will reveal processes and conditions previously unsuspected, as in the recognition of the role of small particulates. Yet, the effort to estimate impacts will be pursued. The sobering prospect is that most of the major public decisions about resource use and environmental management will be made in the face of large uncertainty deriving from ignorance of physical and biological systems and from evolving techniques and social values.

While emphasis during the 1970s was on reducing pollution and preserving wilderness, some of the earlier development programs received new, critical attention. Water resources projects were reevaluated in the light of sterner efficiency criteria and their impacts on natural systems. Concepts of multiple-purpose management in forests embracing use of clear-cutting techniques were reexamined. Fire was recognized as playing a possibly constructive role in ecosystem maintenance when all impacts, including those on air quality, are taken into account. It was ironic, however, that while much effort was expended on wilderness preservation virtually no attention was paid to establishing scientific grounds for managing the huge proportion of the land surface remaining in the public domain and for the basic ecological survey to assist that belated program.

When one seeks to peer ahead, some of the prospective interactions of science and technology with environmental policy can be projected by extrapolating recent trends. Others may constitute sharper breaks with the past. In the environmental field over the years, however, there have been relatively few shifts in the course of events as rapid or dramatic as those of 1933 to 1938 or 1969 to 1972. Perhaps the only three technological changes bringing about sharp discontinuities in development during recent decades were the invention of nuclear power generation, the refinement of techniques for measuring very small quantities of substances in air, water, soil, food, and organic tissue before and after waste discharge, and the provision of computing capacity to handle very large models of natural systems and the data from extensive sampling.

Now appearing prominent on the scene are at least six trends in scientific thinking and action. Others might be identified, but these seem to point to significant problems and opportunities. (1) More serious efforts are made to examine environments in a holistic framework. (2) Greater attention is given to studying processes within the basic life support systems. (3) There is expanded emphasis on canvassing the theoretical range of alternatives in resource management. (4) Numerous efforts are under way to refine methods for the assessment of risk. (5) Techniques for monitoring changes in environmental systems are being streamlined. And (6), increasingly, all this is happening within a global framework. Each of these advances in perspective and method cuts across the traditional divisions of environmental problems according to natural media of air, water, soil, vegetation, and animals. Each also calls for unconventional collaboration across disciplinary lines.

Environment in a Holistic Framework

Much of the activity devoted to enhancement of environmental quality faces a task of integration. The most conspicuous move in that direction was achieved over the past half-century by investigators and planners involved in water management, but only in part. Notwithstanding pious affirmations of the unity of drainage basins and the social efficiency of multiple-purpose management through mixes of single and multiobjective measures, it took a long time to organize genuinely integrated investigations in which diverse but clearly related topics such as sediment flow, floodplain use, aquatic ecosystems, upstream urbanization, and irrigation water application patterns could be attacked in unified fashion. The performance still falls short of aspiration, but the conceptual groundwork and the administrative machinery have begun to emerge.[18] Genuinely national, interdisciplinary programs of research on surface and groundwaters took shape in the 1960s, and although they were laid aside in the 1970s, they will undoubtedly receive fresh and relatively coherent attention in the future under the stimulus of questions as to supplies of water for synthetic fuel and increased crop production.

Not so with the array of studies and regulations initiated during the 1970s. Although the Environmental Protection Agency provides an administrative umbrella for programs on clean air, clean water,

solid waste disposal, and control of toxic substances, the programs
remain disparate activities by virtue of independent congressional
authorization. To take only two sectors, the studies on water supply
and waste treatment under Section 208 of the Federal Water Pollution
Control Act of 1972 are independent of studies to determine air
quality standards and emission limitations under the Clean Air Act
as amended in 1977. It may be argued that the two groups of pollutants are essentially different and that in most respects they operate
with little effect on each other. However, the populations and manufacturing activities involved are the same, locational decisions on
highways and land use are similar for both, and nonpoint sources
are major contributors to water pollution. Genuine pollution control
has the potentiality of what Reilly designates as a "quiet revolution"
in land use.[19] It, furthermore, is abundantly evident that the interchange of particulates and pollutants such as sulfur dioxide and
nitrogen oxides and organic compounds is significant. It will in time
be necessary to examine regularly the interrelationship of flows of
all polluting materials in any given area, and the theoretical framework for doing so is taking shape.[20] The EPA has moved toward
coordination of the investigations it supports, and has begun to establish centers for anticipatory research, but the demands of justifying and monitoring regulatory procedures tend to take precedence
over holistic approaches, and the contracts and grants follow suit.
Lawyers operating within statutory compartments and scientists with
circumscribed expertise still tend to dictate the immediate study
agenda.

No aspect of environmental management in the United States is
more neglected for political reasons than land use and its planning.
The term "land-use planning" with its connotation of encroachment
on private property rights became anathema to powerful sectors of
the electorate, with the result that its direct support at the federal
level is muted. Indirectly, many of the federal programs have changed
the mix of property rights and social mechanisms affecting land use.
National encouragement of planning is still given chiefly for coastal
zones and urbanized areas, and in those instances it is frequently
almost obliterated by the generous use of exceptions. Yet most decisions influencing the location and character of industries generating
waste, or the quality of wild areas and open space, or the transportation of toxic substances are made at the level of determining
how particular pieces of land will be used. Wise action requires
intimate knowledge of local conditions. As long as the process of
making those decisions remains essentially in the hands of private

landowners and local governments, and the federal government treats
them with cautious detachment, the stimulation of research to supply
necessary data will remain weak.

A more comprehensive view of environmental interactions is developing in the field of toxicology. Much of the earlier, fundamental
research on effects of exposure to toxic materials, such as arsenic
and lead, came from studies in the human workplace. Observations,
chiefly clinical and epidemiological, demonstrated pathological effects of various dosages on people producing or using substances
containing the suspected materials. For some of them permissible
levels of exposure were established for purposes of occupational
health and safety.

During the past three decades two technological changes and two
shifts in scientific orientation have altered the situation in a fashion
that profoundly affects public policy. On the technological side, the
commercial production of chemical compounds grew rapidly for
twenty-five years, as noted above, while the capacity to detect and
measure minute quantities of many potentially hazardous substances
increased by several orders of magnitude. On the scientific side, the
concept of threshhold limits of toxicity was severely challenged, and
the view of humans as the major targets for toxic materials was
enlarged to include other constituent organisms in ecosystems.

During the period of growth in chemical production the means of
identifying toxic chemicals in air, water, soil, food, and tissues were
advanced by the refinement, chiefly after 1950, of mass spectrometry, neutron activation analysis, electrochemical techniques, atomic
spectroscopy, and gas and liquid chromatography.[21] Whereas in 1950
polychlorinated biphenyls, dioxin, and nitrosamines—to name only
a few—could not be measured in trace amounts, in the 1970s they
could be detected routinely in amounts of parts per million.[22] The
sensitivity of analysis in some cases, as for chromium, increased to
parts per billion in a few years and dioxin to parts per trillion.[23]
Chloroform was not known to be present in minute quantities in
drinking water as recently as 1960, but its presence at the part per
million level became a target of public concern within a few years
after it was recognized as a carcinogen.[24] Some aspects of analytical
chemistry, previously considered routine, were suddenly on the fore-
front of environmental study.

The implications of these advances in testing and associated problem solving are widespread. Producers as well as regulators of known
or suspected toxic substances find the task of dealing with environmental effects facilitated by rapid and accurate determinations. At

the same time the task is complicated by detecting new ramifications and subtleties of relationships; numerous elements, for example, are found to be essential to the functioning of organisms in very small amounts although they are toxic at higher levels. Further refinements are certain to heighten the complications, and the limit on precision of sensitivity will be set by the capacity to remove residual impurities in reagents and solvents.[25] Thus it is argued that isotope dilution mass spectrometry makes it possible, combined with ultra clean analytical methods, to identify concentrations of lead and other materials from anthropogenic sources previously regarded as natural background.[26]

The subject of toxicity becomes more intricate when the focus shifts from human targets to other organisms in an ecosystem. We are moving into a period when ecotoxicology will claim increasing attention and will address itself to such questions as the synergistic effects of substances, the health of nonhuman species, and the consequences of altering the reproduction or nutrition of selected populations.[27]

Basic Life Support Systems

The health of the land and its associated plants and animals is basic to preservation of the resource base, yet it is rarely investigated in a fashion that permits recognition of the interactions that affect its net ability to support life. Reductionist research prevails and is encouraged by the organization of most universities. We know a great deal about such matters as crop yields per acre and rates of soil loss in test plots and about the effects of specified pesticides on a few insect and bird populations, but we have no fully adequate means of estimating the capacity of land areas as units to permanently sustain a diverse population of species under prevailing techniques. Such appraisal becomes increasingly vital as the number and variety of technological innovations multiply—fertilizers, pesticides, herbicides, fungicides, tillage techniques, narrow genetic strains, urbanization—so that a short-term gain in return from one source may mask long-term deterioration of the producing base. For the world as a whole the volume of production of nitrogen in commercial fertilizer will soon equal the volume made available by biological fixation.[28] With alterations of this magnitude under way in the face of the continuing growth in world population, the time is ripe for a searching examination of what is happening to the fundamental resources.

Only recently has the investigation of drainage areas reached a point at which it is possible, in addition to analyzing stream flow, to trace the flows of nutrients into and out of a small forested area, as at Hubbard Brook, New Hampshire, to measure the effects of different management practices, and to estimate the consequences for stream flora and fauna.[29] The comprehensive view of lakes as unified systems reflecting the interaction of many biological, physical, and social factors has exemplified a long-term goal.[30] This is but a beginning in what promises to be more extensive and penetrating research into the vital processes by which the land is maintained and changed.

Theoretical Range of Management Choice

As reflected in the experience with water management in the United States and many other countries, the trend in handling resource problems is slowly moving away from single-purpose to multiple-purpose programs. Of greater significance, the tendency is to rely less on a single technological fix and more on a multiple mix of various adjustments suited to the unique characteristics of each physical, biological, and social configuration.[31] An earlier example was that of management choices in dealing with floods.[32] These choices have come to encompass upstream and floodplain land use measures, techniques to reduce vulnerabililty of structures, and insurance. Similarly, greater attention is being paid to a wider range of choice and to the balance of gains and losses in coping with pests, waste disposal, and air pollution.[33]

The implications of these management policies for research and technological development are immense. Flood control research expands from corrective works to include modifications in building design, the niceties of zoning ordinances, human response to warnings, and restraints on use of hydrologic data by lending institutions in writing mortgages. When water pollution abatement, instead of concentrating on one type of measure such as municipal waste treatment, reaches out to alternative measures such as irrigation, land disposal, self-contained water and waste recycling, and revisions in industrial processing technology, it makes a diversified set of claims on research and development, and suggests improved procedures for setting research priorities.

Refining the Assessment of Risk

As more attention turns to studying complete natural systems and to alternative methods and strategies for managing them, and as the number and effects of environmental alterations expand, the way of presenting the risks and gains to the public prior to decision calls for greater sophistication. Decisions as to standards setting deserve careful refinement of the method of estimating the likely effects of interventions, but they also demand better understanding of how different publics perceive and are helped to evaluate the consequences.[34]

Much public evaluation of environmental measures today ranges between two positions. The conventional benefit-cost analysis embodied in justifications for public works projects attempts to tally up and compare, for a specified time horizon and discount rate, the future flows of gains and losses. Few fields of research developed more rapidly during the 1950s and 1960s than the economic analysis of resources management measures.[35] The early emphasis on the economic efficiency of firm or nation has expanded to include impacts on environment and regional welfare. This has tended to become routinized and has been subject to criticism on numerous grounds, among them that estimates have been manipulated, that long-term costs are underestimated, and that the full impacts on ecosystems and quality of human life are neglected.[36] At the other extreme certain of the major efforts to preserve environmental quality have been based on broad assumptions that it is in the public interest to achieve a specified standard, with little explicit analysis leading to its setting. The water pollution abatement goal of making all streams swimmable and fishable is one such. The air quality standards for auto emissions specify thresholds in terms of health effects but with limited data to support the precise figure used. Whatever the methods used, the public decision is likely to consider factors broader than those figuring in the risk assessment.

Two recent reports by National Research Council units—those on saccharin and nitrates—illustrate the complications.[37] They explicate uncertainties attaching to effects of the substances in question, present much of their evidence in probabilistic terms, and venture estimates of constructive as well as deleterious consequences of using the material.

It has become dramatically apparent that values attached to these probable flows of gains and losses by expert groups of engineers and scientists may not coincide with the values assigned by various publics concerned.[38] The divergence is nowhere more extreme than in the realm of energy options where the estimated risks of using coal, solar, or nuclear energy not only divide the experts but provoke strong responses from different publics. Nuclear power is seen by certain citizen groups as carrying threats to life ten times those of other environmental hazards, while a National Research Council committee places the risks to health from nuclear power far lower than coal.[39]

The steps under way in the National Research Council and the Royal Society and in various other groups to improve methods of risk assessment and of interpreting the results fairly to interested publics may be expected to be followed by further, wider ranging investigations that will sharpen the choices and clarify the essential part played by value judgments.[40]

Streamlining Monitoring Techniques

Judgment about what is happening in natural systems depends in considerable measure on the accuracy and scope of monitoring observations. To appraise the effectiveness of regulatory measures for air pollution control as well as to support them, extensive networks have been organized to observe air quality. Accompanying them, but rarely in an integrated fashion, are statistics on human morbidity and mortality. These measures provide a point of departure for inferences as to the effects of air quality on health, but the strength of assertions about those relationships has not grown conspicuously.[41]

An integral component of monitoring is the measurement of the perception and valuation by different publics of environmental parameters. The methods have advanced notably in recent decades.

Increasingly it is recognized that there is no static natural system, let alone one which is fully understood, and that management interventions, as with pest control or dam construction, are likely to provoke unexpected results and to require dynamic alteration.[42] It is not a matter of simply taking action after a one-time environmental impact statement: monitoring is needed to provide a ground for revising policy and techniques, perhaps repeatedly. The grand design is flexible and sensitive to new knowledge and technology.

Monitoring of substance or effect or perception of effect may be fruitless or ineffective unless the basic processes are understood or hypothesized and unless the data collection is accompanied by continuing critical assessment of the monitoring strategy and the accuracy of the data set.

As illustrated by the United Nations surveillance of atomic radiation, the greater the precision of comprehension of the diffusion and transport processes involved, the smaller the number of observations required. Accurate and reliable models based on understanding of the underlying processes permit broad description from a small set of established points. But this is not yet true for many environmental parameters, and the task of monitoring rests partly on improving basic knowledge of the system, as in the case of water,[43] partly on improved instrumentation, and partly on innovative methods of using small bits of data to describe large systems. Major advances in electronic and integrated systems for data processing now permit simultaneous determinations of many components in a substance.

At the regional level, a stocktaking is under way in those countries in the Organization for Economic Cooperation and Development.[44] On a global scale, the United Nations Environment Programme has begun to lay the groundwork for detecting changes in major environmental parameters, but a genuine world strategy is just beginning to take shape and faces the difficulty of sifting out the background fluctuations.[45] It will require new approaches like the "mussel watch" which measures changes in coastal water quality through analysis of selected accumulator organisms appropriate to the conditions under surveillance.[46]

Global Perspective

Approaches to investigation of environmental problems at the local or national level gain in strength from the global perspective. A great variety of problems have international implications in that they are common to many countries, or in that what happens in one country affects one or more other countries through physical or economic linkages.[47] There is a good deal of experience with international river basins and a growing disposition to cope with transnational pollution, as in the case of acid rain in North America and Western Europe.

The global perspective is gaining rapidly in scientific work in at least two environmental sectors. One is the accelerated networking of experience in coping with common questions of environmental deterioration or enhancement. Through the United Nations Environment Programme, the Organization for Economic Cooperation and Development, the United Nations Educational, Scientific, and Cultural Organization, bilateral collaboration, and scientific unions a more concerted effort is being made to exchange and bring to bear the available knowledge and to stimulate new research on such questions as desertification and waste disposal. The threat to the resource base from soil depletion in some areas is well established,[48] yet concerted means for coping with it are still feeble.

A second, momentous development during the past decade has been the recognition that a few environmental alterations have attained a magnitude that promises or may already have brought about change in global systems. The convention for the protection of endangered species on a world scale was executed in 1977 with a view to preserving unique genetic resources. Intergovernmental action was taken to control ocean dumping in 1972.

At an international conference in 1977, the prospect of depletion of the ozone layer was first confronted.[49] The World Climate Conference in 1979 saw serious discussion of the probability, as yet unspecified, that mounting concentrations of carbon dioxide in the atmosphere would trigger climatic change.[50] The pace of action is accelerating, and other, perhaps more significant, alterations may be in the wings. But for the first time the world scientific and political communities are systematically contemplating globally significant transformations.[51]

An Illustrative Case: Carbon Dioxide and Biogeochemical Cycles

To a remarkable degree the developments described above are merged in the current attacks on our ignorance concerning the great global biogeochemical cycles. The prevailing efforts stem from two sources. Of long duration is the series of attempts to describe the cycles. The first such investigation was undertaken by Vernadsky;[52] one of the most recent was sponsored by the Scientific Committee on Problems of the Environment (SCOPE).[53] Of shorter duration are the growing expressions of concern over ozone depletion, acid rain, and the mounting levels of carbon dioxide. Interest in reduction in the ozone layer in the stratosphere with its consequent effects on the incidence of skin cancer as a result of augmented ultraviolet radiation stems from discovery of a few of the photochemical processes induced by releases of fluorocarbons.[54] The acid rain problem, which first came into public view in the train of observations of decreases in fish

populations in Scandinavian lakes and of increasing acidity in surface
waters and precipitation, has broader linkages in the sulphur, nitrogen, and phosphorous cycles.[55]

The carbon dioxide concern stems from speculation about global
warming accompanied by observations at Mauna Loa and the Ant-
arctic of an increase of approximately 5 percent in atmospheric
carbon dioxide since 1957 and of 13 percent since before 1850.[56] As
noted by Revelle,[57] the augmentation, while linked with fossil fuel
generation, may also be affected by rates of vegetation burning,
forest destruction, and other interventions in the carbon cycle, as
outlined in figure 29.3 It is symptomatic of the stance commonly
taken by government agencies that they focus on what appear to be
immediate issues of policy. Thus, the carbon dioxide investigations
received strong impetus from debates over plans for increased use
of coal and synthetic fuels in relation to nuclear sources.

The alternative approach, which is slowly taking shape, examines
the problem in the context of the interactions among the cycles of
carbon, nitrogen, phosphorus, and sulphur as affected by other elements, including trace metals. This alternative promises greater returns in the long run by attempting to understand the cycles
themselves instead of trying to solve the carbon dioxide problem, a
goal impossible to reach without making large gains in the former
direction. The holistic framework therefore is essential to finding
solutions. In considering ways of coping with the hazards and opportunities involved in cycle changes—assuming that reasonably
accurate models can be devised to predict direction, magnitude, and
spatial pattern of climatic change—it will be important to look carefully at all theoretical avenues of action, including readjustments in
land use, changes in industrial and agricultural techniques, and the
like. Choices among the various possibilities will require far more
sophisticated methods of assessing the risks of alternative responses
for each of the human populations and habitats affected. These methods will have to take account of the intricacies of physical, biological,
and social systems. They will need to be supplemented by discriminating deployment of new monitoring observations, particularly in
the oceans. All this must take place in an international context if
the full assets of the world scientific community are to be brought
to bear. Collaboration at that scale also will be required if consensus
as to process and consequences is to be reached as a necessary
condition for determining whatever intergovernmental action, if any,
may be suitable. Different areas and populations probably will be
affected quite differently, and there will be pressure to make policy

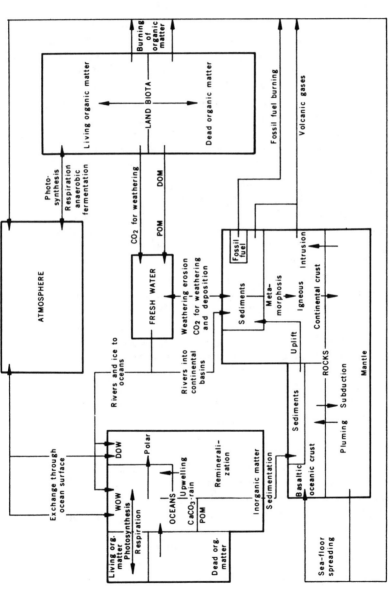

Figure 29.3 Principal reservoirs and fluxes in the carbon cycle. Abbreviations: *WOW*, warm ocean water; *COW*, cold ocean water; *POM*, particulate organic matter; *DOM*, dissolved oceanic matter. From B. Bolin, E. T. Degens, S. Kempe, and P. Ketner, eds., *The Global Carbon Cycle*, SCOPE Report 13 (Chichester: John Wiley and Sons, 1979).

choices about energy generation and vegetation management while neither the basic cycles nor the likely impacts of changing them is well understood. Policy decisions should await the advance of basic research.

The net effect of a widespread effort of the sort shared by many individual investigators and teams dealing with biogeochemical cycles would be the achievement in roughest form of something approaching what Bolin likes to call a "global ecology."[58] Similar undertakings must be expected in other sectors. Nothing less will do in meeting the human family's ultimate need to preserve and use sanely its only base of support.

Broader Bases for Choice: The Next Key Move

Wherever resources conservation and development is discussed the final and sometimes interminable topic is who should do what—what organizations should be devised or strengthened to help us achieve our aims in resources development. My contribution cannot be that of one deep in administration of resources activities or who delves into the political mysteries; it is, rather, that of a person who is trying to see and understand the impression of human organization upon the American landscape of rock, soil, water, vegetation, and people.

(1958d), 205

Notes

1. J. W. Powell, *Report on the Lands of the Arid Region of the United States, with a More Detailed Account of the Lands of Utah*, 45th Cong., 2d sess., H.R. Exec. Doc. 73 (Washington, D.C., 1878).

2. For example, U.S. 33d Cong., 2d Sess., House of Representatives, *Reports of Explorations and Surveys to Ascertain the Most Practicable and Economical Route for a Railroad from the Mississippi River to the Pacific Ocean* (Washington, D.C., 1857), seven volumes.

3. E. Reclus, *Nouvelle Géographie Universelle: La Terre et les Hommes* (Hachette, Paris, 1876–94).

4. G. P. Marsh, *The Earth as Modified by Human Nature*, A Last Revision of *Man and Nature* (New York: Scribner, 1885).

5. *Proceedings of a Conference of Governors in the White House, Washington, D.C., May 13–15, 1908* (Washington, D.C.: Government Printing Office, 1909).

6. S. Hays, *Conservation and the Gospel of Efficiency* (Cambridge: Harvard University Press, 1959).

7. National Academy of Sciences, *Science and Technology: A Five Year Outlook* (San Francisco: Freeman, 1979), 442–45.

8. For example, W. Vogt, *The Road to Survival* (New York: William Sloane Associates, 1948); F. Osborne, *Our Plundered Planet* (Boston: Little, Brown, 1948); H. Brown, *Challenge of Man's Future* (New York: Viking, 1954).

9. D. H. Meadows et al., *The Limits to Growth* (Washington, D.C.: Potomac Associates, 1972).

10. R. Carson, *Silent Spring* (Boston: Houghton Mifflin, 1962).

11. National Academy of Sciences-National Research Council, Committee on Natural Resources, *Report* (Washington, D.C.: National Academy of Sciences, 1960).

12. B. Fischhoff, C. Hohenemser, R. E. Kasperson, R. W. Kates, *Environment 20* (no. 7), 16 (1978).

13. E. Lawless, *Technology and Social Shock—100 Cases of Public Concern Over Technology* (Kansas City: Midwest Research Institute, 1974).

14. E. T. Haefele, *Representative Government and Environmental Management* (Baltimore: Johns Hopkins Press, 1973).

15. R. Beauvais, *Les Tartufes de l'Écologie* (Paris: Fayard, 1978).

16. *Public Interest Law: Five Years Later* (New York: Ford Foundation and American Bar Association, 1976), 17.

17. P. R. Portney, ed., *Current Issues in U.S. Environmental Policy* (Washington, D.C.: Resources for the Future, 1978).

18. National Water Commission, *New Directions in U.S. Water Policy* (Washington, D.C.: Government Printing Office, 1971).

19. W. K. Reilly, "National Land Use Policy," in *Federal Environmental Law*, E. L. Dolgin and T. G. P. Giulbert, eds. (St. Paul, MN: West Publishing, 1974), 1430–36.

20. A. V. Kneese and B. T. Bower, *Environmental Quality and Residuals Management* (Baltimore: Johns Hopkins Press, 1979).

21. G. W. Ewing, "Analytical Chemistry: The past 100 years," in *Chemical and Environmental News* (6 April 1976), 28.

22. National Academy of Sciences-National Research Council, Committee on the Assessment of Polychlorinated Biphenyls in the Environment, *Polychlorinated Biphenyls* (Washington, D.C.: National Academy of Sciences, 1979).

23. R. E. Sievers and J. E. Sadlowski, *Science* 201, 217 (1978); L. L. Lamparski, T. Y. Nestrick, R. M. Stehl, *Analytical Chemistry* 51 (1979), 1453.

24. C. S. Russell, ed., *Safe Drinking Water: Current and Future Problems* (Research Paper R-12) (Washington, D.C.: Resources for the Future, 1978).

25. G. W. Ewing, "Analytical Chemistry," 130 (see note 21 above).

26. National Academy of Sciences-National Research Council, Committee on Lead in the Human Environment, *Lead in the Human Environment* (Washington, D.C.: National Academy of Sciences, 1979), 276–82.

27. G. C. Butler, ed., *Principles of Ecotoxicology* (Chichester: Wiley, 1978), 3–9 and 333–35.

28. B. H. Svensson and R. Söderlund, eds., *Nitrogen, Phosphorus and Sulphur—Global Cycles,* SCOPE Report 7 (Stockholm: Swedish Natural Science Research Council, 1976), 66.

29. F. H. Bormann and G. E. Likens, *Pattern and Process in a Forested Ecosystem* (New York: Springer-Verlag, 1979).

30. G. E. Hutchinson, *A Treatise on Limnology* (New York: Wiley, 1957).

31. National Academy of Sciences-National Research Council, Committee on Water, *Alternatives in Water Management* (Washington, D.C.: National Academy of Sciences, 1966).

32. U.S. 89th Cong., 2d sess., House of Representatives, *Report of the Task Force on Federal Flood Control Policy*, House Document 465 (Washington, D.C.: 1966).

33. M. W. Holdgate, *A Perspective of Environmental Pollution* (London: Cambridge University Press, 1979).

34. National Academy of Sciences-National Research Council, Committee on Environmental Decision Making, *Decision Making in the Environmental Protection Agency* (Washington, D.C.: National Academy of Sciences, 1977), vol. 2.

35. It is best developed in water planning. See C. W. Howe, *Benefit-Cost Analysis for Water Systems Planning* (Baltimore: Publication Press, 1971).

36. National Academy of Sciences-National Research Council, Committee on Principles of Decision Making for Regulating Chemicals in the Environment, *Decision Making for Regulating Chemicals in the Environment* (Washington, D.C.: National Academy of Sciences, 1975), 39–43.

37. National Academy of Sciences-National Research Council, Committee for a Study on Saccharin and Food Safety Policy, *Saccharin: Technical Assessment of Risks and Benefits* (Washington, D.C.: National Academy of Sciences, 1978); Committee for Scientific and Technical Assessments of Environmental Pollutants, *Ni-*

trates: An Environmental Assessment (Washington, D.C.: National Academy of Sciences, 1978).

38. S. Lichtenstein, P. Slovic, B. Fischhoff, M. Layman, B. Combs, *Perceived Frequency of Lethal Events* (Eugene, OR: Decision Research, 1978).

39. National Academy of Sciences-National Research Council, Committee on Nuclear and Alternative Energy Systems, *Energy in Transition, 1985–2010. Final Report* (San Francisco: Freeman, 1979).

40. R. W. Kates, *Risk Assessment of Environmental Hazard*, SCOPE Report 8 (Chichester: Wiley, 1979); A. V. Whyte and I. Burton, *Environmental Risk Assessment*, SCOPE Report 15 (Chichester: Wiley, 1980); W. W. Lowrance, *Of Acceptable Risk* (Los Altos, CA: Kaufmann, 1976); P. Handler, "Some Comments on Risk Assessment," in *The National Research Council in 1979* (Washington, D.C.: National Academy of Sciences, 1979), 3–19; W. D. Rowe, *An "Anatomy" of Risk* (Washington, D.C.: Environmental Protection Agency, 1975).

41. L. B. Lave and E. P. Seskin, *Air Pollution and Human Health* (Washington, D.C.: Resources for the Future, 1977).

42. C. S. Hollings, ed., *Adaptive Environmental Assessment and Management*, IIASA International Series on Applied Systems Analysis 3 (Chichester: Wiley, 1978), 1–21.

43. M. G. Wolman, *Science* 174 (1971), 905.

44. *The State of the Environment in OECD Member Countries* (Paris: Organization for Economic Cooperation and Development, 1979).

45. National Academy of Sciences-National Research Council, International Environmental Programs Committee, *Early Action on the Global Environmental Monitoring System* (Washington, D.C.: National Academy of Sciences, 1976).

46. E. D. Goldberg et al., *Environmental Conservation 5* (1978), 101.

47. C. S. Russell and H. H. Landsberg, *Science* 172 (1971), 1307.

48. *Report of the Conference*, United Nations Conference on Desertification, Nairobi, Kenya (1977); K. A. Hare, *Climate and Desertification*, United Nations Conference on Desertification, Nairobi, Kenya (1977); R. W. Kates, D. L. Johnson, K. Johnson, *Population, Society, and Desertification*, United Nations Conference on Desertification, Nairobi, Kenya (1977); A. Warren and J. K. Maizels, *Ecological Change and Desertification*, United Nations Conference on Desertification, Nairobi, Kenya (1977).

49. United Nations Environment Programme, *Ozone Layer Bulletin*, Issue no. 1 (Kenya: United Nations Environment Programme, January 1978).

50. World Meteorological Organization, *World Climate Conference* (World Meteorological Organization, 1979).

51. L. K. Caldwell, *In Defense of Earth* (Bloomington: Indiana University Press, 1972); D. Pirages, *Global Ecopolitics* (North Scituate, MA: Duxbury, 1978).

52. W. I. Vernadsky, *Problems of Biogeochemistry* (Hamden, CT: Shoestring Press, 1944), vol. 2; W. I. Vernadsky, *La Biosphère* (Paris, 1930).

53. B. H. Svensson and R. Söderlund, eds., *Nitrogen, Phosphorus, and Sulphur* (see note 28 above).

54. National Academy of Sciences-National Research Council, Committee on Impacts of Stratospheric Change, *Stratospheric Ozone Depletion by Halocarbons: Chemistry and Transport* (Washington, D.C.: National Academy of Sciences, 1979).

55. J. N. Galloway, E. B. Cowling, E. Gorham, W. W. McFee, *A National Program for Assessing the Problem of Atmospheric Deposition, A Report to the Council on Environmental Quality,* National Atmospheric Deposition Program, (Fort Collins: Colorado State University, 1978).

56. B. Bolin, E. T. Degens, S. Kempe, P. Ketner, eds., *The Global Carbon Cycle,* SCOPE Report 13 (Chichester: Wiley, 1979), 3.

57. R. Revelle, *Science* 209 (1980), 164.

58. B. Bolin, "Global Ecology and Man," in *World Climate Conference* (Geneva: World Meteorological Organization, 1979), 24.

Gilbert F. White
Published Works, 1935–84

1935   "Shortage of Public Water Supplies in the United States
            During 1934." *Journal of the American Water Works
            Association* 27, no. 7 (July): 841–54.
1936a   "The Limit of Economic Justification for Flood Protec-
            tion." *Journal of Land and Public Utility Economics* 12
            (May): 133–48.
1936b   "Notes on Flood Protection and Land-use Planning."
            Planners Journal 3, no. 3 (May-June): 57–61.
1937a   Prepared with the National Resources Committee. "The
            Broadening Scope of Planning." In *Proceedings of the
            National Zoning Conference* (Chicago, IL, December
            12–13), 25–28.
1937b   "Economic Justification for Flood Protection." *Civil
            Engineering* 7, no. 5 (May): 345–48.
1939   "Economic Aspects of Flood-forecasting." In *Trans-
            actions of 1939 of the American Geophysical Union,*
            218–33. Washington, D.C.
1940   "State Regulation of Floodplain Use." *Journal of Land
            and Public Utility Economics* 16, no. 3 (August): 352–
            57.
1945   *Human Adjustment to Floods.* University of Chicago
            Department of Geography Research Papers, no. 29. Chi-
            cago. Published in limited edition in 1942.
1948   "Water Limits to Human Activity in the United States."
            In *Proceedings of the InterAmerican Conference on
            Conservation of Renewable Natural Resources,* sec. 3,
            317–21. Washington, D.C.: Department of State.
1949a   "National Resources: Progress and Poverty." Univer-
            sity of Chicago *Round Table,* no. 569. NBC radio dis-
            cussion with Edward Ackerman and William Vogt,
            February 13.
1949b   Prepared with the Natural Resources Committee. *Task
            Force Report on Natural Resources,* Appendix L. Pre-
            pared for the Commission on Organization of the Ex-
            ecutive Branch of the Government. Washington, D.C.:
            Government Printing Office, January.

1949c   "Toward an Appraisal of World Resources." *Geographical Review* 39, no. 4: 625–39.

1950a   With the Committee. *A Water Policy for the American People,* Report of the President's Water Policy Commission. Washington, D.C.: Government Printing Office.

1950b   "National Executive Organization for Water Resources." *American Political Science Review* 44, no. 3 (September): 593–610.

1950c   "Reorganization of Federal Agencies for Natural Resources Development." *Journal of the American Water Works Association* 42: 611–14.

1953   "A New Stage in Resources History." *Journal of Soil and Water Conservation* 8, no. 5 (September): 228–32, 248.

1954   "Work of the UNESCO Advisory Committee on Arid Zone Research." *Science* 120, no. 3105 (July 2): 15.

1955a   With the Committee. *The Goals of Student Exchange.* New York: Committee on Educational Interchange Policy, January.

1955b   With the Committee. *Geographic Distribution in Exchange Programs.* New York: Committee on Educational Interchange Policy, January.

1955c   With Peter C. Duisberg. "International Arid Lands Meetings in New Mexico." *Scientific Monthly* 192 (March).

1955d   "Symposium on the Future of the Arid Lands." *Geographical Review* 45, no. 3: 434–35.

1956a   "International Cooperation in Arid Zone Research." *Science* 123, no. 3196 (March 30): 537–38.

1956b   Editor. *The Future of Arid Lands.* Washington, D.C.: American Association for the Advancement of Science. (In English and Russian.)

1956c   "Discussion." In *Water for Industry,* edited by Jack B. Graham and Meredith F. Burrill, 121–24. Washington, D.C.: American Association for the Advancement of Science.

1957a   Prepared with the Special Committee on Renewable Natural Resources, Division of Biology and Agriculture. *The Need for Basic Research with Respect to Renewable Natural Resources.* Washington, D.C.: National Academy of Sciences-National Research Council.

1957b   "A Perspective of River Basin Development." *Law and Contemporary Problems* 22, no. 2 (Spring): 157–84. (In English and Japanese.)

1958a   "Introductory Graduate Work for Geographers." *Professional Geographer* 10, no. 2 (March): 6–8.

1958b   With the Committee. *Integrated River Basin Development.* New York: United Nations Department of Economic and Social Affairs. (In English, French and Spanish.)

1958c   With Wesley C. Calef, James W. Hudson, Harold M. Mayer, John R. Shaeffer, and Donald J. Volk. *Changes in Urban Occupance of Flood Plains in the United States.* University of Chicago Department of Geography Research Papers, no. 57. Chicago.

1958d   "Broader Bases for Choice: The Next Key Move." In *Perspective on Conservation,* edited by Henry Jarrett, 205–26. Baltimore: Johns Hopkins University Press.

1958e   "The Facts About Our Water Supply." *Harvard Business Review* 36, no. 2 (March-April): 87–94.

1958f   "River Basin Development." *News of the U.N.* 2, no. 2 (May): 1, 4. New York: Friends General Conference.

1958g   "Emerging Needs in Development of the World's Rivers." *WMO Bulletin* (July): 108–11.

1959a   "A New Attack on Flood Losses." *State Government* 32, no. 2 (Spring): 121–26.

1959b   "The Changing Dimensions of the World Community." In *The High School in a New Era,* edited by Francis S. Chase and Harold Anderson. Chicago: University of Chicago Press. Also in *Journal of Geography* 59, no. 4 (April 1960): 165–70.

1960a   "Strategic Aspects of Urban Flood Plain Occupance." *Journal of the Hydraulics Division, Proceedings of the American Society of Civil Engineers* 86, no. HY2 (February): 89–102.

1960b   "A Geographer's View of the Problems of the South Platte." In *Resource Development: Frontiers for Research,* Papers of the Western Resources Conference, 1959, 75–83. Boulder: University of Colorado Press.

1960c   "Industrial Water Use: A Review." *Geographical Review* 50, no. 3: 412–30.

1960d   "Alternative Uses of Limited Water Supplies." *Impact* 10, no. 4: 243–63.

1960e   *Science and the Future of Arid Lands.* Paris: United Nations Educational, Scientific, and Cultural Organization. (In English, French and Spanish.)

1960f   "The Changing Role of Water in Arid Lands." *University of Arizona Bulletin Series* 32, no. 2 (November). Also in *Arizona Review* 16, no. 3 (March 1967): 1–8.

1961a   "The Control and Development of Flood Plain Areas." In *Proceedings of the 1960 Institute on Planning and Zoning,* 93–107. Dallas: Southwestern Legal Foundation.

1961b   Consultant. U.S. Congress. Senate. "Report of the Select Committee on National Water Resources." 87th Cong., 1st sess., S. Rept. 29.

1961c   "Introduction: The Strategy of Using Flood Plains." In *Papers on Flood Problems,* edited by Gilbert F. White, 1–4. University of Chicago Department of Geography Research Papers, no. 70. Chicago.

1961d   With Robert W. Kates. "Flood Hazard Evaluation." In *Papers on Flood Problems,* edited by Gilbert F. White, 135–47. University of Chicago Department of Geography Research Papers, no. 70. Chicago.

1961e   Editor. *Papers on Flood Problems.* University of Chicago Department of Geography Research Papers, no. 70. Chicago.

1961f   "Water: A Growing Crisis on the Flood Front." In *Planning,* 132–36. Chicago: American Society of Planning Officials.

1961g   "The Choice of Use in Resource Management." *Natural Resources Journal* 1 (March): 23–40.

1961h   "A Joint Effort to Improve High School Geography." *Journal of Geography* 60, no. 8 (November): 357–60.

1961i   "Water Pollution Control and Its Challenge to Political Economic Research: Discussion." In *Proceedings of the National Conference on Water Pollution,* U.S. Department of Health, Education and Welfare, 485–97. Washington, D.C.: Government Printing Office.

1961j   With Charles C. Colby. "Harlan H. Barrows, 1877–1960." *Annals, Association of American Geographers* 51, no. 4 (December): 395–400.

1962a   "Report of the Past President." *Professional Geographer* 14, no. 4 (July): 13–14.

1962b   "The Lower Mekong Development: Its Meaning and Extent." *Resources,* no. 112 (May): 19–26. (In English and Japanese.)

1962c   With Egbert de Vries, Harold B. Dunkerley, and John V. Krutilla. "Economic and Social Aspects of Lower Mekong Development." In *A Report to the Committee for Coordination of Investigations of the Lower Mekong Basin,* January. (In English and French.)

1962d   "Critical Issues Concerning Geography in the Public Service—Introduction." *Annals, Association of American Geographers* 52, no. 3 (September): 279–80.

1962e   Prepared with the National Research Council Committee on Natural Resources. *Natural Resources: A Summary Report to the President of the United States.* Washington, D.C.: National Academy of Sciences, Publication 1000.

1962f   "Social and Economic Aspects of Natural Resources." *A Report to the Committee on Natural Resources.* Washington, D.C.: National Academy of Sciences, Publication 1000-G.

1962g   "Thirsty Lands—Past and Future." *UNESCO Courier* (May): 4–17.

1962h   "Uncharted Intellect: Our Unworldly Ignorance." *Chicago Sunday Sun-Times*, June 10, sec. 2, 1–2.

1962i   As told to Jack Star. "We're Talking Ourselves into a Water Crisis." *Look* (September 11): 61–64.

1963a   Review of *Man, Mind and Land: A Theory of Resource Use*, by Walter Firey. *Economic Geography* 39, no. 4 (October): 373–75.

1963b   "Contributions of Geographical Analysis to River Basin Development." *Geographical Journal* 129, pt. 4 (December): 412–36.

1963c   "The Mekong River Plan." *Scientific American* 208, no. 4 (April): 49–59. Also in *Arms Control,* edited by Herbert F. York, 318–29. San Francisco: W. H. Freeman and Co., 1973.

1963d   With Howard Cook. "Making Wise Use of Flood Plains." *Science, Technology and Development* 1, U.S. papers prepared for U.N. conference, 343–59. Washington, D.C.: Government Printing Office.

1963e   Review of *Gewässer und Wasserhaushalt des Festlands,* by Reiner Keller. *Geographical Review* 53, no. 4: 628–29.

1963f   Prepared with the National Research Council, Ad Hoc Committee on International Programs in Atmospheric Sciences and Hydrology. "An Outline of International Programs in the Atmospheric Sciences." Washington, D.C.: National Academy of Sciences, Publication 1085.

1963g   "United Nations Conference on the Application of Science and Technology for the Benefit of the Less Developed Areas." *Geographical Review* 53 (October): 608–9.

1964a   "Maps in Our Changing World." *Current Events* 63, no. 21: 163–66.

1964b   "Floodplain Adjustment and Regulations." In *Handbook of Applied Hydrology*, sec. 25-V, edited by Ven Ta Chow. New York: McGraw-Hill.

1964c   *Choice of Adjustment to Floods*. University of Chicago Department of Geography Research Papers, no. 93. Chicago.

1964d   "Rivers of International Concord." *UNESCO Courier*, no. 32 (July-August): 36–37.

1964e   "Vietnam: The Fourth Course." *Bulletin of the Atomic Scientists* 20, no. 10 (December): 6–10.

1964f   "Neglected Alternatives to Flood Protection." In *1964 Annual Report*, Resources for the Future, Inc., 15–24. Washington, D.C.

1965a   "Rediscovering the Earth." *American Education* 1, no. 2 (February): 8–11. Also in *Bulletin of the National Association of Secondary School Principals* 51, no. 316 (February 1967): 1–9.

1965b   "Geography in Liberal Education." In *Geography in Undergraduate Liberal Education: A Report of the Geography in Liberal Education Project*, 13–24. Washington, D.C.: Association of American Geographers.

1965c   "Water Development as Part of a Development Aid Policy." In *Proceedings of the International Conference on Water Development in Less Developed Areas, Berlin, 1963*, 44–56. Berlin: Duncker and Humblot.

1966a   With W. R. D. Sewell. "The Lower Mekong." *International Conciliation*, no. 558 (May): 1–63.

1966b   With the Committee. *Alternatives in Water Management: Report of the Committee on Water, Division of Earth Sciences*. Washington, D.C.: National Academy of Sciences.

1966c   With the Commission. *Weather and Climate Modification: Report of the Special Commission on Weather Modification*. Washington, D.C.: National Science Foundation.

1966d   With the Task Force. *A Unified National Program for Managing Flood Losses: Report by the Task Force on Federal Flood Control Policy*. Washington, D.C.: Bureau of the Budget. Also, U.S. Congress. House. 89th Cong., 2nd sess., H. Doc. 465.

1966e   "Approaches to Weather Modification." In *Human Dimensions of Weather Modification*, 19–23. University of Chicago Department of Geography Research Papers, no. 105. Chicago.

1966f   "Optimal Flood Damage Management: Retrospect and Prospect." In *Water Research,* edited by Allen V. Kneese and Stephen C. Smith, 251–69. Baltimore: Johns Hopkins University Press. Also in *Selected Works in Water Resources,* edited by Asit Biswas, 201–19. Champaign, IL: International Water Resources Association, 1975.

1966g   "Formation and Role of Public Attitudes." In *Environmental Quality in a Growing Economy: Essays from the Sixth RFF Forum*, 105–27. Baltimore: Johns Hopkins University Press.

1966h   "The World's Arid Areas." In *Arid Lands: A Geographical Appraisal,* edited by E. S. Hills, 15–30. London: Methuen and Co.; Paris: UNESCO.

1966i   "Deserts as Producing Regions Today." In *Arid Lands: A Geographical Appraisal,* edited by E. S. Hills, 421–37. London: Methuen and Co.; Paris: UNESCO.

1966j   "Arid Lands." In *Future Environments of North America,* edited by F. Fraser Darling and John P. Milton, 172–84. Garden City, NY: The Natural History Press.

1967a   With the Committee. *African and American Universities Program, 1958–1966: A Summary.* University of Chicago.

1967b   "Flood Plain Safeguards: A Community Concern." In *Outdoors U.S.A.: The Yearbook of Agriculture,* U.S. Department of Agriculture, 133–36. Washington, D.C.: Government Printing Office.

1967c   "River Basin Planning and Peace: The Lower Mekong." In *Problems and Trends in American Geography,* edited by Saul B. Cohen, 187–99. New York: Basic Books.

1967d   "Task Force Report on Federal Flood Control Policy." *Conference Preprint 550,* American Society of Civil Engineers National Meeting on Water Resource Engineering, New York, October 16–20.

1967–68   "Background for Better Teaching: The Uses of New Geography." *Professional Growth for Teachers, Social Studies,* Second Quarter Issue, 2–3.

1968a   With James A. Harder. "The Mekong River Project." In *1968 World Book Yearbook: The Annual Supplement to the World Book Encyclopedia,* 79–95. Chicago: Field Enterprises Educational Corporation.

1968b   "Federal Flood Control Policy." *Civil Engineering* (American Society of Civil Engineers) 38, no. 8: 60–62.

1968c   Prepared with the National Research Council Committee on Water. *Water and Choice in the Colorado Basin: An Example of Alternatives in Water Management.* Washington, D.C.: National Academy of Sciences, Publication 1689.

1969a   *Strategies of American Water Management.* Ann Arbor: University of Michigan Press. (In English and Russian.)

1969b   Editor. *Water, Health and Society: Selected Papers by Abel Wolman.* Bloomington: Indiana University Press.

1969c   "Flood Damage Prevention Policies." In *United Nations Interregional Seminar on Flood Damage Prevention: Measures and Management,* Tbilisi, USSR, September 25–October 15, 18–37. New York: United Nations.

1969d   "Scientific Dimensions of Water Planning." *Earth Sciences Newsletter,* no. 5 (December).

1970a   Prepared with the Commission on Environmental Studies. "The University's Response to Environmental Crisis." Boulder: University of Colorado.

1970b   "Preface." In *Integrated River Basin Development,* rev. ed., ix-xiii. New York: United Nations Department of Economic and Social Affairs. (Orig. ed., 1958.)

1970c   "Flood Losses: A Global Perspective." *Water Spectrum* 2, no. 1 (Spring): 20–23.

1970d   "The University and National Water Policy." In *The University's Role in National Water Policy,* edited by J. Ernest Flack, 1–9 (Proceedings of a conference held July 27–28, 1970). Blacksburg, VA: Universities Council on Water Resources.

1970e   "Recent Developments in Flood Plain Research." *Geographical Review* 60, no. 3 (July): 440–43.

1970f   "Unresolved Issues." In *Arid Lands in Transition,* edited by Harold E. Dregne, 481–91. Washington, D.C.: American Association for the Advancement of Science.

1970g   "Flood-loss Reduction: The Integrated Approach." *Journal of Soil and Water Conservation* 25, no. 5 (September-October): 172–76.

1970h   With the Committee. *From Geographic Discipline to Inquiring Student: Final Report on the High School Geography Project,* edited by Donald J. Patton, 70–72. Washington, D.C.: Association of American Geographers.

1972a   "Collaboration in Natural Hazards Research." *Geographical Review* 62, no. 2: 280–81.

1972b   With David J. Bradley and Anne U. White. *Drawers of Water: Domestic Water Use in East Africa.* Chicago: University of Chicago Press.

1972c   "International Cooperation Among Geographers: The IGU Commission on Man and Environment." *Geoforum: Journal of Physical, Human, and Regional Geosciences* 3, no. 10: 98–99.

1972d   "International Dimensions." In *High School Geography Project: Legacy for the Seventies,* edited by Angus M. Gunn, 37–41. Prepared for the International Geographical Union Commission on the Teaching of Geography. Montreal: Educatif et Culturel.

1972e   "Organizing Scientific Investigations to Deal with Environmental Impacts." In *The Careless Technology,* edited by M. Taghi Farvar and John P. Milton, 914–26. Garden City, NY: Natural History Press.

1972f   *Man-made Lakes as Modified Ecosystems.* Prepared with the SCOPE Working Group on Man-made Lakes. Paris: International Council of Scientific Unions, SCOPE Report No. 2.

1972g   "Geography and Public Policy." *Professional Geographer* 24, no. 2: 101–4.

1972h   "Environmental Impact Statements." *Professional Geographer* 24, no. 3: 302–9.

1972i   "Human Response to Natural Hazard." In *Perspectives on Benefit-risk Decision Making,* 43–49. Report of a colloquium conducted by the Committee on Public Engineering Policy, April. Washington, D.C.: National Academy of Engineering.

1972j   "History of Fire in North America." In *Fire in the Environment: Symposium Proceedings* (December), 3–11. (Published in cooperation with the fire services of Canada, Mexico, and the United States—members of the Fire Management Study Group, North American Forestry Commission, Food and Agriculture Organization.) Washington, D.C.: Forest Service, Department of Agriculture.

1972k   "Regional Alternatives." In *The High Plains: Problems of Semiarid Environments,* edited by Donald D. MacPhail, 97–100. Fort Collins: Colorado State University. 48th annual meeting of Southwestern and Rocky Mountain Division of the American Association for the Advancement of Science, 26–29 April.

1972l   Prepared with the National Advisory Committee on
             Oceans and Atmosphere. *The Agnes Floods*. Report for
             the Administrator of the National Oceanographic and
             Atmospheric Administration, November 22. Washing-
             ton, D.C.: Government Printing Office.

1972m   "Response to the Presentation of the Iben Award." *Water
             Resources Bulletin* 8, no. 6 (December): 1287–89.

1972n   With the Committee. *Experiment Without Precedent*.
             Report of an American Friends Service Committee Del-
             egation's Visit to China, May.

1973a   With William C. Ackermann and E. B. Worthington, ed-
             itors. *Man-made Lakes: Their Problems and Environ-
             mental Effects*. Geophysical Monograph 17. Washington,
             D.C.: American Geophysical Union.

1973b   "Public Opinion in Planning Water Development." In
             Environmental Quality and Water Development, edited
             by Charles R. Goldman, James McEvoy, II, and Peter J.
             Richerson. San Francisco: W. H. Freeman and Co.

1973c   "Prospering with Uncertainty." In *Transfer of Water
             Resources Knowledge,* edited by Evan Vlachos. (Pro-
             ceedings of the First International Conference on Trans-
             fer of Water Resources Knowledge), 22–28. Fort Collins,
             CO: Water Resources Publications. Also in *Floods and
             Droughts,* edited by E. F. Schulz, V. A. Koelzer, and
             Khalid Mahmood (Proceedings of the Second Interna-
             tional Symposium in Hydrology), 9–15. Fort Collins,
             CO: Water Resources Publications.

1973d   With Daya U. Hewapathirane. "Obstacles to Consid-
             eration of Resources Management Alternatives: South
             Asian Experience." In *Transfer of Water Resources
             Knowledge,* edited by Evan Vlachos (Proceedings of the
             First International Conference on Transfer of Water Re-
             sources Knowledge), 252–61. Fort Collins, CO: Water
             Resources Publications.

1973e   "Preface." In *Man, Materials, and Environment,* 7–10.
             Report for the National Commission on Materials Policy
             by the Study Committee on Environmental Aspects of
             a National Materials Policy, Committee for International
             Environmental Programs. Washington, D.C.: National
             Academy of Sciences-National Academy of Engineering.

1973f   "Natural Hazards Research." In *Directions in Geog-
             raphy,* edited by Richard J. Chorley, 193–216. London:
             Methuen and Co.

1973g "The Changing Role of Water in Arid Lands." In *Coastal Deserts: Their Natural and Human Environments*, edited by David H. K. Amiran and Andrew W. Wilson, 37–43. Tucson: University of Arizona Press, adapted from the Riecker Lecture. Also in *Arizona Review* 16, no. 3 (March 1967): 1–8.

1973h "The Last Settler's Syndrome." In *Exploring Options for the Future: A Study of Growth in Boulder County*, vol. 8: *Social and Humanistic Aspects*, 80–85. Boulder, CO: Boulder Area Growth Study Commission, November.

1974a "Domestic Water Supply: Right or Good?" In *Proceedings of Ciba Foundation Symposium on Human Rights in Health, July 4–6, 1973, London*, 35–59. Amsterdam-London-New York: Associated Scientific Publishers.

1974b "Comparative Field Observations on Natural Hazards." In *Man and Environment*, edited by Marton Pecsi and Ferenc Probald, 73–79. Budapest: Research Institute of Geography, Hungarian Academy of Sciences.

1974c "Role of Geography in Water Resources Management." In *Man and Water*, edited by L. Douglas James, 102–21. Lexington: University Press of Kentucky.

1974d "Environmental Threats to Man." In *Energy, Environment, Productivity*, Proceedings of the First Symposium on RANN: Research Applied to National Needs, November 18–20, 1973, 86–93. Washington, D.C.: National Science Foundation, May.

1974e "Edward A. Ackerman, 1911–1973." *Annals, Association of American Geographers* 64, no. 2 (June): 297–309.

1974f "Natural Hazards Research: Concepts, Methods, and Policy Implications." In *Natural Hazards: Local, National, Global*, edited by Gilbert F. White, 3–16. New York: Oxford University Press.

1974g With Paul Slovic and Howard Kunreuther. "Decision Processes, Rationality, and Adjustment to Natural Hazards." In *Natural Hazards: Local, National, Global*, edited by Gilbert F. White, 187–205. New York: Oxford University Press.

1974h Editor. *Natural Hazards: Local, National, Global*. New York: Oxford University Press. (In English and Russian.)

1974i "Mekong River." *Encyclopedia Britannica*, 15th ed., 860–63.

1974j   "Comments." In *A Time to Choose: America's Energy Future,* Final Report by the Energy Policy Project of the Ford Foundation, 410–11. Cambridge, MA: Ballinger Publishing Co.

1974k   With Miriam Orleans, eds. *Carbon Monoxide and the People of Denver,* University of Colorado Environmental Council. Boulder: University of Colorado, Institute of Behavioral Science.

1975a   With J. Eugene Haas. "The Next Time." *Landscape Architecture* 66 (April): 169.

1975b   "National Perspective." In *The Water's Edge, The National Forum on the Future of the Flood Plain,* U.S. Department of the Interior, Minneapolis, MN, September 17–19.

1975c   With J. Eugene Haas. *Assessment of Research on Natural Hazards.* Cambridge, MA: MIT Press.

1975d   "Interdisciplinary Studies of Large Reservoirs in Africa," and "Discussion." In *Proceedings of the Conference on Interdisciplinary Analysis of Water Resource Systems,* edited by J. Ernest Flack (University of Colorado, Boulder, CO, June 19–22, 1973), 63–85, 102–4. New York: American Society of Civil Engineers.

1975e   With G. O. Lang. "Community Mobilization for Adaptation to Change in Rapid Growth Areas." In *Energy Development in the Rocky Mountain Region: Goals and Concerns,* 76–81. Denver, CO: Federation of Rocky Mountain States, Inc.

1975f   With W. A. R. Brinkmann, Harold C. Cochrane, and Neil J. Ericksen. *Flood Hazard in the United States: A Research Assessment.* Boulder: University of Colorado, Institute of Behavioral Science.

1976a   June 24 testimony before the U.S. House of Representatives, Committee on Science and Technology, Subcommittee on Science, Research and Technology, "Earthquake Hearings." Washington, D.C.

1976b   "The IGU Commission on Man and Environment." *Geoforum: Journal of Physical, Human, and Regional Geosciences* 7, no. 2: 143–47.

1976c   "Scientific Capacity and Global Environmental Problems." In *Science: A Resource for Humankind* (Proceedings of the National Academy of Sciences Bicentennial Symposium, Washington, D.C., October 10–14), 37–40. Washington, D.C.: National Academy of Sciences.

1976d   Prepared with the International Environmental Programs Committee, Environmental Studies Board, Commission on Natural Resources, National Research Council. *Early Action on the Global Environmental Monitoring System.* Washington, D.C.: National Academy of Sciences.

1976e   Prepared with the International Institute for Environment and Development, Earthscan. "Water for All." Statement for the Symposium on Water (in preparation for the United Nations Conference, Mar Del Plata, Argentina, March 14–25, 1977). Washington, D.C., December 9–11. (In English, Spanish, French and Arabic.)

1977a   "Environmental Health in Developing Countries." *Geographia Polonica* 36: 226–37.

1977b   "Water Supply Service for the Urban Poor: Issues." In *Energy, Water and Telecommunications Department Public Utilities Notes.* International Bank for Reconstruction and Development, International Development Association, PU Report no. PUN 31, August. Also in *Water Supply and Management* 2 (1978): 425–54.

1977c   "Comparative Analysis of Complex River Development." In *Environmental Effects of Complex River Development: International Experience,* edited by Gilbert F. White, 1–21. Boulder, CO: Westview Press.

1977d   Editor. *Environmental Effects of Complex River Development: International Experience.* Boulder, CO: Westview Press.

1977e   With Martin W. Holdgate, editors. *Environmental Issues: SCOPE Report 10.* Chichester, John Wiley & Sons.

1977f   "Natural Hazards Management in the Coastal Zone." In *Proceedings of the American Shore and Beach Preservation Association, Washington, D.C., October 17–19, 1977.* Berkeley, American Shore and Beach Preservation Association.

1978a   Consultant. "Resources and Needs: Assessment of the World Water Situation." In *Water Development and Management: Proceedings of the United Nations Water Conference, Mar Del Plata, Argentina, March, 1977,* pt. 1, 1–46. New York: Pergamon Press.

1978b   "The Hazards of Wetlands Use." In *Proceedings of the National Wetland Protection Symposium* (Reston, VA, June 6–8, 1977, U.S. Department of the Interior, Fish and Wildlife Service), 3–5. Washington, D.C.: Government Printing Office.

1978c   As told to Henry Spall. "Gilbert White Talks about Natural Hazards." *Earthquake Information Bulletin* 10, no. 1 (January-February): 16–25.

1978d   "Natural Hazards Data Needs." *Environmental Data Service* (May): 3.

1978e   With Ian Burton and Robert W. Kates. *The Environment as Hazard.* New York: Oxford University Press.

1978f   "Stewardship of the Earth." In *Right Sharing of World Resources: Basic Concerns,* 15–23. Committee on Right Sharing of World Resources of Friends World Committee for Consultation, March.

1978g   "Domestic Water Supply in the Third World." *Progress in Water Technology* 11, nos. 1–2: 13–19.

1978h   "Foreword." In *Principles of Ecotoxicology: SCOPE 12,* edited by G. C. Butler, v-vi. Chichester: John Wiley & Sons.

1978i   "Introduction." In *Water in a Developing World,* edited by Albert E. Utton and Ludwik Teclaff, 1–5. Boulder, CO: Westview Press. Also in *Natural Resources Journal* 16 (October 1976): 737–41.

1978j   "Advising on the Environment." In *The National Research Council in 1978,* 209–19. Washington, D.C.: National Academy of Sciences.

1978k   With Eve C. Gruntfest and Thomas E. Downing. "Big Thompson Flood Exposes Need for Better Flood Reaction System to Save Lives." *Civil Engineering* (American Society of Civil Engineers) 48 (February): 72–73.

1978l   Editor. *Environmental Effects of Arid Land Irrigation in Developing Countries.* MAB Technical Notes 8. Paris: UNESCO.

1978m   "Natural Hazards and the Third World—A Reply." *Human Ecology* 6, no. 2: 229–31.

1978n   "Foreword." In *Saharan Dust: SCOPE 14,* edited by C. Morales, v-vi. Chichester: John Wiley & Sons.

1979a   With Anne U. White. "Behavioral Factors in Selection of Technologies." In *Appropriate Technology in Water Supply and Waste Disposal,* edited by Charles G. Gunnerson and John M. Kalbermatten, 31–51. New York: American Society of Civil Engineers.

1979b   With Mostafa K. Tolba. "Global Life Support Systems." *United Nations Environment Programme Information/47* (Nairobi, June 5), 1–4.

1979c   "Natural Hazards Policy and Research Issues." In *Natural Hazards in Australia,* edited by R. L. Heathcote and B. G. Thom, 15–24. Canberra: Australian Academy of Science.

1979d   "Presentation(s)" and "Discussion(s)." In *Water: Resource Policies Appropriate to Pervasive Uncertainty* (Seminars with Gilbert White), edited by K. Denike. Center for Human Settlements Occasional Paper no. 8. Vancouver: University of British Columbia.

1979e   "Problems of Communication Between Scientists and Decision Makers in Resource Management." In *Samdene International Workshop,* p. 79. Alexandria: Alexandria University.

1979f   *Nonstructural Floodplain Management Study: Overview.* Prepared for the U.S. Water Resources Council, October. Washington, D.C.: Government Printing Office.

1979g   "International Exploration of the Global Environment." In *The National Research Council in 1979,* 189–99. Washington, D.C.: National Academy of Sciences.

1980a   "The Environmental Movement at a Turning Point." *PITT* 34, no. 3 (February): 7–10.

1980b   "Overview of the Flood Insurance Program." Statement to the Senate Committee on Banking, Housing and Urban Affairs, February 28.

1980c   With John H. Sorensen. "Natural Hazards: A Cross-cultural Perspective." In *Human Behavior and Environment* 4, edited by I. Altman, A. Rapoport, and J. F. Wohlwill, 279–318. New York: Plenum Press.

1980d   "What is Enough Information About Earthquake Prediction?" In *Proceedings of the Conference on Earthquake Prediction Information* (Los Angeles, California, January 28–30), 57–64. Menlo Park, CA: USGS, Open-File Report 80–843.

1980e   "Environment." In *SCIENCE Centennial Review,* edited by Philip H. Abelson and Ruth Kulstad, 165–72. Washington, D.C.: American Association for the Advancement of Science. Also in *Science* 209, no. 4452 (July 4): 183–90.

1981a   "National Water Issues: Growing Opportunities." *Water Resources Specialty Group Newsletter* 1, no. 1 (March): 2.

1981b   With Ian Burton and Robert W. Kates. "The Future of Hazard Research: A Reply to William I. Torry." *Canadian Geographer* 25, no. 3: 286–89.

1981c   With Edwin Kessler. "Thunderstorms in a Social Context." *Thunderstorms: A Social, Scientific, and Technological Documentary* 1, (September): 1–22.
Washington, D.C.: Department of Commerce, NOAA.

1981d   "Geographic Contributions to Analysis of Global Environmental Problems." In *The Environment: Chinese
and American Views,* edited by Laurence J. C. Ma and
Allen G. Noble, 385–93. New York: Methuen & Co.

1981e   Review of *A Perspective of Environmental Pollution,* by
M. W. Holdgate. *Environmental Conservation* 8, no. 1
(Spring): 84.

1981f   With Anne U. White. "Reappraisal and Response to
Changing Service Levels." In *Project Monitoring and
Reappraisal in the International Drinking Water Supply
and Sanitation Decade,* edited by Charles G. Gunner-
son and John M. Kalbermatten, 1–21. Papers presented
at the American Society of Civil Engineers International
Convention, New York, New York, May 11–15. New
York: American Society of Civil Engineers.

1981g   "Reservoir Systems—A Socioeconomic Perspective."
In *Proceedings of the National Workshop on Reservoir
Systems Operations,* 550–59. Workshop held in Boulder, Colorado, August 13–17. New York: American Society of Civil Engineers.

1981h   "Opening Remarks: Day Two," and "Closing Remarks." In *The Impact of Interventions in Water Supply
and Sanitation in Developing Countries,* edited by
James D. Lindstrom, 91–92, 157–58. (Proceedings of a
seminar held at the Pan American Health Organization,
Washington, D.C., March 25–26, 1980.) Washington,
D.C.: Agency for International Development.

1981i   "Natural Hazards Reduction—Will it Change its
Course?" *Natural Hazards Observer* 6, no. 1 (September): 1–2.

1982a   Chapter coordinator and joint author. "Water Resources." In *Outlook for Science and Technology: The
Next Five Years* (Report of the National Research Council), 255–85. San Francisco: W. H. Freeman and Co.

1982b   With Martin W. Holdgate and Mohammed Kassas, editors. *The World Environment, 1972–1982.* Dublin: Ty-
cooly International Publishing.

1982c   "The State of the World Environment." *Transition* 12,
no. 3: 2–12. Also in hearings before the Subcommittee
on Human Rights and International Organizations,

Committee on Foreign Affairs, U.S. House of Representatives, Washington, D.C., April 1, 1982.

1982d "Epilogue." In *Regional Conflict and National Policy,* edited by Kent A. Price, 126–31. Washington, D.C.: Resources for the Future.

1982e With Martin W. Holdgate and Mohammed Kassas. "World Environmental Trends Between 1972 and 1982." *Environmental Conservation* 9, no. 1: 11–29.

1982f "Special Human Behavioral Response Associated With Volcanic Hazard Conditions." In *Status of Volcanic Prediction and Emergency Response Capabilities in Volcanic Hazard Zones of California,* edited by Roger C. Martin and James F. Davis, 147–51. Special Publication No. 63. Sacramento: California Department of Conservation.

1983a "Water Resource Adequacy: Illusion and Reality." *Natural Resources Forum* 7, no. 1: 11–21.

1983b "Foreword." In *Floodplain Management: The TVA Experience,* i-ii. Knoxville: Tennessee Valley Authority.

1984a "Environmental Perception and Its Uses: A Commentary." In *Environmental Perception and Behavior: An Inventory and Prospect,* edited by Thomas F. Saarinen, David Seamon, and James L. Sell, 93–96. University of Chicago Department of Geography Research Papers, no. 209. Chicago.

1984b "Notes on Geographers and the Threat of Nuclear War." *Transition* 14, no. 1: 2–4.

1984c With Julius London, editors. *The Environmental Effects of Nuclear War.* Boulder, CO: Westview Press. Also with London, Chapters 1 and 6.

1984d "A Meeting of Minds on Global Resources." *Science* 226, no. 4674 (November 2): 495.

Index

Abbot, H. L., 44
Acid rain, 434
Ackerman, Edward, 106, 111, 146–47, 167, 169
Adjustments. *See* Floods and flood hazard, range of adjustments to; Natural hazards research: alternative adjustments
Agricultural Research Service, USDA, rural floodplain studies of the, 334–35
Agriculture, real costs in, 32
Agriculture, U.S. Department of: flood control surveys, 13; flood policy of, 337; Program Surveys in the, 228; and watershed protection, 61
American Friends Service Committee, 11, 377
American Society of Civil Engineers, 49
American Water Works Association, 1–2
Amiran, David, 273
Arey, D., 338
Arid lands: amenities in, 274; attitudes toward water in, 132, 140–41; changes in, 267–70; decision making in, 127, 139–40; development in, 270–71; 273–74; grazing industry in, 137; industrialization of, 129, 133–34; investment in, 137–38; irrigation in, 132–33, 135, 272–73; land use changes in, 130–31; municipal water pricing systems in, 135; mystique of, 266–72; nomadism in, 128; recreational use of, 129, 135–36; as regional foci, 125; research in, xi, 126, 141–42, 271, 274–75; salt accumulation in, 132–33; social processes in, 126–29, 139–41, 174–75, 271; technology in, 141; uniqueness of, 127; urbanization of, 129–30, 268; use and supply of water

in, 125, 131–32; waterlogging in, 132–33; water quality standards in, 130; water rights in, 135; world linkages in, 275–76
Arkansas River, 62
Aschmann, H., 167, 169
Association of American Geographers (AAG), 166, 169, 185, 219; Committee on Geography and Afro-America, 319; Council, 320; high school geography units, 195; responsibility for public policy, 317–18, 320–22; symposium on perception, 334
Aswan reservoir, 284
Attitudes, 219, 223–24, 291; and personality attributes, 234–36; role of, in decision making, 222; role of information in, 236

Balchin, W. G. V., 110, 117
Banfield, E., 146
Barrows, Harlan H., 1, 10, 57, 328
Benefits-cost analysis. *See* Floodplain management, benefit-cost analysis; Resource management, economic analysis of; Water resources management, economic analysis of
Bennett, Hugh, 60
Biogeochemical cycles, 434–37
Bird, Isabella, 350
Blake, N. M., 114
Bolin, B., 437
Borland, Hal, *High, Wide and Lonesome*, 351
Boulder, CO: decision making in, 220–21, 227; settlement of, 349–55
Boulder Area Growth Study Commission, 353–54
Boulder Canyon Project Act of 1928, 46
Bradley, David J., 291, 361
Bristol, TN, 205
Brown, Harrison, *The Challenge of Man's Future*, 29